AF676652

Qualitative-Environmental Aspects to Improve Materials and Processes of Their Formation

Qualitative-Environmental Aspects to Improve Materials and Processes of Their Formation

Guest Editors

Andrzej Pacana
Dominika Siwiec

MDPI

Basel • Beijing • Wuhan • Barcelona • Belgrade • Novi Sad • Cluj • Manchester

Guest Editors
Andrzej Pacana
Faculty of Mechanical
Engineering and Aeronautics
Rzeszow University of
Technology
Rzeszow
Poland

Dominika Siwiec
Faculty of Mechanical
Engineering and Aeronautics
Rzeszow University of
Technology
Rzeszow
Poland

Editorial Office
MDPI AG
Grosspeteranlage 5
4052 Basel, Switzerland

This is a reprint of the Special Issue, published open access by the journal *Materials* (ISSN 1996-1944), freely accessible at: https://www.mdpi.com/journal/materials/special_issues/materials_quality.

For citation purposes, cite each article independently as indicated on the article page online and as indicated below:

Lastname, A.A.; Lastname, B.B. Article Title. *Journal Name* **Year**, *Volume Number*, Page Range.

ISBN 978-3-7258-5257-4 (Hbk)
ISBN 978-3-7258-5258-1 (PDF)
https://doi.org/10.3390/books978-3-7258-5258-1

© 2025 by the authors. Articles in this book are Open Access and distributed under the Creative Commons Attribution (CC BY) license. The book as a whole is distributed by MDPI under the terms and conditions of the Creative Commons Attribution-NonCommercial-NoDerivs (CC BY-NC-ND) license (https://creativecommons.org/licenses/by-nc-nd/4.0/).

Contents

About the Editors

Andrzej Pacana

Andrzej Pacana is Full Professor of the Rzeszow University of Technology and Vice-Dean for Education at the Faculty of Mechanical Engineering and Aviation of the Rzeszow University of Technology (Poland). He is the author or co-author of over 660 scientific publications, including 75 books (monographs and academic textbooks) devoted to production engineering issues, and his H-index is 28. His research focuses on process improvement in production engineering. He is listed on Standford's ranking of the World's Top 2% Scientists.

Dominika Siwiec

Dominika Siwiec is Assistant Professor at the Department of Machine Technology and Production Engineering at the Rzeszow University of Technology (Poland). Her research focuses on predicting product quality and conducting life cycle sustainability assessments. Her achievements include over 180 publications (including monographs and academic textbooks), and her H-index is 19. She is listed on Stanford's ranking of the World's Top 2% Scientists.

materials

MDPI

Article

New Insights into a Vanadium Precipitation Process by Hydrolysis Based on Manganese Salt Pretreatment

Mengxia Liu [2], Tao Jiang [2,*], Jing Wen [2], Zibi Fu [1], Tangxia Yu [2], Guangdong Yang [2], Sanyuan Xia [2] and Hao Xiao [2]

[1] State Key Laboratory of Vanadium and Titanium Resources Comprehensive Utilization, Panzhihua 617000, China; 13982377105@163.com

[2] School of Metallurgy, Northeastern University, Shenyang 110819, China; 15265373295@163.com (M.L.); wenjing@smm.neu.edu.cn (J.W.); 2210665@stu.neu.edu.cn (T.Y.); 2110566@stu.neu.edu.cn (G.Y.); 13739801061@163.com (S.X.); 15101851894@163.com (H.X.)

* Correspondence: jiangt@smm.neu.edu.cn

Abstract: Vanadium precipitation is the key step in producing vanadium products from vanadium solution. The sustainable development of the vanadium industry requires new environmentally friendly processes for vanadium precipitation. In this study, $NaVO_3$ solution was pretreated with manganese salt to preliminarily separate the vanadium and sodium components. The product of vanadium extraction by manganese salt was dissolved by acid to produce manganese vanadate solution. After vanadium precipitation by hydrolysis, manganese removal, and calcination, the target product V_2O_5 was obtained. Scanning electron microscopy (SEM), X-ray diffraction (XRD), inductively coupled plasma emission spectrometry (ICP-OES), and atomic absorption spectrometry (AAS) were used to perform the characterization and analyses. The results showed that vanadium and manganese have a strong binding ability. The rate of vanadium extraction by manganese salt reached 99.75%, and the product of vanadium extraction by manganese salt was $Mn_2V_2O_7$, with a sodium content of only 0.089%, confirming the effective separation of vanadium and sodium. The acid dissolution rate of the vanadium extraction product reached 99.95%, and the rate of vanadium precipitation by hydrolysis reached 97.87%. After manganese removal and calcination, the purity of the V_2O_5 product reached 98.92%. In addition, the recyclability of manganese sulfate and ammonium sulfate was analyzed. The process reduced the production of ammonia–nitrogen wastewater, laying a foundation for researching new technologies for extracting vanadium from vanadium slag.

Keywords: sodium metavanadate solution; manganese vanadate; acid dissolution; vanadium precipitation by hydrolysis

1. Introduction

Vanadium is an indispensable nonferrous metal, an important strategic resource in modern national defense and industry [1,2]. Vanadium slag is obtained by smelting vanadium–titanium magnetite in a blast furnace and blowing vanadium in a converter. V_2O_5 is prepared by extracting vanadium from vanadium slag, which is widely used in steel, the chemical industry, medical treatments, aerospace, new energy, and other fields [3–10]. This wide usage necessitates establishing an industrial chain for the clean and efficient preparation of V_2O_5 products for the utilization of vanadium resources and the development of the vanadium industry.

The roasting and leaching process group of vanadium slag includes sodium roasting–water leaching, calcification roasting, and manganese roasting–acid leaching [11–13]. The sodium roasting–water leaching process is widely used because of its high leaching rate. $NaVO_3$ [14–16] is the main vanadium-containing phase in the sodium roasting clinker of vanadium slag, with its leaching solution obtained through water immersion. Vanadium pentoxide is mainly prepared from a vanadium-containing leaching solution through chemical precipitation, ion exchange, and extraction [17–21]. However, the ion exchange and extraction methods are relatively complex and costly, making them more suitable for experimental studies on vanadium extraction rather than large-scale industrial applications. Hence, the chemical precipitation method is widely used because of its short process, simple operation, and low production cost.

The chemical precipitation method mainly includes the ammonium salt precipitation and hydrolysis precipitation methods [22–28]. Xiong [29] studied a vanadium-containing solution with a vanadium concentration of about 18 g/L and a sodium concentration of about 23 g/L. Vanadium was precipitated by ammonium salt at a pH of 8.0, and the precipitation rate was 99.52%. The sodium impurity in the V_2O_5 product was high, leading to a purity of only 92.37%. At a pH of 2.0, reaction temperature of 90 °C, and reaction time of 60 min, the vanadium precipitation rate and product purity could exceed 99%. However, the ammonium salt precipitation method cannot separate ammonium and sodium, leading to the generation of a considerable amount of refractory ammonia–nitrogen wastewater, which pollutes the environment and affects the ecological balance [30–33]. The hydrolysis precipitation method is more environmentally friendly than the ammonium salt precipitation method, but it has a low vanadium precipitation rate and produces low-purity vanadium products, which is due to the sodium ion concentration in solution. Wu [34] studied the hydrolysis and precipitation of the leaching solution of sodium roasting of vanadium–chromium slag. The vanadium concentration of the solution was 17.86 g/L, and the sodium concentration was 54.80 g/L. At a reaction pH of 1.8, reaction temperature of 95 °C, and reaction time of 60 min, the vanadium precipitation rate reached 91.61%, and the sodium content in the vanadium products was approximately 7.71%. Meanwhile, the process of sodium ion concentration in the solution from 35.2 to 90 g/L was studied, and the vanadium precipitation rate showed an obvious downward trend. Evidently, an excessively high sodium concentration adversely affects both the vanadium precipitation rate and the product purity in the processes of alkaline ammonium salt precipitation and vanadium precipitation by hydrolysis.

Therefore, if an additive is added to the sodium vanadate solution to separate vanadium and sodium, it would be crucial to improve the vanadium precipitation rate and enhance the product purity during the vanadium precipitation by hydrolysis. Wen [35] found that manganese has a strong binding capacity with vanadium. Under specific roasting conditions, vanadium and manganese combine to form $Mn_2V_2O_7$, which dissolves in a sulfuric acid medium to yield a vanadium-containing solution, transforming vanadium from solid to liquid. This process establishes a foundation for the clean and efficient preparation of V_2O_5 products through the hydrolysis and precipitation of vanadium.

In this study, we used $NaVO_3$ solution instead of a leaching solution of sodium-roasted vanadium slag to eliminate the influence of other impurity ions. We studied the effects of vanadium extraction conditions on the vanadium extraction rate of manganese salt, acid dissolution conditions on vanadium dissolution rate, and vanadium precipitation conditions on vanadium precipitation rate. The vanadium precipitation product was washed to remove manganese and calcined to obtain V_2O_5, and the purity of the related products was analyzed.

2. Experiment

2.1. Experimental Materials

$NaVO_3$ (Macklin Bio-Chemical Co., Ltd., Shanghai, China, analytical purity > 99%) was used as the experimental raw material. $MnSO_4$ (Fucheng (Tianjin) Chemical Reagents Co., Ltd., Tianjin, China, analytical purity > 99%) was used for vanadium extraction by manganese salt. H_2SO_4 (Sinopharm Chemical Reagent Co., Ltd., Beijing, China, analytical purity > 98%) was the acid dissolution leaching medium and was also used to adjust the pH during vanadium precipitation by hydrolysis. $(NH_4)_2SO_4$ (Sinopharm Chemical Reagent Co., Ltd., Beijing, China, analytical purity > 99%) was used to wash and remove impurities in the product.

Other chemical reagents were used to titrate vanadium with standard ammonium sulfate. H_3PO_4 (Sinopharm Chemical Reagent Co., Ltd., Beijing, China, analytical purity > 85%) solution (50% volume fraction) was used to acidify the vanadium solution. $KMnO_4$ (Sinopharm Chemical Reagent Co., Ltd., analytical purity > 99%) solution (20 g/L) was used to oxidize vanadium ions. $NaNO_2$ (Sinopharm Chemical Reagent Co., Ltd., analytical purity > 99%) solution (20 g/L) and urea (Sinopharm Chemical Reagent Co., Ltd., analytical purity > 99%) solution (100 g/L) were used to reduce excess potassium permanganate. N-phenylanthranilic acid (Sinopharm Chemical Reagent Co., Ltd., analytical purity > 99%) solution (2 g/L) was used as a vanadium indicator. $(NH4)_2Fe(SO_4)_2$ (Sinopharm Chemical Reagent Co., Ltd., analytical purity > 99.5%) solution (50 g/L) was used to reduce vanadium and other high-valent ions, and $(NH_4)_2Fe(SO_4)_2$ standard titration solution (0.005 mol/L) was used to titrate vanadium until the titration ended.

2.2. *Experimental Process*

We prepared a $NaVO_3$ solution with a specific concentration and pH that are consistent with those of an actual vanadium slag sodium roasting leaching solution. NaOH solution was used to adjust the pH of the solution to 11 during the preparation.

2.2.1. Vanadium Extraction by Manganese Salt

A 30 mL aliquot of the $NaVO_3$ solution was placed in a conical flask, and a certain amount of $MnSO_4$ was added. According to the experimental conditions outlined in Table 1, the content was placed in a constant-temperature water bath for the reaction to occur. After the reaction was completed, the resultant mixture was vacuum-filtered. The solid obtained from this filtration step was designated as the vanadium extraction product by manganese salt, which was used to perform characterization analysis and the subsequent acid dissolution process. A specified volume of the filtrate was designated as the waste liquor from the vanadium extraction by manganese salt; this volume was used to calculate the rate.

Table 1. Experimental parameter design.

Parameter Setting	pH	T/°C	t/min	n(Mn)/n(V)
Vanadium extraction by manganese salt	11	20, 30, 50, 70, 90	5, 15, 30, 45, 60	0.5, 0.75, 1.0, 1.25, 1.5
Acid dissolution	1.9, 2.2, 2.5, 2.8, 3.1, 3.4, 3.7	60	5, 10, 20, 40, 60	—
Vanadium precipitation by hydrolysis	1.4, 1.6, 1.8, 2.0, 2.2	40, 60, 80, 90, 95	10, 30, 45, 60, 90	—

2.2.2. Acid Dissolution

We placed 2 g of the vanadium extraction product by manganese salt into a three-necked flask and added 20 mL of deionized water. In accordance with the reaction conditions outlined in Table 1, the three-necked flask was positioned within a constant-temperature water bath pot, and H_2SO_4 was gradually added in a dropwise manner to adjust the pH of the solution. Upon completion of the reaction, vacuum filtration was performed on the mixture to obtain a specific volume of liquid, which was designated as the manganese vanadate solution and subsequently used in the vanadium precipitation by hydrolysis. Furthermore, the insoluble solid was dried to a constant mass at 110 °C for 24 h, and the acid solubility of the product was calculated by determining the difference in solid mass before and after the reaction.

2.2.3. Vanadium Precipitation by Hydrolysis

We placed 40 mL of manganese vanadate solution, with a concentration of 17.29 g/L, into a three-necked flask. In accordance with the reaction conditions detailed in Table 1, the flask was positioned within a constant-temperature water bath to facilitate the reaction, and H_2SO_4 was gradually added to adjust the pH of the solution. Upon completion of the reaction, vacuum filtration was performed. The solid obtained was designated as the vanadium precipitation product by hydrolysis, which was subjected to characterization and calcination processes to yield the target product. The liquid obtained was designated as the waste liquor from vanadium precipitation by hydrolysis and used for calculating the rate of vanadium precipitation by hydrolysis.

The hydrolyzed vanadium precipitate was washed with $(NH_4)_2SO_4$ solution in a conical flask [36]. The product was calcined in a muffle furnace at 550 °C for 2 h to obtain the V_2O_5 product. The experimental flow chart of this study is shown in Figure 1. The vanadium content in the solution is calculated using Equation (1), while the vanadium precipitation or dissolution rate is calculated using Equation (2).

$$m = \frac{n \times C_V \times V_V \times 50.94}{1000} \tag{1}$$

$$L = \frac{m_1 - m_2}{m_1} \times 100\% \tag{2}$$

where m is the mass of vanadium in the solution, g; n is the volume of solution, mL; C_V is the concentration of $(NH_4)_2Fe(SO_4)_2$ standard titration solution, mol/L; V_V is the volume of $(NH_4)_2Fe(SO_4)_2$ standard titration solution consumed by 1 mL solution; 50.94 is the molar mass of vanadium, g/mol; L is the rate, %; m_1 is the mass of vanadium before precipitation or dissolution, g; m_2 the mass of vanadium after precipitation or dissolution, g.

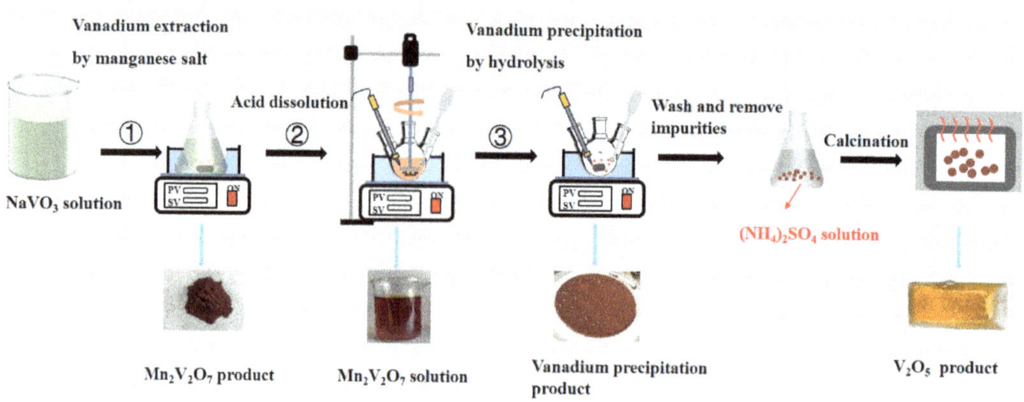

Figure 1. Experimental flowchart. (① Vanadium extraction by manganese salt; ② Acid dissolution; ③ Vanadium precipitation by hydrolysis).

2.3. Characterization

The concentration of vanadium in the liquid was determined through ammonium ferrous sulfate titration. The contents of manganese and vanadium in the solid samples were determined by inductively coupled plasma emission spectrometry (ICP-OES, Perkin Elmer Optima-4300 DV, Perkin Elmer Enterprise Management (Shanghai) Co., Ltd., Shanghai, China). The phases of the vanadium precipitation product obtained by manganese salt, vanadium precipitation products obtained by hydrolysis, and V_2O_5 were characterized using a powder X-ray diffraction system (XRD, X Pertpro, PANalytical B.V., Almelo, The Netherlands). The sodium content was quantified using an atomic absorption spectrophotometer (AAS, TAS-990, Beijing Purkinje general Instrument Co., Ltd., Beijing, China). The micro-morphology and distribution of the product were observed using a scanning electron microscope (Ultra Plus, Zeiss, Germany).

3. Experimental Results and Analysis

3.1. Vanadium Extraction by Manganese Salt

Figure 2a depicts the impact of various reaction temperatures on the rate of vanadium extraction by manganese salt at the n(Mn)/n(V) ratio of 1.5 and a reaction time of 30 min. The results indicate that the rate of vanadium extraction by manganese salt steadily increased as the reaction temperature rose from 20 °C to 70 °C. This is because vanadium extraction by manganese salt is an endothermic reaction [37], meaning a rise in temperature will increase the molecular kinetic energy and the number of activated molecules, thus speeding up the reaction. The extraction rate of vanadium was 99.73% at 70 °C. The reaction temperature rose until it reached 90 °C, and the vanadium extraction rate remained basically unchanged. This result indicates that a reaction temperature of 70 °C is optimal for this process. Thus, vanadium extraction by manganese salt has a very high efficiency, proving the feasibility of its application in vanadium purification. Figure 2b illustrates the impact of varying reaction times on the vanadium extraction rate using manganese salt, maintaining the n(Mn)/n(V) ratio at 1.5 at a reaction temperature of 70 °C. The results indicate a gradual increase in the vanadium extraction rate from 97.60% to 99.73% as the reaction time extends from 5 to 30 min. This result further indicates that vanadium extraction from manganese salt has a very high efficiency. Extending the reaction time to 60 min does not further enhance the rate of vanadium extraction, which accelerates notably within the first 15 min and stabilizing at approximately 30 min. This stability behavior suggests that the optimal reaction duration is approximately 30 min. Figure 2c depicts how various n(Mn)/n(V) ratios affect the vanadium extraction rate when using manganese salt at a reaction temperature of 70 °C and at a duration of 30 min. The results represent the n(Mn)/n(V) ratio from 0.5 to 1.0; the vanadium extraction rate increases, indicating that the amount of manganese is insufficient and the vanadium in the solution cannot be completely extracted. As the n(Mn)/n(V) ratio increases from 1.0 to 1.25, the extraction rate also increases from 93.20% to 99.75%. However, when the ratio of n(Mn)/n(V) is 1.5, no further increase in speed is obse n(Mn)/n(V)rved. Therefore, the optimal ratio of n(Mn)/n(V) is 1.25.

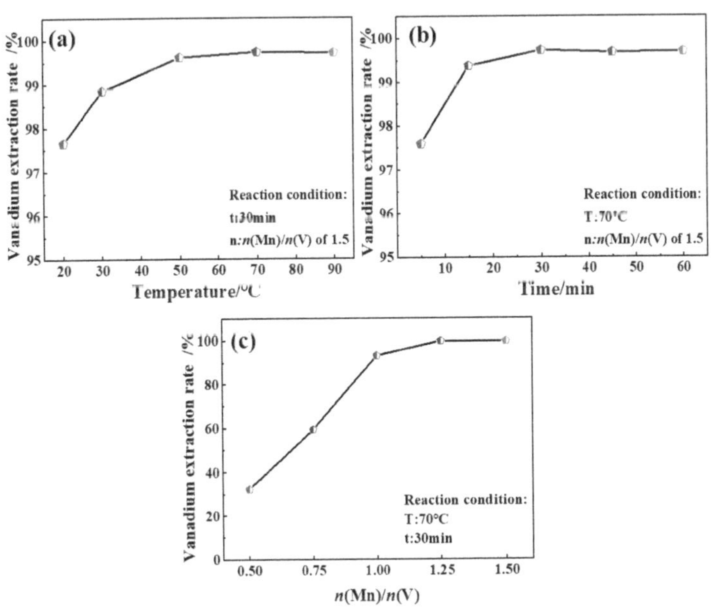

Figure 2. Effect of reaction conditions on rate of vanadium extraction by manganese salt. (**a**) Temperature; (**b**) Time; (**c**) n(Mn)/n(V).

The XRD diffraction peak of the vanadium extraction product obtained by manganese salt is depicted in Figure 3I, and the crystallinity of the product is poor. To improve the crystallinity and determine the phase composition of the product, the temperature was raised to 500 °C under an N_2 atmosphere and maintained at that level for 30 min. This process helps to avoid other reactions involving the manganese vanadate products [37]. The XRD diffraction peak of the product after high-temperature crystallization is shown in Figure 3II. It closely aligns with the standard JCPDS card number 00-022-0436, confirming that the product is $Mn_2V_2O_7$. No diffraction peaks of sodium-related compounds are observed in the figure: thus, vanadium remains in the solid phase in the form of $Mn_2V_2O_7$, whereas sodium remains in the liquid phase in the form of Na_2SO_4, verifying the separation of vanadium and sodium. The reaction mechanism for vanadium extraction by manganese salt is expressed in Equation (3).

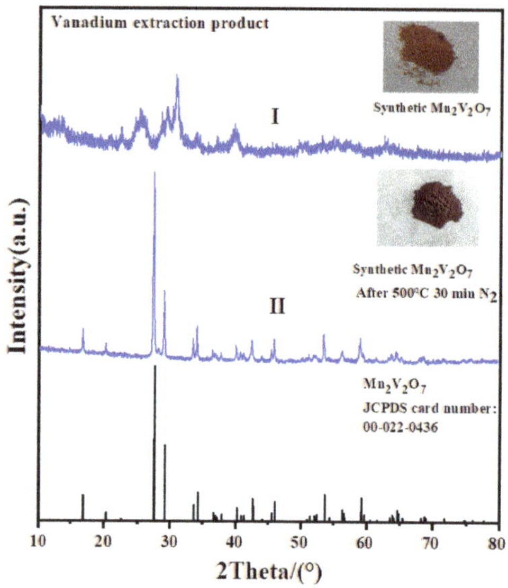

Figure 3. XRD characterization of vanadium extraction product by manganese salt.

Figure 4a,b show SEM images of vanadium extraction products by manganese salts at different magnifications. This product has a loose, flaky microstructure. Figure 4c,d depict the vanadium extraction products by manganese salts after being enhanced by crystallization at 500 °C in an N_2 atmosphere for 30 min. This process prevents other reactions between vanadium and manganese [38]. The vanadium precipitation product resulting from the manganese salt process has a compact, massive microstructure. A comparative analysis reveals that the product's degree of crystallization is significantly improved. Figure 4e shows that vanadium is distributed in various regions of the product. Figure 4f shows that manganese is distributed in all regions of the product, which is consistent with the distribution region of vanadium. Figure 4g shows that the distribution regions of oxygen components and vanadium–manganese components coincide. Figure 4h shows that the sodium component area is dark, with obvious changes in brightness observed. This result indicates that sodium mainly adheres to the product because of incomplete washing during the suction filtration process. Figure 4i shows the X-ray energy spectrum of the product, which indicates three main energy spectrum peaks of vanadium, manganese, and oxygen; the energy spectrum peaks of sodium are few, and the peak intensity is low. Figure 4j is the EDS analysis of point A in Figure 4b, which shows that the mass fractions and atomic percentages of vanadium, oxygen, and manganese are similar to those of the three elements in $Mn_2V_2O_7$; this observation is consistent with the XRD analysis results. Therefore, the product of vanadium precipitation by manganese salt is $Mn_2V_2O_7$, confirming the separation of vanadium and sodium.

$$2MnSO_4 + 2NaVO_3 + 2NaOH = Mn_2V_2O_7\downarrow + 2Na_2SO_4 + H_2O \quad (3)$$

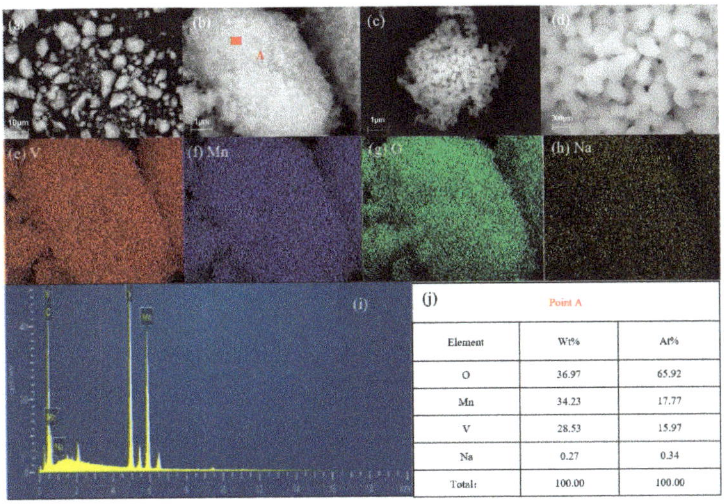

Figure 4. SEM and EDS characterization of vanadium extraction product by manganese salt. (**a**,**b**) Different magnification images of the product before crystallization; (**c**,**d**) Different magnification images of the product after crystallization; (**e**–**h**) elements mappings; (**i**,**j**) EDS analysis.

Table 2 describes the chemical composition of the product. The product shows a mass ratio $m(Mn)/m(V)$ of 1.055, which is equal to the mass ratio $m(Mn)/m(V)$ of $Mn_2V_2O_7$ of 1.058. The sodium content in the product is only 0.089%, indicating an efficient separation of vanadium and sodium. This extraction method effectively achieves the purification of vanadium in the $NaVO_3$ solution and also lays a foundation for the subsequent vanadium precipitation by hydrolysis.

Table 2. Chemical composition of vanadium extraction product by manganese salt (wt%).

Element	V	Mn	Na
content	30.8	32.5	0.089

3.2. *Acid Dissolution*

Figure 5 shows that the acid solubility of vanadium extraction products by manganese salt gradually decreases with a decrease in the pH value in the range of 1.9–2.5. This behavior occurs because as the pH decreases, the vanadium solution obtained through acid dissolution undergoes hydrolysis and precipitation, transforming vanadium from a liquid to a solid state. In the pH range of 2.8–3.4, the acid solubility of the vanadium extraction product by manganese salt is relatively stable above 99.60%. However, as the pH rises continuously, the time required for the pH of the solution to stabilize during acid dissolution (the time required from the start of the reaction to the slight changes in the pH of the solution) is prolonged. Meanwhile, the acid solubility of the vanadium extraction product is reduced at the same reaction time. For example, when the reaction time is 60 min, the acid solubility of the vanadium extraction product from manganese salt is only 81.32% at pH 3.7, which is because the leaching acidity decreases with the increase in pH, and the reaction rate decreases. Therefore, pH 2.8 is a suitable acid dissolution condition, considering that the vanadium solution will be hydrolyzed when the pH is too low. Moreover, the reaction rate will be reduced and the reaction time will be prolonged when the pH is too high. The acid dissolution mechanism of this product is expressed as Equation (4).

$$Mn_2V_2O_7 + 6H^+ = (VO_2)_2^{2+} + 2Mn^{2+} + 3H_2O \quad (4)$$

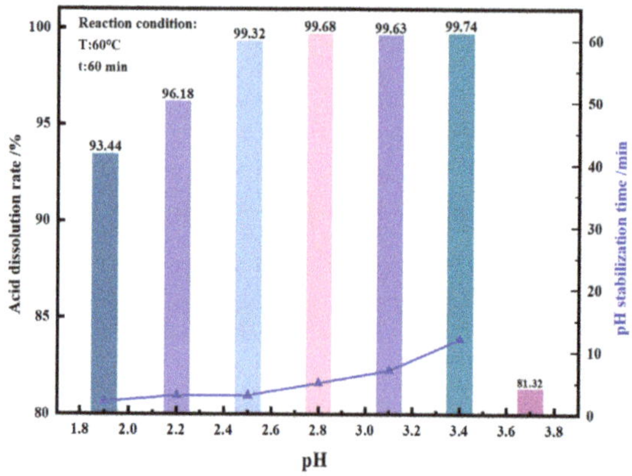

Figure 5. Effect of pH on acid dissolution rate of vanadium extraction product by manganese salt.

Figure 6 illustrates the effect of reaction time on the acid solubility of the vanadium extraction product by manganese salt. This investigation was conducted at a reaction temperature of 60 °C and a reaction pH of 2.8. Within a reaction time of 5 min, 99.84% of the product was dissolved in the acid solution, and when the reaction time was extended to 20 min, 99.95% of the product was dissolved in the acid solution. As the reaction time further increased, the acid solubility decreased slightly, and the product solubility decreased to 99.78% in 40 min and 99.68% in 60 min. This is because as the reaction time is extended, weak hydrolysis and precipitation reactions occur in the vanadium solution, resulting in the transformation of liquid manganese vanadate solution after acid dissolution into solid hydrolysis and precipitation products, and the acid solubility decreases. To sum up, if the reaction pH is too low or the reaction time is too long, vanadium will undergo hydrolysis and precipitation, reducing the acid solubility of the vanadium extraction products from manganese salts. Therefore, the optimal dissolution conditions are a pH of 2.8 and a reaction time of 20 min, under which 99.95% of the product is dissolved in the acid solution.

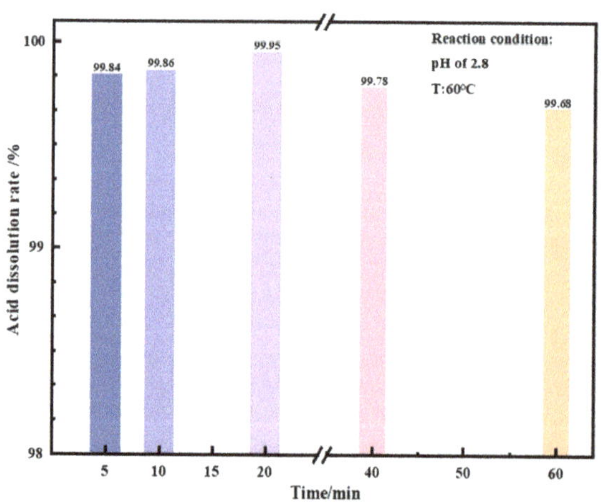

Figure 6. Effect of time on acid dissolution rate of vanadium extraction product by manganese salt.

3.3. Hydrolyzing and Precipitating Vanadium: Washing and Calcination

As depicted in Figure 7a, with the increase in pH, the rate of hydrolysis and precipitation of vanadium first increases and then decreases. At a pH of 2.0, the rate peaks at 97.87%. However, as the pH decreases to 1.6, the vanadium precipitation product begins to dissolve, decreasing the rate of vanadium precipitation by hydrolysis to 95.99%. When the pH is reduced to 1.2, a notable decrease is observed in the rate of vanadium precipitation by hydrolysis, reaching only 91.44%. This result shows that the vanadium precipitation product becomes soluble at high acidity. At a pH of 2.2, the hydrolysis precipitation rate of vanadium is only 72.51%, suggesting that a further increase in the pH becomes unfavorable for the hydrolysis precipitation of vanadium. It can be seen from Figure 7a that the separation of vanadium and manganese is relatively stable, and the change in pH has little effect on the concentration of manganese ions in the vanadium precipitation waste liquid. These manganese ions in the waste liquid can be recovered in the form of manganese sulfate by evaporation and crystallization. As shown in Figure 7b, when the reaction temperature rises from 40 °C to 80 °C, the precipitation rate of vanadium by hydrolysis increases significantly from 42.37% to 97.42%. From a reaction temperature of 80 °C to 95 °C, the hydrolysis and precipitation rate of vanadium increases from 97.42% to 97.87%, indicating that the hydrolysis and precipitation effect of vanadium is optimal in this temperature range. Further, with a rise in reaction temperature, the concentration of manganese ions in the waste liquid of vanadium precipitation increase slightly, indicating that more manganese ions could be recovered from the waste liquid. An evident increase is observed in the rate of vanadium precipitation by hydrolysis in Figure 7c, rising from 45.83% to 95.88% when the reaction time is extended from 10 min to 30 min. Extending the reaction time from 30 min to 60 min yields a relatively small increase in the rate of vanadium precipitation by hydrolysis, from 95.88% to 97.87%. However, as the reaction time is prolonged to 90 min, the rate of vanadium precipitation through hydrolysis decreases to 95.49%. This result shows that if the reaction time is too long, the precipitate obtained by hydrolysis will dissolve, decreasing the vanadium precipitation rate by hydrolysis. Meanwhile, Figure 7c shows that the reaction time has little effect on the concentration of manganese ions in the vanadium precipitation waste liquid. To sum up, the optimum conditions for vanadium precipitation by hydrolysis are a pH of 2.0, a reaction temperature of 95 °C, and a reaction time of 60 min.

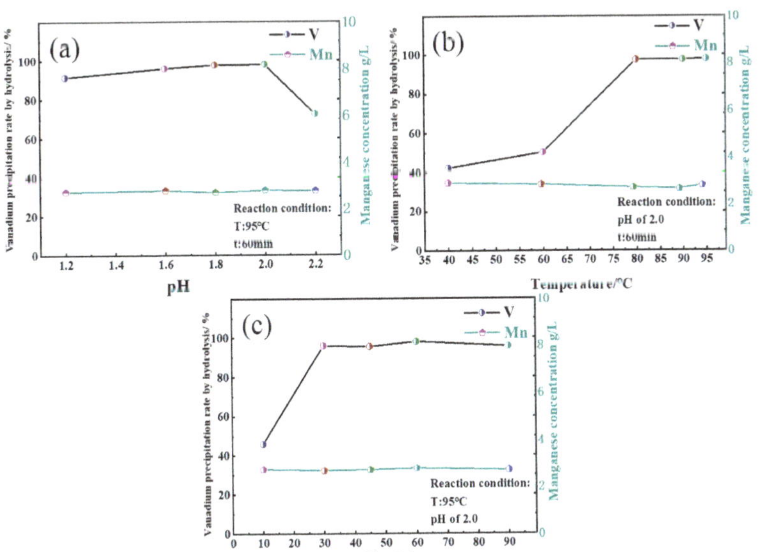

Figure 7. Effect of reaction conditions on vanadium precipitation rate by hydrolysis. (**a**) pH; (**b**) Temperature; (**c**) Time.

Figure 8a shows the XRD pattern of the product of hydrolysis and vanadium precipitation under optimum conditions: a pH of 2.0, reaction temperature of 95 °C, and reaction time of 60 min. The diffraction peak is consistent with the standard peak of JCPDS card number 00-047-0146, indicating that the product of vanadium precipitation by hydrolysis is a compound of $MnV_{12}O_{31}$ $10H_2O$. As observed in Figure 9, at a pH of 2.0, the vanadium solution of 0.3394 mol/L (17.29 g/L) mainly exists in the form of $H_2V_{10}O_{28}{}^{4-}$, and the hydrolysis and vanadium precipitation reactions are expressed in Equation (5). However, at this pH, the H^+ in the polymer is replaced by Mn^{2+}, as illustrated in Equation (6). The results show that under these reaction conditions, a small amount of Mn^{2+} will replace H^+ to participate in the hydrolysis and precipitation process of vanadium. Figure 8b shows the XRD pattern of the product obtained by calcining the hydrolyzed vanadium precipitation product at 550 °C for 2 h, and its diffraction peak is consistent with the standard JCPDS card number 01-077-2418. These results demonstrate that under these reaction conditions, a small amount of Mn^{2+} replaces H^+ to participate in the hydrolysis and precipitation process of vanadium. However, there are also low-intensity diffraction peaks corresponding to MnV_2O_6. This is because manganese in the hydrolysis precipitation product, after calcination, exists in the form of MnV_2O_6.

$$6H_2V_{10}O_{28}{}^{4-} + 24H^+ = 5H_2V_{12}O_{31}\downarrow + 13H_2O \quad (5)$$

$$H_2V_{12}O_{31} + Mn^{2+} = 5MnV_{12}O_{31} + 2H^+ \quad (6)$$

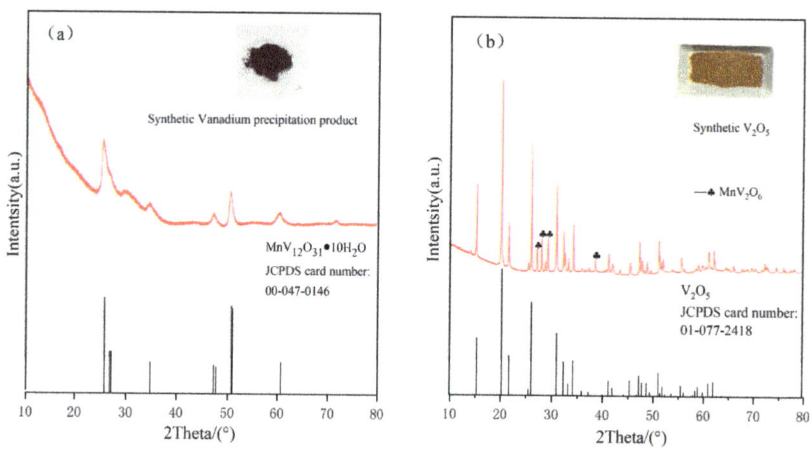

Figure 8. XRD images of the hydrolyzed product and calcined product. (**a**) Vanadium precipitation product; (**b**) V_2O_5.

Figure 10 shows SEM and EDS images of hydrolyzed vanadium precipitation products. Figure 10a,b are images of vanadium precipitation products at different magnifications, which reveal a network structure. Figure 10c,d show the surface scanning distribution of each element in the product. The product is mainly composed of vanadium, oxygen, and a small amount of manganese. Figure 10f shows the energy spectrum diagram of the product, demonstrating pronounced energy spectrum peaks for vanadium and oxygen, whereas those for manganese are comparatively weak. Figure 10g, on the other hand, shows the EDS results for point B in Figure 10b. The results indicate that the product is mainly composed of vanadium and oxygen, as well as a small amount of manganese, which is consistent with the XRD result.

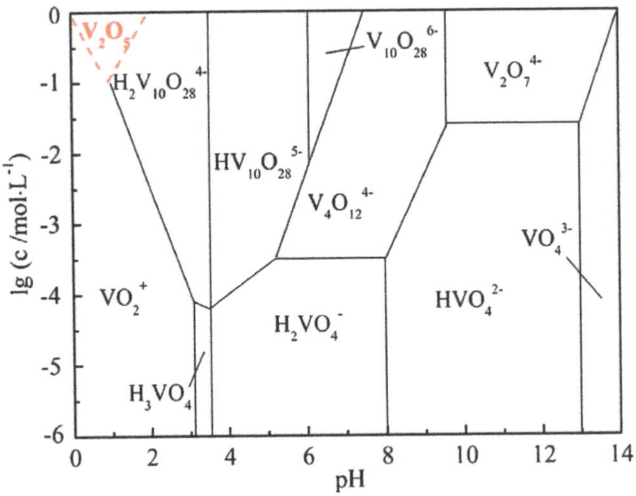

Figure 9. Existing form of vanadium at different pH values and concentrations [28].

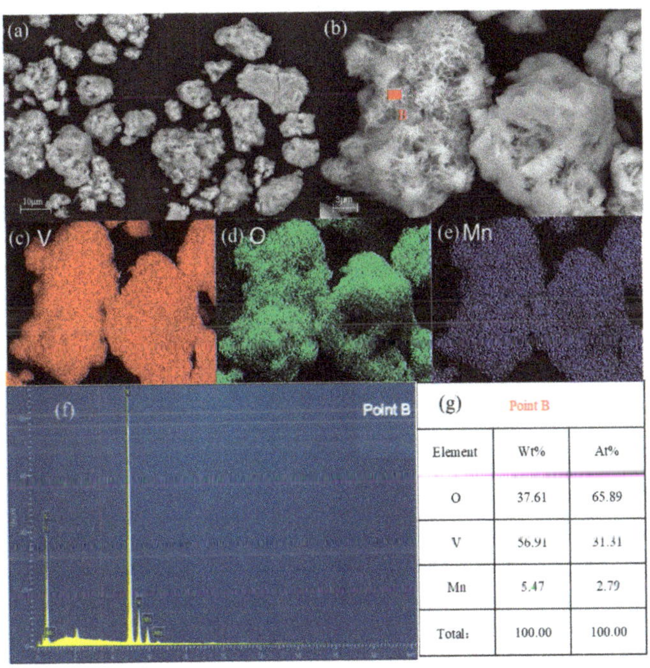

Figure 10. SEM and EDS images of a hydrolyzed product. (**a**,**b**) Different magnifications; (**c**–**e**) elements mappings; (**f**,**g**) EDS analysis.

Figure 11 presents the SEM and EDS diagrams of V_2O_5 obtained by calcining the hydrolysis vanadium precipitation product. Figure 11a,b V_2O_5 are images at different magnifications, which reveal a rod structure. Figure 11c,e show the surface scanning distribution of each element in the product. The images show that the product is primarily composed of vanadium and oxygen, including a small amount of manganese. Figure 11f shows the energy spectrum of the product, with obvious peaks of vanadium and oxygen, whereas the peaks of manganese are relatively weak. Figure 11g shows the EDS results of point C in Figure 11b. The product is mainly composed of vanadium and oxygen, including a small amount of manganese, which is consistent with the XRD results in Figure 8b.

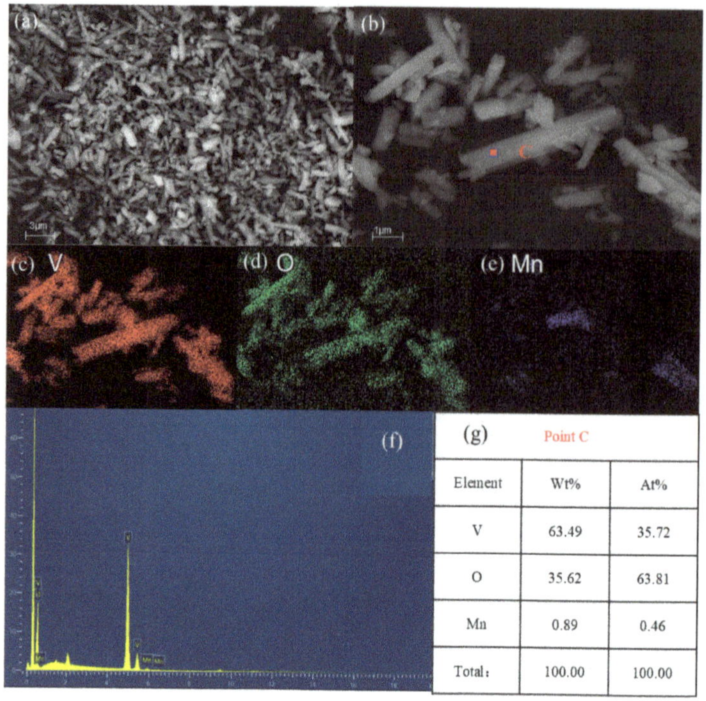

Figure 11. SEM and EDS images of a calcined product. **(a,b)** Different magnifications; **(c–e)** elements mappings; **(f,g)** EDS analysis.

Table 3 presents the chemical composition of V_2O_5 obtained by calcining hydrolysis and vanadium precipitation products. According to the previous analysis, manganese inevitably enters the hydrolysis solution during the vanadium precipitation by hydrolysis. Therefore, the V_2O_5 product contains 5.930% manganese and 0.048% sodium without washing with $(NH_4)_2SO_4$ solution, and the purity of the V_2O_5 product is 92.21%. At the $n(NH_4^+)/V_2O_5)$ ratio of 2, reaction temperature of 65 °C, reaction time of 5 min, and using the $(NH_4)_2SO_4$ solution to wash twice to hydrolyze and precipitate vanadium, the V_2O_5 product contains 0.818% manganese and 0.006% sodium; its purity after manganese removal is 98.92%. NH_4^+ can replace Mn^{2+} and Na+ in the product of vanadium precipitation by hydrolysis. The pH of the $(NH_4)_2SO_4$ solution after manganese removal is adjusted with ammonia water to precipitate Mn^{2+} as $Mn(OH)_2$, and it is further adjusted with sulfuric acid for recycling. This process greatly reduces the production of ammonia nitrogen wastewater compared with the ammonium salt, direct vanadium precipitation process.

Table 3. Chemical composition of products before and after impurity removal (wt%).

Sample	V_2O_5	Mn	Na
V_2O_5 before manganese removal	92.21	5.930	0.048
V_2O_5 after manganese removal	98.92	0.818	0.006

4. Conclusions

This study investigated the manganese salt pretreatment of $NaVO_3$ solution, which yields vanadium extraction product by manganese salt and separates vanadium from sodium. This process establishes a foundation for the subsequent vanadium precipitation by hydrolysis from the vanadium solution. The manganese vanadate products undergo acid dissolution, facilitating the transformation of vanadium from solid to liquid, resulting in a manganese vanadate solution. The vanadium solution then undergoes precipitation by hydrolysis, followed by impurity removal and calcination to obtain the target product, V_2O_5. Through theoretical analysis and characterization, we draw the following conclusions:

(1) The optimum technological conditions for extracting vanadium from manganese salt of the $NaVO_3$ solution are as follows: reaction temperature of 70 °C, reaction time of 30 min, and $n(Mn)/n(V)$ ratio of 1.25, with a 99.75% vanadium extraction efficiency. The final product of vanadium extraction mediated by manganese salt is $Mn_2V_2O_7$, which contains only 0.089% sodium, effectively completing the purification of vanadium.

(2) The most favorable conditions for the acid dissolution of $Mn_2V_2O_7$ products are a pH of 2.8 and reaction time of 20 min, yielding a dissolution rate of 99.95%. $Mn_2V_2O_7$ shows excellent solubility in a sulfuric acid medium.

(3) The optimum conditions for vanadium precipitation by hydrolysis are a pH of 2.0, reaction temperature of 95 °C, and reaction time of 60 min, yielding a vanadium precipitation rate of 97.87%. The purity of V_2O_5 after washing, manganese removal, and calcination reaches 98.92%. In addition, $MnSO_4$ can be recovered by evaporation and crystallization. Meanwhile, the $(NH_4)_2SO_4$ solution can also be recycled, greatly reducing the production of ammonia–nitrogen wastewater.

Author Contributions: M.L.: scheme design, experiment development, writing—original draft. T.J.: experimental adjustment, writing—revision. J.W.: experimental supervision, writing—review and editing. Z.F.: project help. T.Y.: experimental help. G.Y.: experimental assistance. S.X.: experimental help. H.X.: experimental help. All authors have read and agreed to the published version of the manuscript.

Funding: This work was financially supported by the National Natural Science Foundation of China (Nos. 52174277, 52374300, and 52204309) and the Opening Foundation of State Key Laboratory of Vanadium and Titanium Resources Comprehensive Utilization, China (No. 2022P4FZG11A).

Institutional Review Board Statement: Not applicable.

Informed Consent Statement: Not applicable.

Data Availability Statement: The original contributions presented in this study are included in the article. Further inquiries can be directed to the corresponding author.

Conflicts of Interest: The authors declare no conflict of interest.

References

1. Yang, Q.-Q.; Fang, Y.-W.; Mu, Y.-W.; Zhang, S.-H.; Zhao, Y.-S.; Li, Y.-G.; Liu, J.-T.; Wu, Z.-J. Summary of the application of vanadium. *Appl. Mech. Mater.* **2014**, *598*, 55–59. [CrossRef]
2. Gao, F.; Olayiwola, A.U.; Liu, B.; Wang, S.-N.; Du, H.; Li, J.-Z.; Wang, X.-D.; Chen, D.-H.; Zhang, Y. Review of vanadium production part I: Primary resources. *Miner. Process. Extr. Metall. Rev.* **2022**, *43*, 466–488. [CrossRef]
3. Capdevila, C.; Garcia-Mateo, C.; Chao, J.; Caballero, F.G. Advanced vanadium alloyed steel for heavy product applications. *Mater. Sci. Technol.* **2009**, *25*, 1383–1386. [CrossRef]
4. Cai, Z.; Mao, X.-P.; Bao, S.-Q.; Zhao, G.; Hu, X. Influence of vanadium microalloying on isothermal transformation behavior of eutectoid steel. *Phys. Met. Metallogr.* **2019**, *120*, 936–942.
5. Semiz, S. Vanadium as potential therapeutic agent for COVID-19: A focus on its antiviral, antiinflamatory, and antihyperglycemic effects. *Emir. J. Trace Elem. Med. Biol.* **2022**, *69*, 126887. [CrossRef]
6. Mukherjee, R.; Donnay, E.G.; Radomski, M.A.; Miller, C.; Redfern, D.A.; Gericke, A.; Damron, D.S.; Brasch, N.E. Vanadium-vitamin B (12) bioconjugates as potential therapeutics for treating diabetes. *Chem. Commun.* **2008**, *238*, 3783–3785. [CrossRef]

7. Amante, C.I.; Aureliano, M.; De Sousa-Coelho, A.L. Vanadium compounds therapeutic effects in melanoma. *Eur. J. Public Health* **2021**, *11*, 828. [CrossRef]
8. Chen, D.; Lu, M.-J.; Wang, B.-R.; Chai, R.-Q.; Li, L.; Cai, D.; Yang, H.; Liu, B.-K.; Zhang, Y.-P.; Han, W. Uncover the mystery of high-performance aqueous zinc-ion batteries constructed by oxygen-doped vanadium nitride cathode: Cationic conversion reaction works. *Energy Storage Mater.* **2021**, *35*, 679–686. [CrossRef]
9. Wan, J.; Wang, C.; Tang, Q.; Gu, X.; He, M.-Q. First-principles study of vanadium carbides as electrocatalysts for hydrogen and oxygen evolution reactions. *RSC Adv.* **2019**, *9*, 37467–37473. [CrossRef]
10. Shao, S.-X.; Liu, B.-Y.; Zhang, M.; Yin, J.-L.; Gao, Y.-Y.; Ye, K.; Wang, G.-L.; Zhu, K.; Cao, D.-X. Synthesis and electrochemical performance of $LiVO_3$ anode materials for full vanadium-based lithium-ion batteries. *J. Energy Storage* **2021**, *35*, 102254. [CrossRef]
11. Li, H.-Y.; Fang, H.-X.; Wang, K.-Z.; Zhou, W.; Yang, Z.; Yan, X.-M.; Ge, W.-S.; Li, Q.-W.; Xie, B. Asynchronous extraction of vanadium and chromium from vanadium slag by stepwise sodium roasting-water leaching. *Hydrometallurgy* **2015**, *156*, 123–135. [CrossRef]
12. Yu, T.-X.; Jiang, T.; Wen, J.; Sun, H.-Y.; Li, M.; Peng, Y. Effect of chemical composition on the element distribution, phase composition and calcification roasting process of vanadium slag. *Int. J. Miner. Metall. Mater.* **2022**, *29*, 2144–2151. [CrossRef]
13. Wen, J.; Jiang, T.; Yu, T.-X.; Chen, B.-J.; Li, L. Clean and efficient extraction of vanadium from vanadium slag: Effect of manganese on the phase composition and vanadium extraction process. *J. Clean. Prod.* **2022**, *367*, 133077. [CrossRef]
14. Qin, X.-M. Study on decomposition mechanism of vanadium slag by Na_2CO_3 Roasting. *Chem. Ind.* **2021**, *50*, 15–16.
15. Li, W.; Zheng, H.-Y.; Shen, F.-M. Effect of roasting characteristics of vanadium-rich slag on its vanadium leaching behavior. *Met. Mater. Soc.* **2019**, *71*, 3190–3195. [CrossRef]
16. Li, C.-Q.; Zhang, H.-C.; Tao, M.; Wang, X.-F.; Li, H.; Li, Y.-G.; Tian, Y. Efficient Separation of silicon and vanadium by sodium roasting-water leaching from vanadium slag and CaV_2O_6 preparation. *Crystals* **2021**, *11*, 255. [CrossRef]
17. Yan, H.; Du, H.; Wang, S.-N.; Zheng, S.-L.; Zhang, Y. Solubility data in the ternary NH_4HCO_3-NH_4VO_3-H_2O and $(NH_4)_2CO_3$-NH_4VO_3-H_2O systems at (40 and 70 °C). *J. Chem. Eng. Data* **2016**, *61*, 2346–2352. [CrossRef]
18. Feng, M.; Wenzel, M.; Wang, S.-N.; Du, H.; Zhang, Y.; Weigand, J.J. Separation of Na_3VO_4 and Na_2CrO_4 from high alkalinity solutions by solvent extraction. *Sep. Purif. Technol.* **2020**, *255*, 117282. [CrossRef]
19. Guo, X.-M.; Wang, S.-N.; Du, H.; Feng, M.; Zheng, S.-L.; Zhang, Y. Cooling crystallization of ammonium metavanadate from ammonium bicarbonate solution. *Chem. Ind. Eng. Prog.* **2018**, *37*, 853–860.
20. Zeng, L.; Li, Q.-G.; Xiao, L.-S.; Zhang, Q.-X. A study of the vanadium species in an acid leach solution of stone coal using ion exchange resin. *Hydrometallurgy* **2010**, *105*, 176–178. [CrossRef]
21. Zhang, Y.; Zhang, T.-A.; Lv, G.-Z.; Zhang, G.-Q.; Liu, Y.; Zhang, W.-G. Synergistic extraction of vanadium (IV) in sulfuric acid media using a mixture of D2EHPA and EHEHPA. *Hydrometallurgy* **2016**, *166*, 87–93. [CrossRef]
22. Zhang, J.-H.; Zhang, W.; Zhang, L.; Gu, S.-Q. A critical review of technology for selective recovery of vanadium from leaching solution in V_2O_5 production. *Solvent Extr. Ion Exch.* **2014**, *32*, 221–248. [CrossRef]
23. Rong, Y.-Y.; Ye, G.-H.; Kang, X.-X.; Zhu, S.-Q.; Liang, X.-Y.; Xiang, X.-Y.; Zhang, Y.; Song, C.-X. Research progress in preparation of V_2O_5 from vanadium-rich solution. *Min. Metall.* **2023**, *32*, 78–86.
24. Wu, Z.-X. Research status of preparation of vanadium pentoxide from vanadium-containing solution. *Iron Steel Vanadium Titan.* **2023**, *44*, 9–19.
25. Li, M.; Jiang, L.; Wu, Z.-X. Investigation on separation of vanadium and chromium in alkaline solution and preparation of V_2O_5 and Cr_2O_3 products. *Iron Steel Vanadium Titan.* **2018**, *39*, 30–35+47.
26. Wang, M.-Y.; Chen, B.-F.; Huang, S.; Wang, X.-W.; Liu, B.; Ge, Q.; Xie, S.-S. A novel technology for vanadium and chromium recovery from V-Cr-bearing reducing slag. *Hydrometallurgy* **2017**, *171*, 116–122. [CrossRef]
27. Xiong, P.; Zhang, Y.-M.; Huang, J.; Bao, S.-X.; Yang, X.; Shen, C. High-efficient and selective extraction of vanadium (V) with N235-P507 synergistic extraction system. *Chem. Eng. Res. Des.* **2017**, *120*, 284–290. [CrossRef]
28. Jiang, L.; He, W.; Du, G.; Zheng, H.; Peng, Y. Implication of hydrolysis on vanadium precipitation with acidic ammonium salt from high concentration of alkaline vanadium solution. *Korean Inst. Chem. Eng.* **2023**, *40*, 2513–2519. [CrossRef]
29. Xiong, P.; Zhang, Y.-M.; Bao, S.-X.; Huang, J. Precipitation of vanadium using ammonium salt in alkaline and acidic media and the effect of sodium and phosphorus. *Hydrometallurgy* **2018**, *180*, 113–120. [CrossRef]
30. Chen, B.; Bao, S.-X.; Zhang, Y.-M.; Ren, L.-Y. A novel and sustainable technique to precipitate vanadium from vanadium-rich solutions via efficient ultrasound irradiation. *J. Clean. Prod.* **2022**, *339*, 130755. [CrossRef]
31. Zhang, J.-H.; Zhang, W.; Xue, Z.-L. An environment-friendly process featuring calcified roasting and precipitation purification to prepare vanadium pentoxide from the converter vanadium slag. *Metals* **2018**, *9*, 21. [CrossRef]
32. Wen, J.; Jiang, T.; Xu, Y.-Z.; Liu, J.-Y.; Xue, X.-X. Efficient separation and extraction of vanadium and chromium in high chromium vanadium slag by selective two-stage roasting-leaching. *Metall. Mater. Trans. B* **2018**, *49*, 1471–1481. [CrossRef]
33. Xu, Z.-K.; Cao, J.; Qin, X.-M.; Qiu, W.-Q.; Mei, J.; Xie, J. Toxic effects on bioaccumulation, hematological parameters, oxidative stress, immune responses and tissue structure in fish exposed to ammonia nitrogen: A review. *Animals* **2021**, *11*, 3304. [CrossRef] [PubMed]
34. Wu, Z.-X.; Jiang, L. Study on vanadium precipitation by hydrolysis of chromium-vanadium solution. *Iron Steel Vanadium Titan.* **2020**, *41*, 22–26.

35. Wen, J.; Jiang, T.; Sun, H.-Y.; Yu, T.-X. Novel understanding of simultaneous extraction of vanadium and manganese from vanadium slag and low-grade pyrolusite based on selective oxidation-reduction roasting. *Acs Sustain. Chem. Eng.* **2020**, *8*, 5927–5936. [CrossRef]
36. Lin, G.-C. Study on Technique to Prepare Vanadium Pentoxide via Dynamic Continuous Purification. Master's Thesis, Wuhan University of Science and Technology, Wuhan, China, 2022.
37. Fu, Z.-B.; Jiang, L.; Rao, Y.-Z. Separation of Vanadium and Chromium from Vanadium-chromium Solution by Calcium Salt Precipitation. *Iron Steel Vanadium Titan.* **2020**, *41*, 5.
38. Wen, J.; Jiang, T.; Wang, J.-P.; Lu, L.-G.; Sun, H.-Y. Cleaner extraction of vanadium from vanadium-chromium slag based on MnO_2 roasting and manganese recycle. *J. Clean. Prod.* **2020**, *261*, 121205. [CrossRef]

Disclaimer/Publisher's Note: The statements, opinions and data contained in all publications are solely those of the individual author(s) and contributor(s) and not of MDPI and/or the editor(s). MDPI and/or the editor(s) disclaim responsibility for any injury to people or property resulting from any ideas, methods, instructions or products referred to in the content.

materials

MDPI

Article

Intelligent Optimization and Impact Analysis of Energy Efficiency and Carbon Reduction in the High-Temperature Sintered Ore Production Process

Yuxing Yuan [1,2,3], Jingchao Sun [1,2,3], Lei Zhang [1,2,3], Su Yan [1,2,3], Tao Du [1,2,3,*] and Hongming Na [1,2,3,*]

1 SEP Key Laboratory of Eco-Industry, Northeastern University, Shenyang 110819, China
2 Engineering Research Center of Frontier Technologies for Low-Carbon Steelmaking, Ministry of Education, Northeastern University, Shenyang 110819, China
3 Frontiers Science Center for Industrial Intelligence and Systems Optimization, Northeastern University, Shenyang 110819, China
* Correspondence: dut@smm.neu.edu.cn (T.D.); nahm@mail.neu.edu.cn (H.N.)

Abstract: The coordinated optimization of energy conservation, efficiency improvement, and pollution reduction in the sintering production process is vital for the efficient and sustainable development of the sintering department. However, previous studies have shown shortcomings in the multi-objective collaborative optimization of sintering systems and the quantification of pollutant impacts. To address these, this paper proposes a multi-objective optimization method integrated with the NSGA-III algorithm and establishes an integrated system optimization model for sintered ore production and high-temperature waste heat recovery. The results demonstrate significant improvements: energy utilization efficiency increased by 0.67%, energy consumption decreased by 17.3 MJ/t, production costs were reduced by 11.45 CNY/t, and the emissions of CO_2, SO_2, and NO_x were reduced by 0.464 kg/t, 0.034 kg/t, and 0.008 kg/t, respectively. Additionally, the study identified optimal configuration parameters and analyzed the quantitative impact of several key factors on multiple indicators. The results also show that reducing the water content of the mixture, decreasing the middling coal content in the fuel, and increasing the thickness of the material layer are effective strategies to reduce energy consumption and pollutant emissions in the sintering process. Overall, implementing these optimizations can bring significant economic and environmental benefits to the steel industry.

Keywords: sintering process; multi-objective optimization; energy saving; improving efficiency; reducing pollution emissions

1. Introduction

As a typical representative of heavy industry, the steel industry has continually faced challenges such as high energy consumption, significant pollution, and economic fluctuations. In 2022, global steel production reached 1.890 billion tons, with energy consumption per ton of steel exceeding 20.99 GJ [1]. In the complex production chain of steel, sintered ore is the most important raw material in the ironmaking process, and its production scale has exceeded 1.078 billion tons in 2022 [2]. Notably, it is worth stating that the sintering process accounts for the second-largest proportion of energy consumption and carbon emissions in the steel manufacturing process, accounting for 6–10% and 5.7–7.8% [1,2], respectively. Sintering is also the process with the highest pollutant emissions in the entire steel smelting process, with SO_2, NOx, and particulate matter accounting for 64.84%, 78.72%, and 56.40% of the total emissions in the blast furnace-converter steelmaking route [3,4], respectively. Therefore, optimizing sintering operations and process parameters to achieve high efficiency, energy conservation, environmental protection, and cost reduction has become crucial for enhancing the competitiveness and sustainable development of steel enterprises.

At present, significant progress has been made in energy conservation and pollution reduction in the sintering process within the steel industry. These advancements can be roughly divided into the following categories. Firstly, fuel substitution involves using cleaner and more efficient energy sources [5,6], such as semi-coke, biomass, and natural gas, to replace coke powder and coal [7]. Secondly, efficient combustion and material ratio optimization involve the rational combination of fuel components [8], the optimization of low-carbon batching schemes, and the precise monitoring and control of key parameters. Thirdly, comprehensive treatment of air leakage and air volume control technology for large-scale sintering machines and systems has been adopted, reducing the air leakage rate to below 35% [2]. Fourthly, there is full recovery and use of the waste heat generated during the sintering process [9–11]. This involves constructing and optimizing systems to recover waste heat from flue gases and sensible heat from sintered ore [12]. Fifth, flue gas purification and resource recycling have reduced emissions of SO_2, NO_X, and particulate matter by nearly 80% and have increased the recycling of iron-containing waste materials [6,13]. These technological advancements have resulted in significant energy savings and a substantial reduction in environmental impact.

Among these advancements, optimization plays a crucial role in industrial process control and technological development. Previous studies have found that optimization research for sintering primarily focuses on two aspects: optimization of the sintering thermal process and optimization of waste heat recovery. In terms of the optimization of the thermal process, Liu et al. [14,15] proposed a sintering ore ratio optimization model based on mathematical programming, which improved the energy-saving and emission reduction effect by 4.92%. Yuan et al. [16] achieved a cost reduction of 8.46% in the overall optimization of the steel manufacturing process while meeting the requirements for ore blending in the production process. In addition, when aiming for a comprehensive carbon ratio in the sintering process, Hu et al. [17] developed an integrated prediction model for the comprehensive coke ratio in multiple modes for real-time predicting and optimizing of carbon efficiency during the process [18]. Chen et al. [19] used backpropagation neural network (BPNN) modeling and the particle swarm optimization (PSO) algorithm to reduce the comprehensive carbon ratio by 2.40 kg/t.

On the other hand, in optimizing the recovery of high-temperature sintered ore waste heat, Feng et al. [20] aimed to maximize the exercise output. They determined the optimal fabric height of the sintered ore layer in the circular cooler, resulting in a 22.80% increase in the exercise output. When optimizing the waste heat recovery and exergy destruction quantity of the sintering vertical cooling furnace, they increased the waste heat recovery rate to 83.02% by determining the optimal structural and operating parameters [21]. Under the optimal temperature parameter combination, the recovery rate of the low-grade sinter cooling flux gas reached 64.86% [22]. Tian et al. [23] quantified the influence of sintering cooler operating parameters on indicators such as waste heat recovery rate and final sintering temperature. They used the Non-Dominated Sorting Genetic Algorithm II (NSGA-II) to solve the multi-objective optimization model and identify the optimal operating conditions. Liu et al. [24] also used NSGA-II to achieve the dual objectives of waste heat utilization and cost for sintered ore cooling beds, and found the optimal operating conditions of air flow rate to be about 400 kg/s, the cooling bed to be 0.8 m to 1.0 m, and the moving speed to be 0.020 m to 0.022 m. Overall, previous studies have mainly optimized processes [25] and key operating parameters [26,27], such as operating speed, flow rate, temperature, time, pressure, and atmosphere, which can effectively improve energy utilization and recovery rates. However, these studies have primarily focused on optimizing either the thermal process or waste heat recovery individually. There is a lack of comprehensive models that integrate the sintering thermal process with waste heat recovery into a unified system. Additionally, there is insufficient collaborative research addressing the interplay between energy efficiency, cost, energy consumption, and pollutants. The analysis of how sintering process operations and parameter optimization affect pollutant emissions is also lacking, particularly concerning the key factors influencing the synergy between energy conservation and pollutant emission in multi-objective integrated optimization. Therefore, this study addresses these gaps by developing a comprehensive model, conducting multi-objective optimization, and analyzing parameter impacts to advance the coordinated improvement of sintering processes in terms of energy conservation, environmental protection, and efficiency enhancement.

Therefore, based on the intricate principles of metallurgical chemistry and the process mechanisms involved in sintered ore formation, this study establishes a comprehensive system model for both sintered ore production and high-temperature waste heat recovery. The Non-Dominated Sorting Genetic Algorithm III (NSGA-III) optimization algorithm is utilized to perform multi-objective optimization targeting energy efficiency, energy consumption, and economic costs in the sintering process. This approach identifies the Pareto front and determines the optimal solutions for balancing quality, energy conservation, and cost objectives. Additionally, it derives the optimal operating conditions for resource allocation and analyzes the quantitative effects of various key factors on multiple objectives and pollutant emissions throughout the production process. This paper is organized as follows. Section 2 provides the explanation of the methodology, Section 3 describes the optimization algorithms and data sources, Section 4 analyzes and discusses the multi-objective optimization results of the sintering plant, including a comparison of evaluation indicators before and after optimization and an examination of various influencing factors, and Section 5 draws conclusions based on the above analysis and discussion.

2. Materials and Methods

2.1. Research Boundary

This study takes the sintering plant (SP) of a steel enterprise as the research subject, The specific materials and processes are illustrated in Figure 1, which includes two main components: the high-temperature sintering process and the high-temperature waste heat recovery from sintered ore. The sintering plant involves the combination of ore and auxiliary materials at high temperatures, generated by fuel combustion, to form a solid mass. The equipment used in this process includes silos, mixers, igniters, sintering machines, crushers, fans, and flues. The sensible heat recovery process involves a blower directing cooling air onto the ring cooling bed. This cold air undergoes multi-stage gas-solid heat exchange with the high-temperature sintered ore, absorbing heat from the sintered ore. The heated air is then collected by hoods on top of the sintered ore cooling bed and recycled through various heat exchangers. The equipment used in this process includes ring coolers, hoods, pumps, fans, and heat exchangers.

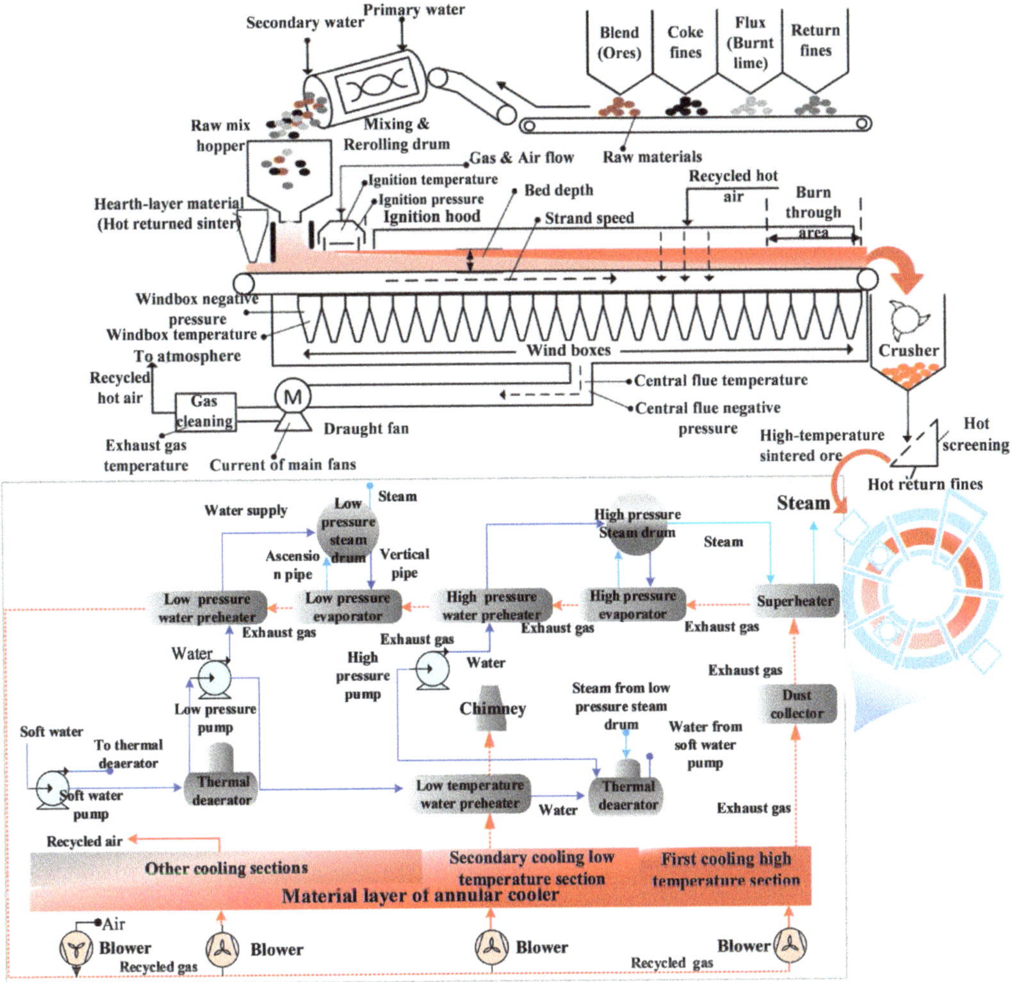

Figure 1. Sintering process and equipment employed in the production cycle at the SP [15,28].

2.2. Model Establishment

The sintering thermal process model involves the coupling of multiple pieces of field information, including mass, heat, momentum, and chemical reactions. Firstly, there is the material conversion process. This process describes the mass energy relationship of SP based on various physical and chemical reaction mechanisms, using the principles of conservation of matter and conservation of energy. The equilibrium relationships of each element in the model can be summarized as follows:

$$\sum_i \left(m_{SP,i,in} \times [X]_{SP,i,in} \right) = \sum_j \left(m_{SP,j,out} \times [X]_{SP,j,out} \right) \quad (1)$$

where the left input item includes various sintering powders, quicklime, limestone, dolomite, return ore, iron slag, coke powder, coal powder, gas, etc. [X] is the composition of element X in the material or energy, and the components such as Fe, FeO, CaO, MgO, SiO_2, Al_2O_3, P, and S in the material have the same equilibrium relationship. The products or by-products of the sintering process include sintered ore, return ore, furnace dust, flue gas, etc.

The ingredients are meticulously produced to meet the specified alkalinity requirements. The required range for the ternary alkalinity of sintered ore is 1.75 to 2.10. The calculation is as follows:

$$R = \frac{\sum_I m_{SP,i,in} \times [\mathrm{CaO}]_{SP,i,in} + m_{SP,i,in} \times [\mathrm{MgO}]_{SP,i,in}}{\sum_I \left(m_{SP,i,in} \times [\mathrm{SiO_2}]_{SP,i,in} \right)} \quad (2)$$

Secondly, the energy consumed and released during the material conversion process, along with the flow and loss of energy, are primarily constrained by the first law of thermodynamics. The overall heat balance equation is as follows:

$$\begin{aligned} &Q_{SP,sen,in} + Q_{SP,oxidi,in} + Q_{SP,air,in} + Q_{SP,circul,in} \\ &\quad = Q_{SP,sen,out} + Q_{SP,decom,out} + Q_{SP,circul,out} + Q_{SP,gas,out} \\ &\quad + Q_{SP,incom,out} + Q_{SP,loss} \end{aligned} \quad (3)$$

where the input heats are the sensible heat of various materials, the heat released by the oxidation of fuel and iron-containing minerals, and the heat brought in by circulation. The outputs are product and by-product sensible heat, reaction decomposition heat, cycle heat, flue gas heat, incomplete combustion loss, and heat loss.

The calculation formulas for the sensible heat carried by materials and gases are:

$$Q_{m,i} = \int m_i \times c_{m,i} \times dt \quad (4)$$

$$Q_{v,i} = \int V_i \times c_{v,i} \times dt \quad (5)$$

The reactions in the sintering process include water evaporation and condensation, carbon combustion, carbonate decomposition, iron oxide decomposition, oxidation and reduction reactions, removal of harmful substances such as sulfur and potassium, and consolidation reactions. The main reactions and reaction heats involved in this model are shown in Table 1.

$$Q_r = \sum_r \left(\sum_i \left(m_{SP,i,in} \times [X]_{SP,i,in} \right) - \sum_j \left(m_{SP,j,out} \times [X]_{SP,j,out} \right) \right) / M_r \times \Delta H_r \quad (6)$$

Table 1. The main chemical reactions and reaction enthalpy reactions in SP.

Reaction Types	Reaction	ΔH/kJ·mol^{-1}
Oxidation reaction	$C + 0.5O_2 = CO$	−110.5
	$C + O_2 = CO_2$	−393.5
	$2Fe_3O_4 + 0.5O_2 = 3Fe_2O_3$	−235.8
	$3FeO + 0.5O_2 = 3Fe_3O_4$	−302.4
	$4FeS_2 + 8O_2 = 2Fe_2O_3 + 6SO_2$	−3310
Reduction reaction	$CO_2 + 2C = 2CO$	+172.5
	$3Fe_2O_3 + CO = 2Fe_3O_4 + CO_2$	−47.2
	$Fe_3O_4 + CO = 3FeO + CO_2$	+19.4
Decomposition reaction	$MeCO_3 = MeO + CO_2$	$FeCO_3$(+85.1), $CaCO_3$(+178.6), $MgCO_3$(+120.9)
	$MeSO_4 = MeO + SO_2 + 0.5O_2$	$FeSO_4$(+340.4), $CaSO_4$(+500.6), $MgSO_4$(+221.6)

After the sintered ore is flipped, the sensible heat of the sintered ore is recovered by a circular cooler. The circulating hot air, after heat exchange, enters the waste heat boiler to generate steam.

$$\begin{aligned} Q_{SP,SCR} &= a_{air} \cdot \eta_{SCR} \cdot m_{SP,sinter,out} \left(\int_{T_{SCR,sinter,out}}^{T_{SCR,sinter,in}} c_{SP,sinter}(T) dT \right) \\ &= \left(\sum S_{SP,s,out} \times h_s \right) \end{aligned} \quad (7)$$

where a_{air} is the comprehensive heat transfer coefficient between sintered ore and air, and η_{SRC} is the efficiency of the waste heat boiler of the ring cooler, %. $T_{SRC,sinter,in}$, and $T_{SRC,sinter,out}$ refer to the temperatures of the sintered ore entering and exiting the sintering ring cooler, respectively, K. Additionally, $S_{SP,s,out}$ and h_s are the s-th steam extraction amount and corresponding enthalpy values, kg/t and kJ/kg, respectively.

2.3. *Evaluation Indexes*

2.3.1. Heat Utilization Efficiency (HUE)

The definition and classification of energy efficiency quantitatively represents the effective energy supply level of thermal equipment for specific purposes [29,30], as defined in Equation (1). Energy efficiency is calculated as the ratio of the effective utilization of energy to the energy input during the manufacturing process.

$$\eta_{SP} = \frac{\sum Q_{useful}}{\sum_i Q_{i,in}} \tag{8}$$

where Q_{useful} is the useful heat required for the material reaction process, the heat carried by the product, and some by-products, kJ. $\sum_i Q_{i,in}$ is the total energy input, which is the heat brought into the equipment by the material and energy and the heat released by the oxidation reaction, kJ. Effective heat refers to the heat required for the material reaction process.

2.3.2. Energy Consumption (EC)

When studying energy conservation issues in steel enterprises or processes, the primary focus is on the energy consumption per ton of physical production. This is defined as the difference between the energy consumed per unit of the product process and auxiliary production during the statistical period and the energy recovered and supplied externally [31,32]. The specific process of energy consumption is as follows:

$$e_{SP} = \frac{\sum_i m_{i,in} \times \varepsilon_{i,in} - \sum_k r_k \times \varepsilon_k}{P} \tag{9}$$

where e_{SP} is the energy consumption of the SP, MJ/t. $m_{i,in}$ is the amount of i-th energy consumption in the SP, kg/t, m^3/t.

2.3.3. Production Costs (PC)

The production process cost index calculates the production cost of each process and procedure based on the different sources of raw materials, solvents, energy, and power, considering both the purchase price and some internal prices. The production cost of each process product includes material cost, energy cost, maintenance cost, labor cost, depreciation cost, and energy recovery income. The specific details are as follows:

$$C_{cost} = \sum_i (c_i \times m_{i,in}) + \left(c_{main} + c_{labor} + c_{dep}\right) - \sum_k (c_k \times r_k) \tag{10}$$

where C_{cost} is the production cost per ton of sintered ore products, CNY/t. c_i is the price coefficient of the i-th material or energy, CNY/t/kg, CNY/t/m^3, or CNY/t/kWh; c_{main}, c_{labor}, and c_{dep} are the unit product maintenance, labor, and depreciation price coefficients, respectively, CNY/t. c_k is the price coefficient for the k-th energy recovered in the process, CNY/t/kg, CNY/t/m^3.

2.3.4. Pollutants and Carbon Emissions

The accounting of pollutant and carbon emissions in SP is as follows:

$$E_{SO_2} = (1 - \beta_{SO_2}) \sum \left(m_{SP,i,in} \times [S]_{SP,i,in} - m_{SP,j,out} \times [S]_{SP,j,out}\right) \times O_{SP,i} \tag{11}$$

$$E_{NO_x} = (1 - \beta_{NO_x}) \sum_{I} \left(m_{SP,i,in} \times [N]_{SP,i,in} \right) \times \varphi_{SP,NO_x,i} \tag{12}$$

$$\begin{aligned} E_{CO_2} = \Sigma\left(m_{SP,i,in} \times [C]_{SP,i,in} - m_{SP,j,out} \times [C]_{SP,j,out} \right) \times \tfrac{44}{12} \\ + \Sigma(m_{SP,i,in} \times f) \end{aligned} \tag{13}$$

where $O_{SP,i}$ is the oxidation rate of sulfur, while β_{SO_2} indicates the desulfurization rate of the desulfurization equipment. $[N]_{SP,i,in}$ is the nitrogen content in the fuel. The coefficient for nitrogen oxide generation is denoted as $\varphi_{SP,NO_x,i}$. Carbon emissions are categorized into direct and indirect emissions. Direct carbon emissions refer to the amount of carbon dioxide directly released from the process, whereas indirect carbon emissions pertain to the emissions resulting from the energy consumed during production. The variable f represents the carbon emission factor.

3. Optimization Algorithms and Data Sources

3.1. Optimization Algorithm

This paper aims to optimize the heat utilization efficiency, energy consumption, and cost of the sintering process. The mathematical formulation of the multi-objective optimization problem is described as follows:

$$\begin{gathered} Min(and\ Max)\ y = [f_1(x), f_2(x), \cdots, f_n(x)](n = 1, 2, \ldots, N) \\ st.h(x) = [h_1(x), h_2(x), ..h_m(x)] = 0, m = 1,2,\ldots M \\ g(x) = [g_1(x), g_2(x), ..g_d(x)] \leq 0, d = 1,2,\ldots D \\ x = [x_1, x_2, \ldots, x_i, \ldots x_I] \\ x^l \leq x \leq x^u \end{gathered} \tag{14}$$

where y is the objective function, and N is the total number of optimization objectives. $f_n(x)$ is the n-th sub objective function. $g(x)$ and $h(x)$ represent the D-term inequality and M-term equality constraint conditions, respectively. These constraints define the feasible domain. Additionally, x^l and x^u are the lower and upper limits of the parameters, respectively. The optimization objective of this paper is to maximize heat utilization efficiency while minimizing energy consumption and carbon emissions during the sintering process. The various constraints in the optimization model include product quality and composition, process temperature, air volume requirements, and oxygen content.

In addition, NSGA-III is employed to solve multi-objective optimization problems by introducing a set of predefined reference points generated using the Das-Dennis method [33,34]. These reference points are evenly distributed within the objective space, guiding the population to evenly cover the entire Pareto front. NSGA-III utilizes non-dominated sorting and reference point-based selection mechanisms to maintain population diversity effectively and prevent the solution set from becoming overly concentrated in certain regions of the Pareto front [35,36]. This approach ensures a well-distributed set of optimal solutions across the entire Pareto front, enhancing the robustness and comprehensiveness of the optimization process. Model building and optimization work using Python language editing and 11th Gen Intel (R) Core (TM) i7-11700 @ 2.50 GHz 2.50 GHz processor, 128 GB memory server.

3.2. Data Sources

The primary data for this study are sourced from an advanced integrated steel plant located in Tangshan, Hebei Province, China. The plant is equipped with state-of-the-art sintering machines and circular cooling units, and utilizes a dual pressure waste heat boiler to capture low-temperature flue gas. With an annual output of approximately 11 million tons, the plant boasts a product qualification rate of 99% and a drum strength exceeding 81.0%. The main material components of the sintering plant as of August 2022 are detailed in Tables A1–A4 of Appendix A.

3.3. Model Validation

In order to ensure the model's accuracy, on-site statistical data from the case enterprise were used for validation. The comparison results between the field data and the model's operation are shown in Table 2. The small errors observed are within acceptable limits, indicating that the model is effective and applicable. Additionally, the model's accuracy was further verified by comparing it with previous research results, as shown in Table 3.

Table 2. Comparison of the simulated results and field data.

Category	Type	Field Data	Simulated Data
Input materials	Mixture materials (kg/t)	1176.030	1176.030
	iron ore (kg/t)	888.378	888.378
	Returned ore (kg/t)	159.457	159.457
	Flux (kg/t)	128.691	128.691
	Coke powder (kg/t)	49.952	49.952
	COG (m^3/t)	4.78	4.78
Output materials	Tfe (%)	56.902	56.994
	FeO of sintered ore (%)	8.590	8.500
	C of sintered ore (%)	0.059	0.043
	CaO of sintered ore (%)	10.100	10.492
	MgO of sintered ore (%)	1.784	2.107
	SiO_2 of sintered ore (%)	4.934	5.512
	Al_2O_3 of sintered ore (%)	1.851	2.006
	TiO_2 of sintered ore (%)	0.108	0.071
	P of sintered ore (%)	0.044	0.057
	S of sintered ore (%)	0.025	0.004
	Recovery of steam (kg/t)	39.62	39.19

Table 3. Comparison of the simulated and previous research results.

Index	Field Data	Simulated	Wu et al. [8]	Sun et al. [32]
Heat utilization efficiency	57.83%	57.98%	57.92%	/
Energy consumption	1474.44 MJ	1464.76 MJ	1471.8 MJ	1402 MJ

4. Results and Discussion

4.1. Multi Objective Optimization Results

This paper is based on the established integrated model of the high-temperature sintering process and high-temperature waste heat recovery in sintering plants. Using the NSGA-III algorithm, we optimized the heat utilization efficiency, energy consumption, and production cost of the sintering model, resulting in a solution set comprising 61 distinct solutions. These solutions illustrate the trade-off relationships between the various objectives, forming a Pareto front. Figure 2 provides a three-dimensional visualization, intuitively showcasing the distribution of the solution set and the interrelationships among the objectives. These solution sets offer a range of choices for sintering decision-makers. Sintering managers typically focus on cost objectives when optimizing the ratio of various ores. To further select multi-objective optimal solutions that prioritize cost, solutions with weights of 0.2, 0.3, and 0.5 for heat utilization efficiency, energy consumption, and production cost objectives, respectively, were chosen from the frontier solution set to identify the optimal solution for the sintering production process. The results before and after optimization are shown in Table 4.

The results showed that the optimized sintering process increased heat utilization efficiency by 0.67%, indicating more efficient energy conversion and utilization under the same production conditions, thereby reducing waste. Energy consumption decreased by 17.3 MJ/t, a reduction of 1.17%, effectively lowering energy use in the production process. Production costs were reduced by 11.45 CNY/t, a decrease of 1.22%. Additionally, CO_2, and SO_2 emissions per ton of sintered ore decreased by 0.464 kg/t and 0.034 kg/t, and NOx emissions decreased by 0.008 kg/t. This not only cuts production costs but also reduces pollutants and carbon emissions, contributing positively to environmental protection. The company currently produces over 34,900 tons of sintered ore per day, with an annual output of approximately 11.74 million tons. This production can reduce energy consumption by 203.12 million MJ, production costs by 134.47 million CNY, CO_2 emissions by 5.45 million kg, SO_2 emissions by 0.399 million kg, and NO_X by 0.094 million kg annually. In summary, the model method proposed in this article effectively optimizes the sintering process. Implementing these optimizations enhances heat utilization efficiency, reduces production costs, and minimizes environmental pollution, while also providing significant economic and environmental benefits to the steel industry.

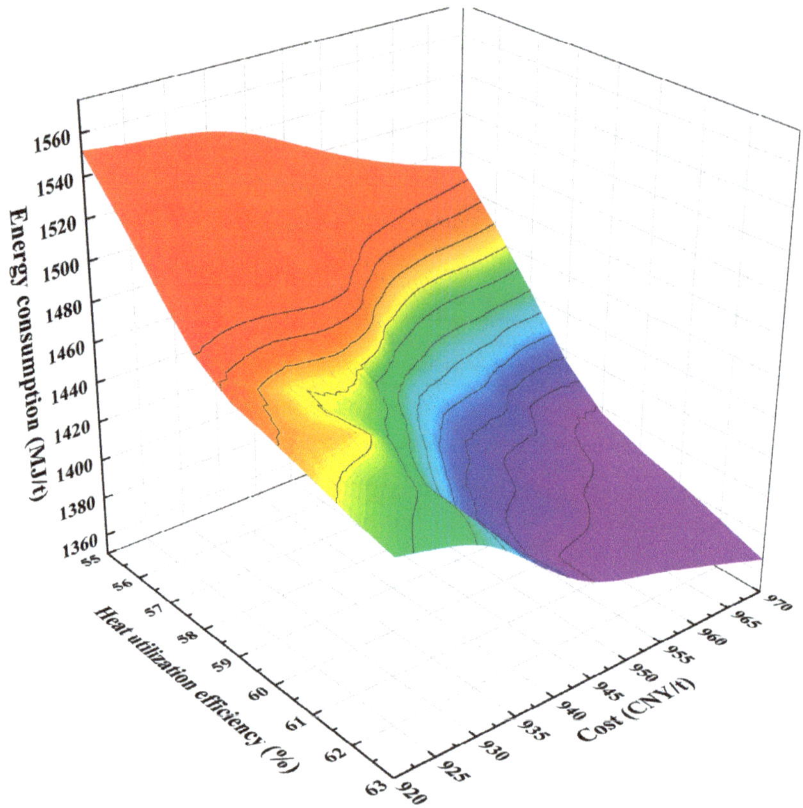

Figure 2. Pareto front of multi-objective optimization results.

Table 4. Comparison of results before and after optimization.

Index	HUE (%)	EC (MJ/t)	PC (CNY/t)	CO_2(kg/t)	SO_2(kg/t)	NOx (kg/t)
Before optimization	57.83	1474.44	942.38	160.040	0.781	0.484
After optimization	58.50	1457.14	930.93	159.576	0.747	0.476

4.2. Parameter and Energy Changes

Various ores and fluxes produce high-temperature sintering through oxidation, reduction, and solidification reactions in the action of fuel, producing exhaust gases containing carbon dioxide, sulfur dioxide, and nitrogen oxides. The material and energy input/output and conversion relationships of various sintering equipment and subsystems, both before and after multi-objective optimization, are illustrated in Figures 3 and 4. These figures provide a detailed display of the flow direction and conversion process of materials and energy. Through comparing different ore ratios, it was discovered that adjusting the mix led to a decrease in the cost of sintered ore products from 942.38 CNY to 930.93 CNY per ton. The highest unit price of ore6 decreased from 128.81 kg/t to 28.26 kg/t, while the lower unit price of ore3 increased from 39.78 kg/t to 108.58 kg/t. This change significantly impacted sintering costs and resulted in a substantial difference in iron content. This was achieved by incorporating better iron ore powders such as ore2, ore8, and ore9, with increases of 29.71 kg/t, 22.17 kg/t, and 37.85 kg/t, respectively. Therefore, through multi-objective optimization under strict production constraints, the ore ratio can be adjusted more effectively to reduce sintering costs in a scientifically sound manner.

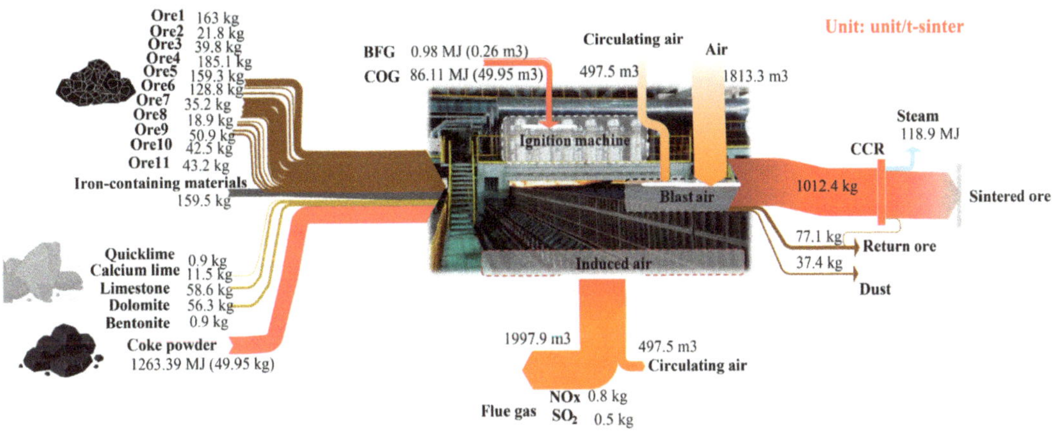

Figure 3. The material and energy parameters of the SP before optimization.

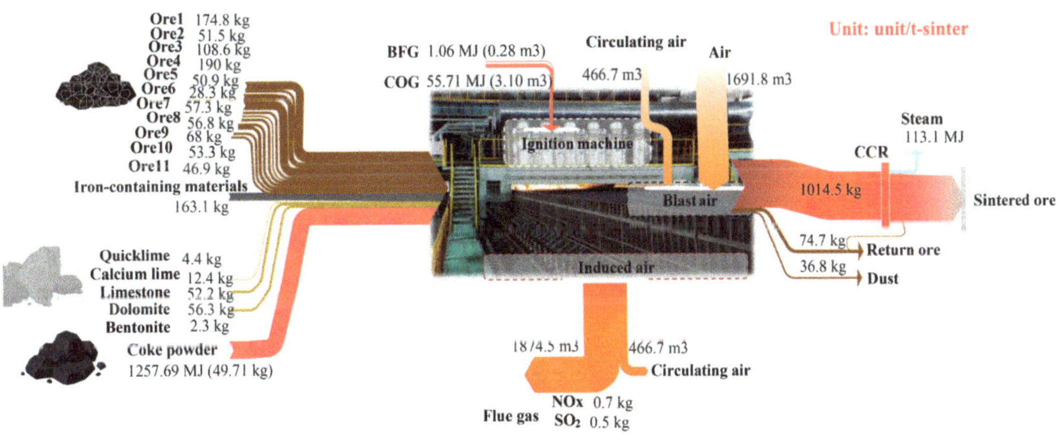

Figure 4. The material and energy parameters of the SP after optimization.

In addition, the amount of air injected during the high-temperature operation of sintering is determined by a comprehensive analysis of factors such as the theoretical air volume needed for fuel combustion, air leakage, circulating air volume, and the limitation of oxygen content in the flue gas. From Figures 3 and 4, it can be clearly seen that after optimization, the air volume injected into the sintering process decreased from 1813.26 m^3/t to 1691.78 m^3/t, and the flue gas volume decreased from 1997.89 m^3/t to 1874.45 m^3/t. As illustrated in Figure 5a,b the optimization of the sintering high-temperature process significantly impacted the proportion of heat input and output. Notably, the reduction in flue gas volume led to a decrease in the proportion of heat loss carried away by the flue gas, reducing it from 21.56% to 20.73%. The reduction in air volume and the decrease in heat loss carried away by flue gas after optimization are the primary reasons for the improvement in heat utilization efficiency in the sintering process. This optimization also resulted in decreased energy consumption, as reflected by a 0.45% reduction in coke powder consumption. Additionally, although the circulating hot air volume in the sintering internal and external circulation decreased from 497 m^3/t before optimization to 466.74 m^3/t, the utilization efficiency of circulating heat slightly reduced. This slight reduction is also attributable to the decrease in the total exhaust air volume.

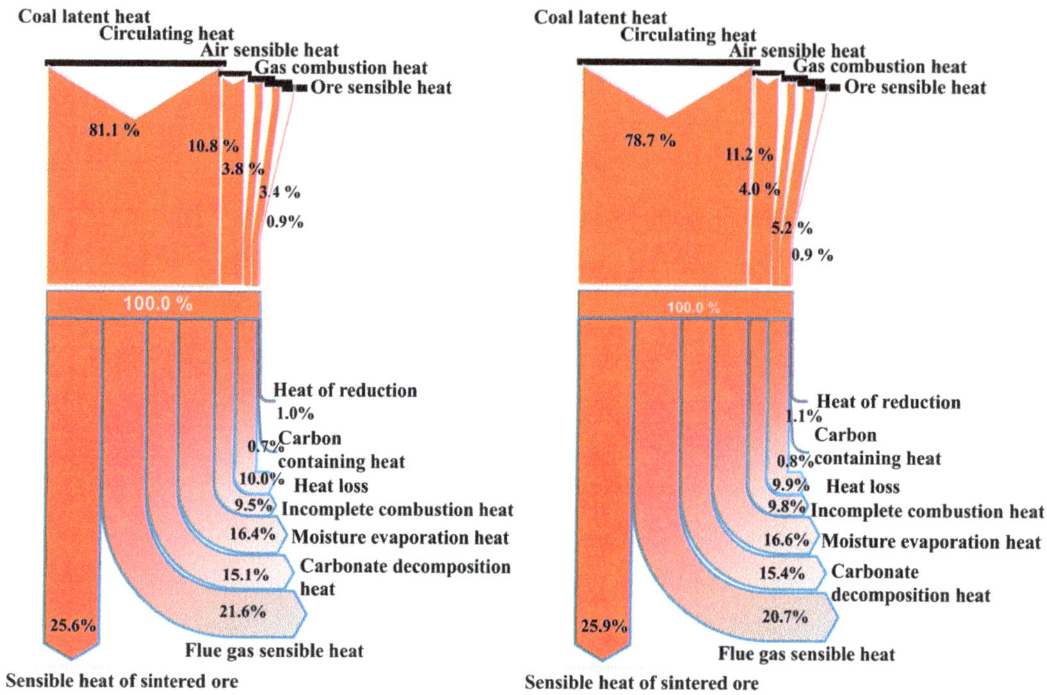

Figure 5. Heat utilization input and output before (**a**) and after (**b**) optimization of sintering high-temperature process.

The comparison of the proportion of heat input and output before and after optimization is shown in Figure 5. Due to the adjustment of the ore type ratio after optimization, the proportion of sensible heat conduction, moisture evaporation heat, and carbonate decomposition heat of sintered ore increased by 0.3%, 0.2%, and 0.3%, respectively. This improvement is due to the optimization of multiple ore ratios, in which increases the grade of the mixed ores and sintered ore. This optimization leads to a slight reduction in the amount of gangue and impurities introduced by the material, thereby decreasing the energy consumption required for gangue melting and discharge. Consequently, more heat is used for effective utilization in the production process. Additionally, the reduction in exhaust volume decreases the proportion of exhaust heat loss by 0.83%. Overall, the optimization of parameters such as ore blending and airflow in the sintering process, has led to a reduction in the cost and energy consumption of the sintering process, as well as an improvement in heat utilization efficiency. It has also reduced the number of pollutants and carbon emissions in the exhaust gas during the sintering process.

4.3. Analysis of Influencing Factors

Based on the in-depth analysis results in the previous section, we have identified the significant impact of raw material parameters and operating parameters on determining the quality, production efficiency, energy utilization efficiency, and emission characteristics of sintered ore. To provide more systematic guidance for sintering production practice and optimize process control, this section combines the basic principles of material layer distribution, temperature gradient, and reaction interval in the sintering thermal process (as shown in Figure 6a,b), and analyzes in detail the specific impact laws of raw material chemical composition (iron ore grade, impurity content), and key operating parameters (material layer thickness, mixture moisture content, and fuel ratio) on improving multiple key sintering indicators.

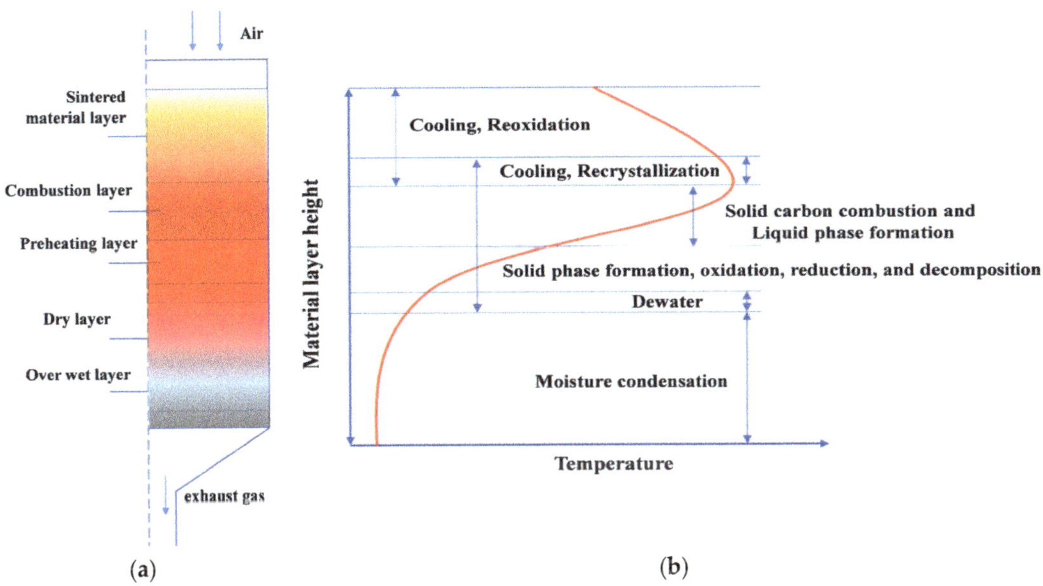

Figure 6. (**a**) Distribution diagram of material layer in SP. (**b**) Temperature changes and reaction processes of each material layer in SP [37].

4.3.1. Changes in Sintered Ore Grade

The quantitative relationship between the change in sintered ore grade and sintering indicators is illustrated in Figure 7. The sintered ore grade has a positive correlation with heat utilization efficiency and production cost of the sintering production process. For every 1% increase in sintered ore grade, the heat utilization efficiency and production cost decrease by 0.81% and 0.19%, respectively. However, the sintering energy consumption increases by 0.15%. The impact of changes in sintered ore grade on sintering pollutants is depicted in Figure 8. As the grade increases, sintering pollutants and carbon emissions are significantly reduced. CO_2, SO_2, and NO_x emissions are reduced by 3.13 kg/t, 0.058 kg/t, and 0.009 kg/t, respectively. This reduction is mainly due to changes in ore powder. Higher ore grades contain fewer impurities, leading to reduced heat consumption during the sintering production process and decreased fuel consumption. Therefore, appropriately increasing the sintered ore grade is beneficial for efficient sintering energy conservation, pollution reduction, and high efficiency.

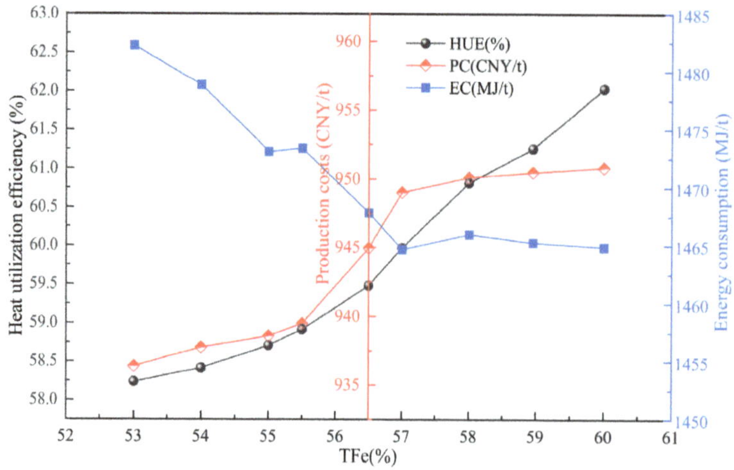

Figure 7. The influence of changes in iron content in sintered ore on sintering heat utilization, energy consumption, and cost.

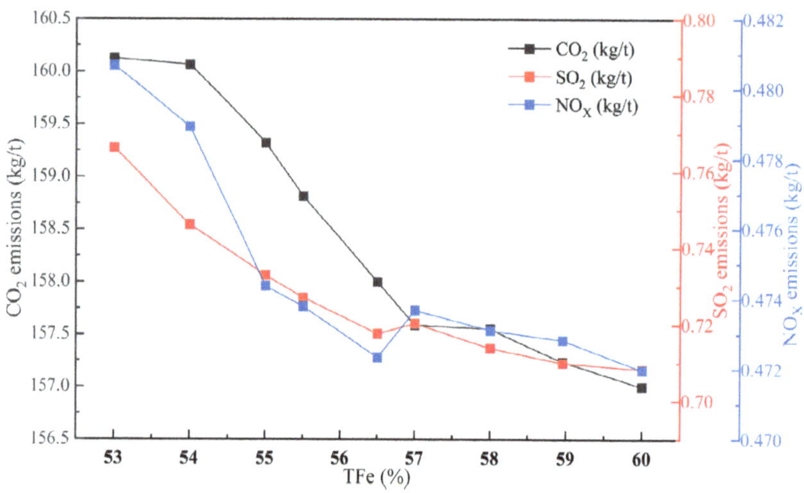

Figure 8. The influence of changes in iron content in sintered ore on pollutant emissions and carbon emissions in sintering plants.

4.3.2. Changes in the Thickness of the Material Layer

By increasing the thickness of the material layer, the energy utilization of the sintering process can be promoted, fully utilizing the self-heat storage effect of the material layer and improving the heat supply of the lower material layer [38,39]. Based on the constructed model, this section briefly discusses and quantifies the impact of material layer thickness changes on various sintering indicators under the condition that technology meets practical production needs. As shown in Figures 9 and 10, the increase in the material layer can greatly improve heat utilization efficiency. When the material layer changes from 820 mm to 950 mm, the production cost and energy consumption decrease by 3.13% and 3.59%, respectively, showing a linear relationship. This is mainly due to the increase in output and the effect of thermal recycling. At the same time, SO_2, NO_X, and CO_2 emissions decrease by more than 2.5%. Overall, increasing the thickness of the sintering material layer is very beneficial for the sustainable development of sintering.

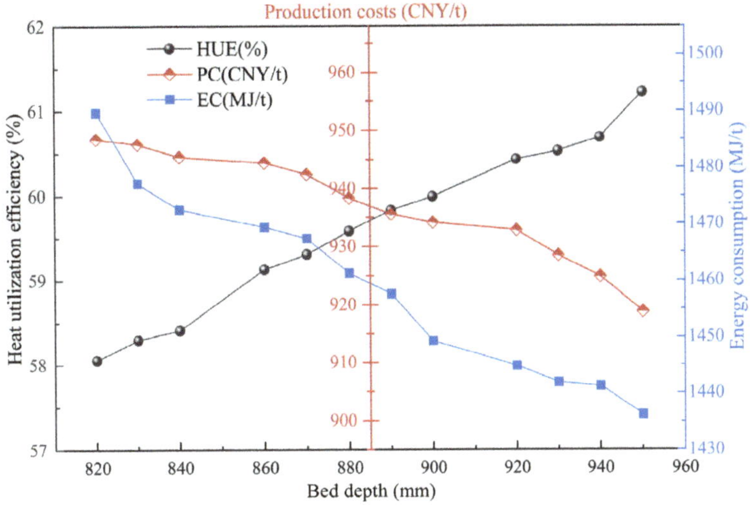

Figure 9. The influence of material layer thickness on sintering energy utilization, energy consumption, and carbon emissions.

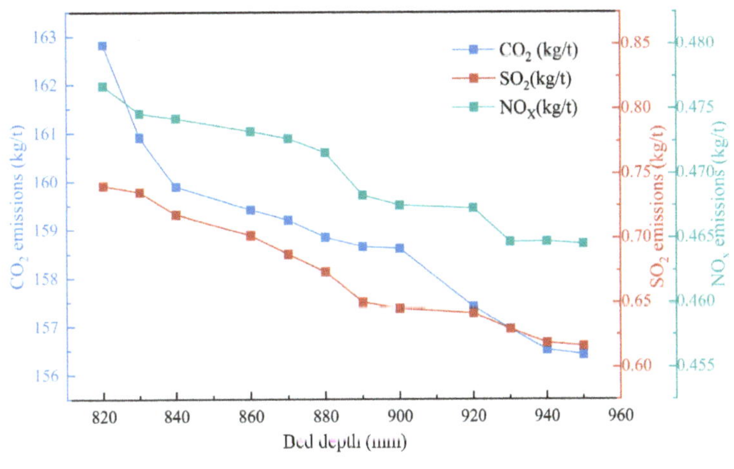

Figure 10. The influence of material layer thickness on sintering pollutants and carbon emissions.

4.3.3. Change in Coal Ratio

Figures 11 and 12 show the quantitative relationship between the change in the proportion of middling coal consumption in the total fuel consumption and the impact on the sintering index. Based on the energy content of the fuel, varying dosage ratios of coal and coke powder are applied. As the proportion of coal increases from 10% to 54%, the cost price of coke powder is lower than the market price of coal, but the calorific value is higher than that of coal. As the coal consumption increases, the production cost and energy consumption of sintering gradually increase. At the same time, with the production of the same product, as the amount of coal used increases, it inevitably leads to an increase in the total amount of coal and coke powder used, as well as an increase in the sulfur content and nitrogen-containing substances in coal, especially the sulfur content. As a result, the emissions of pollutants and carbon during the sintering process increase, with CO_2, SO_2, and NO_x emissions increasing by 5.82 kg/t, 0.432 kg/t, and 0.031 kg/t, respectively, which is not conducive to energy conservation and pollution reduction in the sintering process.

Compared to purchased coal, the coking process has already removed most impurities from coal. Therefore, using coke powder instead of coal in the sintering process offers several advantages.

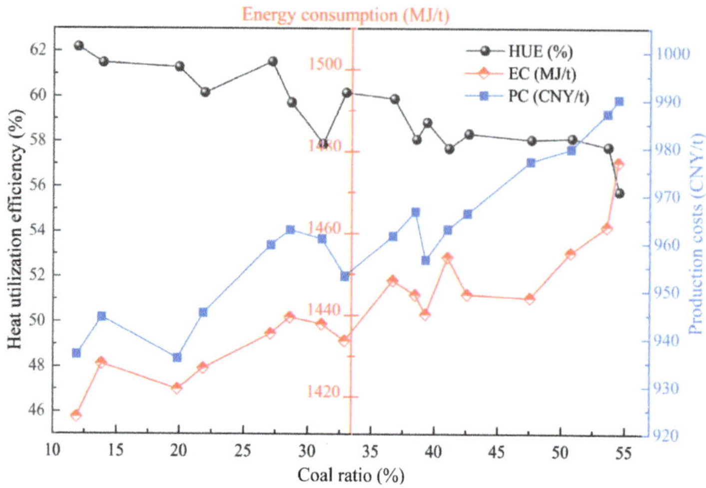

Figure 11. The influence of changes in coal ratio in fuel on sintering heat utilization, energy consumption, and cost.

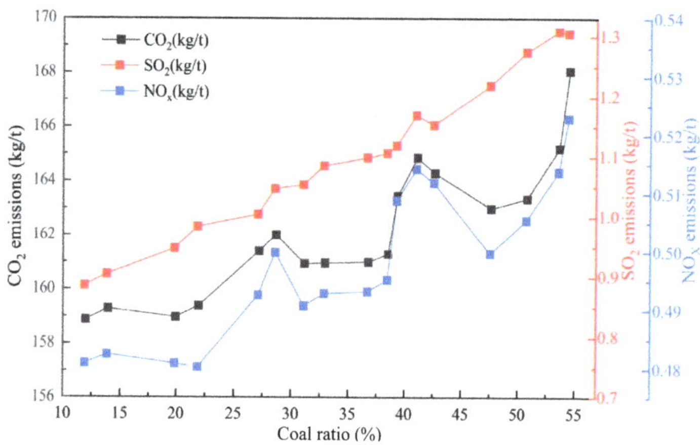

Figure 12. The influence of changes in coal ratio in fuel on pollutant emissions and carbon emissions in sintering plants.

4.3.4. Changes in Moisture Content of Materials

The moisture content of the material is shown in Figure 13. The change in moisture content directly affects the energy consumption and carbon emissions of the sintering process. As the moisture content increases from 5% to 8.5%, the energy consumption and carbon emissions gradually increase. This is because more water evaporation requires excessive heat consumption, which needs to be supplemented by an increase in fuel consumption. Therefore, energy consumption and carbon emissions increase significantly, by 7.84% and 7.91%, respectively, as a result of using coke powder in the sintering process; other pollutant emissions remain unchanged. Therefore, accurately and strictly controlling material moisture can effectively reduce energy consumption in the sintering process.

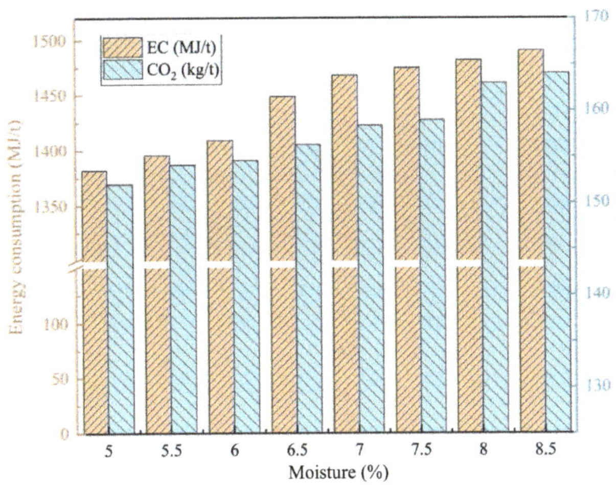

Figure 13. Influence of material moisture content on sintering energy consumption and carbon emissions.

5. Conclusions

This paper is grounded in the complex principles and mechanisms of metallurgical chemistry related to sintered ore formation. Firstly, an integrated system model for sintered ore production and high-temperature waste heat recovery was established. A multi-objective optimization method for the sintering process, combined with the NSGA-III optimization algorithm, was then proposed. This method optimizes the heat utilization efficiency, energy consumption, and production cost of the sintering process. Additionally, it quantifies the influence of several typical factors on various sintering indicators and pollutant emissions. The research has led to the following conclusions:

(1) Using the NSGA-III algorithm, a Pareto front solution set for the multi-objective optimization of the sintering process was obtained, and the optimal solution for balancing quality, energy saving, and cost objectives was found. After optimization, the heat utilization efficiency of the sintering process increased by 0.67%, energy consumption decreased by 17.3 MJ/t, and production costs decreased by 11.45 CNY/t. Additionally, CO_2 and SO_2 emissions per ton of sintered ore decreased by 0.464 kg/t and 0.034 kg/t, and NOx emissions decreased by 0.008 kg/t. This optimization effectively improved the heat utilization efficiency of the sintering process while reducing energy consumption, production costs, pollutants, and carbon emissions.

(2) In addition, optimized operating parameters for the optimal allocation of production resources, such as ore blending, flux, fuel consumption, and various airflow parameters, were obtained. These optimizations are beneficial for enhancing energy savings, improving efficiency, and reducing pollution in the sintering process. By implementing these optimizations, enterprises are expected to reduce energy consumption by 203.12 million MJ, production costs by 134.47 million CNY, CO_2 emissions by 5.45 million kg, SO_2 emissions by 0.399 million kg, and NO_X by 0.094 million kg annually. It has brought significant economic and environmental benefits to the steel industry.

(3) By studying and quantifying the influence of various key factors on multiple targets and pollutant emissions in the production process, especially changes in parameters such as sintered ore grade, material layer thickness, mixed material moisture content, and coal ratio, it was found that for every 1% increase in sintered ore grade, heat utilization efficiency improved by 0.81%, energy consumption decreased by 0.19%, and sintering production cost increased by 0.15%. Additionally, while ensuring production quality requirements are met, reasonably reducing the moisture content in the mixture and minimizing the consumption of coal in the fuel, along with increasing

the material layer thickness, can contribute to lowering sintering energy consumption and reducing pollutant emissions.

Author Contributions: Conceptualization, Y.Y.; Methodology, Y.Y. and J.S.; Software, Y.Y. and J.S.; Validation, Y.Y., J.S. and L.Z.; Investigation, S.Y. and H.N.; Data Curation, Y.Y.; Writing—Original Draft, Y.Y. and H.N.; Writing—Review and Editing, Y.Y. and T.D.; Visualization, J.S. and L.Z.; Supervision, H.N.; Funding Acquisition, T.D. All authors have read and agreed to the published version of the manuscript.

Funding: The authors are grateful to the financial support provided by National Natural Science Foundation of China (No.: 52270177), Young Elite Scientists Sponsorship Program by CAST (No.: 2022QNRC001), the 111Project (B16009), Jianlong Group-Northeastern University Youth Science and Technology Innovation Fund (No.: 2023012600001) and Key Technical Research Project of Shenyang (No.: 22-101-0-28).

Institutional Review Board Statement: Not applicable.

Informed Consent Statement: Not applicable.

Data Availability Statement: The original contributions presented in the study are included in the article, further inquiries can be directed to the corresponding authors.

Conflicts of Interest: The authors declare no conflicts of interest.

Abbreviations

SP	Sintering plant	COG	Coke Oven Gas
SRC	Sintering ring cooler	BFG	Blast Furnace Gas
HUE	Heat utilization efficiency	NSGA-III	Non-dominated Sorting Genetic Algorithm III
EC	Energy consumption	CNY	Chinese Yuan
PC	Production costs		
i	The i-th type of material or energy	$T_{SRC,sinter,in}$	Temperature of sintered ore entering the ring cooler, °C
j	The j-th type of output material or energy	$T_{SRC,sinter,out}$	Temperature of the sintered ore exiting the ring cooler, °C
$m_{SP,i,in}$	The input dosage of the i-th material or energy, kg/t	$S_{SP,s,out}$	Extraction volume of the s-th type of steam, kg/t
$m_{SP,j,out}$	The output dosage of the j-th material or energy, kg/t	h_s	Enthalpy value of the s-th type of steam, kJ/kg
R	Dosage of material in process, kg/t	e_{SP}	energy consumption of the sintering process, MJ/t
$Q_{SP,i,in}$	The i-th type of input heat in SP, kJ/t	r_k	The dosage of k-th energy recovered and supplied externally, kg/t, m^3/t
$Q_{SP,j,out}$	The j-th type of output heat in SP, kJ/t	$\varepsilon_{i,in}$, ε_k	Conversion coefficient of standard coal, MJ/t, MJ/m^3
m_i	The dosage of the i-th material or energy, kg/t	P	The dosage of sintered ore produced during the statistical period, kg/t
$c_{m,i}$	The specific heat capacity of the i-th material or energy, kJ/(kg·K)	C_{cost}	production cost per ton of sintered ore products, CNY/t
V_i	The specific heat capacity of the i-th gas, m^3/t	c_i	Price coefficient of the i-th substance or energy, CNY/kg, CNY/m^3, CNY/kWh
$c_{v,i}$	Specific heat capacity of the i-th gas, kJ/(m^3·K)	c_{main}, c_{labor}, c_{dep}	Price coefficients for maintenance, labor, and depreciation of unit products, CNY/t
r	The r-th chemical reaction	c_k	Price coefficient for the k-th type of energy recovery, CNY/kg, CNY/m^3
M_r	Relative molecular mass of the r-th material	$O_{SP,i}$	Oxidation rate of sulfur, %
ΔH_r	Reaction enthalpy of the r-th reaction, kJ/mol	β_{SO_2}	The desulfurization rate of desulfurization equipment, %
a_{air}	Comprehensive heat transfer coefficient between sintered ore and air, W/(m^2·K)	$\varphi_{SP,NO_x,i}$	Nitrogen oxide generation coefficient
η_{SRC}	Efficiency of waste heat boiler in sintering ring cooler, %	f	Carbon emission factor, kg/t, kg/m^3, kg/kWh

Appendix A

Table A1. The composition and price parameters of raw material in sintering plant.

Material Species	Component (%)									Price (CNY)
	TFe	FeO	SiO_2	CaO	MgO	Al_2O_3	S	P	TiO_2	
Ore1	65.80	0.23	1.42	0.45	0.01	1.22	0.01	0.06	0.09	830.00
Ore2	62.50	0.22	4.20	0.48	0.03	2.30	0.02	0.09	0.10	815.00
Ore3	61.60	0.31	4.35	0.45	0.03	1.88	0.04	0.05	0.11	500.00
Ore4	58.50	0.23	4.16	0.44	0.01	1.42	0.01	0.05	0.09	795.00
Ore5	56.20	0.20	5.53	0.10	0.03	2.22	0.03	0.08	0.11	760.00
Ore6	66.60	26.00	2.50	0.60	0.40	1.30	0.11	0.01	0.17	870.00
Ore7	60.60	11.32	10.33	0.69	0.83	1.14	0.02	0.02	0.00	800.00
Ore8	66.03	26.01	2.08	0.55	1.04	0.90	0.19	0.02	0.00	840.00
Ore9	57.43	0.28	5.57	0.24	0.64	1.20	0.02	0.07	0.06	720.00
Ore10	58.74	0.91	5.16	0.01	0.10	2.61	0.02	0.05	0.00	730.00
Ore11	59.61	0.33	5.51	0.27	0.12	2.81	0.01	0.05	0.00	750.00
Quicklime	0.00	0.00	4.75	56.95	6.25	1.49	0.09	0.00	0.00	680.00
Limestone	0.00	0.00	0.84	51.14	3.01	0.93	0.02	0.00	0.00	500.00
Dolomite	0.00	0.00	1.83	31.09	20.29	0.00	0.02	0.00	0.00	180.00
Bentonite	1.18	0.00	68.40	1.51	2.48	13.95	0.00	0.00	0.08	300.00
Calciumlime	0.00	0.00	0.64	83.96	3.20	0.33	0.01	0.00	0.00	880.00

Table A2. Proportion of gas components in sintering plant.

Gas	H_2	O_2	N_2	CO	CO_2	CH_4	C_2H_6	C_3H_8	C_2H_4	C_2H_2
COG	59.8	0.14	3.54	6.53	1.86	23.60	0.62	0.00	1.35	0.00
BFG	4.535	0.25	49.60	24.86	20.75	0.01	0.00	0.00	0.00	0.00

Table A3. The proportion of each component of fuel coal in sintering plant.

Coal Species	Volatile Content ($V_{daf}/\%$)	Carbon Content (Car/%)	Sulfur Content ($S_{td}/\%$)	Ash Content ($A_d/\%$)
Anthracite	9.45	78.92	0.14	12.50
Coke	0.008752	85.77	0.59	11.95

Table A4. Composition requirements for sintered ore products in sintering plant.

Composition	TFe	FeO	SiO_2	CaO	MgO	Al_2O_3	S	P	TiO_2
Sintered ore	55–58	≤9.00	≤5.80	≤11.00	≤2.50	≤2.20	≤0.03	≤0.09	≤0.25

Table A5. The composition of sintered ore before and after optimization.

Composition	TFe	FeO	SiO_2	CaO	MgO	Al_2O_3	S	P	TiO_2
Before	56.32	8.50	5.75	10.95	2.42	2.09	0.0043	0.0016	0.084
After	56.68	8.40	5.66	10.94	2.47	2.02	0.0041	0.0011	0.068

References

1. World Steel Association. *Sustainability Indicators for the Steel Industry*; World Steel Association: Brussels, Belgium, 2023.
2. 2023 China Steel Yearbook. *China Iron and Steel Industry Yearbook Society*; China Iron and Steel Association: Beijing, China, 2023.
3. Jiang, M.; Guo, M.; Liu, H.; Zhao, S.; Wang, Z. Discussion on optimization technology of whole process control of ultra-low emission of sintering flue gas. *Sinter. Pelletizing* **2023**, *48*, 115–121. [CrossRef]
4. Tang, L.; Jia, M.; Bo, X.; Xue, X. High resolution emission inventory and atmospheric environmental impact research in Chinese iron and steel industry. *China Environ. Sci.* **2019**, *40*, 1493–1506. [CrossRef]
5. Wang, J.; Meng, H.; Zhou, H. Effect of biochar substitution on iron ore sintering characteristics based on optimization of fuel distribution through the bed. *Fuel Process. Technol.* **2023**, *247*, 107817. [CrossRef]

6. Cheng, Z.; Tan, Z.; Guo, Z.; Yang, J.; Wang, Q. Recent progress in sustainable and energy-efficient technologies for sinter production in the iron and steel industry. *Renew. Sustain. Energy Rev.* **2020**, *131*, 110034. [CrossRef]
7. Cavaliere, P. (Ed.) Sintering: Most Efficient Technologies for Greenhouse Emissions Abatement. In *Clean Ironmaking and Steelmaking Processes: Efficient Technologies for Greenhouse Emissions Abatement*; Springer International Publishing: Cham, Switzerland, 2019; pp. 111–165. [CrossRef]
8. Wu, Y.; Gan, M.; Ji, Z.; Fan, X.; Zhao, G.; Zhou, H.; Zheng, H.; Wang, X.; Liu, L.; Li, J. New approach to improve heat energy utilization efficiency in iron ore sintering: Exploration of surface fuel addition. *Process Saf. Environ. Prot.* **2024**, *190*, 125–137. [CrossRef]
9. Jouhara, H.; Khordehgah, N.; Almahmoud, S.; Delpech, B.; Chauhan, A.; Tassou, S.A. Waste heat recovery technologies and applications. *Therm. Sci. Eng. Prog.* **2018**, *6*, 268–289. [CrossRef]
10. Zhang, X.; Chen, Z.; Zhang, J.; Ding, P.; Zhou, J. Simulation and optimization of waste heat recovery in sinter cooling process. *Appl. Therm. Eng.* **2013**, *54*, 7–15. [CrossRef]
11. Dong, H.; Yang, Y.; Jia, F.; Zhao, L.; Cai, J. Thermodynamic analysis of efficient recovery and utilisation of waste heat resources during sintering process. *Int. J. Exergy* **2013**, *12*, 552–569. [CrossRef]
12. Liu, Y.; Yang, J.; Wang, J.; Cheng, Z.; Wang, Q. Energy and exergy analysis for waste heat cascade utilization in sinter cooling bed. *Energy* **2014**, *67*, 370–380. [CrossRef]
13. Cui, L.; Liu, M.; Yuan, X.; Wang, Q.; Ma, Q.; Wang, P.; Hong, J.; Liu, H. Environmental and economic impact assessment of three sintering flue gas treatment technologies in the iron and steel industry. *J. Clean. Prod.* **2021**, *311*, 127703. [CrossRef]
14. Liu, C.; Xie, Z.; Sun, F.; Chen, L. Optimization for sintering proportioning based on energy value. *Appl. Therm. Eng.* **2016**, *103*, 1087–1094. [CrossRef]
15. Wang, J.; Qiao, F.; Zhao, F.; Sutherland, J.W. A Data-Driven Model for Energy Consumption in the Sintering Process. *J. Manuf. Sci. Eng.* **2016**, *138*, 12. [CrossRef]
16. Yuan, Y.; Na, H.; Du, T.; Qiu, Z.; Sun, J.; Yan, T.; Che, Z. Multi-objective optimization and analysis of material and energy flows in a typical steel plant. *Energy* **2023**, *263*, 125874. [CrossRef]
17. Hu, J.; Wu, M.; Chen, X.; Cao, W.; Pedrycz, W. Multi-model ensemble prediction model for carbon efficiency with application to iron ore sintering process. *Control Eng. Pract.* **2019**, *88*, 141–151. [CrossRef]
18. Hu, J.; Wu, M.; Chen, L.; Cao, W.; Pedrycz, W. Real-time dynamic prediction model of carbon efficiency with working condition identification in sintering process. *J. Process Control* **2022**, *111*, 97–105. [CrossRef]
19. Chen, X.; Chen, X.; Wu, M.; She, J. Modeling and optimization method featuring multiple operating modes for improving carbon efficiency of iron ore sintering process. *Control Eng. Pract.* **2016**, *54*, 117–128. [CrossRef]
20. Feng, H.; Chen, L.; Liu, X.; Xie, Z.; Sun, F. Constructal optimization of a sinter cooling process based on exergy output maximization. *Appl. Therm. Eng.* **2016**, *96*, 161–166. [CrossRef]
21. Feng, J.; Yan, Y.; Zhao, L.; Dong, H. Numerical study of gas–solid counterflow heat transfer in sinter vertical cooling furnace based on energy and exergy analysis. *Appl. Therm. Eng.* **2024**, *244*, 122773. [CrossRef]
22. Feng, J.; Cheng, X.; Wang, H.; Zhao, L.; Wang, H.; Dong, H. Performance analysis and multi-objective optimization of organic Rankine cycle for low-grade sinter waste heat recovery. *Case Stud. Therm. Eng.* **2024**, *53*, 103915. [CrossRef]
23. Tian, W.; Jiang, C.; Ni, B.; Wu, Z.; Wang, Q.; Yang, L. Global sensitivity analysis and multi-objective optimization design of temperature field of sinter cooler based on energy value. *Appl. Therm. Eng.* **2018**, *143*, 759–766. [CrossRef]
24. Liu, Y.; Yang, J.; Wang, J.; Ding, X.; Cheng, Z.; Wang, Q. Prediction, parametric analysis and bi-objective optimization of waste heat utilization in sinter cooling bed using evolutionary algorithm. *Energy* **2015**, *90*, 24–35. [CrossRef]
25. Tian, W.; Ni, B.; Jiang, C.; Wu, Z. Uncertainty analysis and optimization of sinter cooling process for waste heat recovery. *Appl. Therm. Eng.* **2019**, *150*, 111–120. [CrossRef]
26. Zhang, X.; Zhao, L.; Dong, H.; Wang, D.; Zhang, J. Numerical investigation of gas-solid heat transfer process and parameter optimization in shaft kiln for high-purity magnesia. *Chem. Eng. Res. Des.* **2023**, *193*, 576–586. [CrossRef]
27. Zhu, S.; Gao, C.; Gao, C.; Guo, Y.; Zhang, X.; Li, X. Exploration of a new path to reduce air pollutant emissions in the sinter plant of steelworks. *J. Clean. Prod.* **2022**, *373*, 133831. [CrossRef]
28. Zhang, L.; Na, H.; Yuan, Y.; Sun, J.; Yang, Y.; Qiu, Z.; Che, Z.; Du, T. Integrated optimization for utilizing iron and steel industry's waste heat with urban heating based on exergy analysis. *Energy Convers. Manag.* **2023**, *295*, 117593. [CrossRef]
29. Yuan, Y.; Na, H.; Chen, C.; Qiu, Z.; Sun, J.; Zhang, L.; Du, T.; Yang, Y. Status, challenges, and prospects of energy efficiency improvement methods in steel production: A multi-perspective review. *Energy* **2024**, *304*, 132047. [CrossRef]
30. Li, M.-J.; Tao, W.-Q. Review of methodologies and polices for evaluation of energy efficiency in high energy-consuming industry. *Appl. Energy* **2017**, *187*, 203–215. [CrossRef]
31. Na, H.; Sun, J.; Qiu, Z.; He, J.; Yuan, Y.; Yan, T.; Du, T. A novel evaluation method for energy efficiency of process industry—A case study of typical iron and steel manufacturing process. *Energy* **2021**, *233*, 121081. [CrossRef]
32. Sun, W.; Wang, Q.; Zheng, Z.; Cai, J. Material–energy–emission nexus in the integrated iron and steel industry. *Energy Convers. Manag.* **2020**, *213*, 112828. [CrossRef]
33. Gu, Z.-M.; Wang, G.-G. Improving NSGA-III algorithms with information feedback models for large-scale many-objective optimization. *Future Gener. Comput. Syst.* **2020**, *107*, 49–69. [CrossRef]

34. Reddy, S.R.; Dulikravich, G.S. Many-objective differential evolution optimization based on reference points: NSDE-R. *Struct. Multidisc. Optim.* **2019**, *60*, 1455–1473. [CrossRef]
35. Ding, R.; Dong, H.; He, J.; Feng, X.; Yu, X.; Li, L. U-NSGA-III: An Improved Evolutionary Many-Objective Optimization Algorithm. In *Bio-Inspired Computing: Theories and Applications*; Springer: Singapore, 2018; pp. 24–35. [CrossRef]
36. Cui, Z.; Chang, Y.; Zhang, J.; Cai, X.; Zhang, W. Improved NSGA-III with selection-and-elimination operator. *Swarm Evol. Comput.* **2019**, *49*, 23–33. [CrossRef]
37. Ding, C.; Jiang, F.; Xue, S.; Chang, R.; Long, H.; Yu, Z.; Ding, X. Prediction model of sintering bed temperature based on lognormal distribution function: Construction and application. *J. Mater. Res. Technol.* **2023**, *26*, 5478–5487. [CrossRef]
38. Wang, Y.-Z.; Zhang, J.-L.; Liu, Z.-J.; Du, C.-B. Recent Advances and Research Status in Energy Conservation of Iron Ore Sintering in China. *J. Miner.* **2017**, *69*, 2404–2411. [CrossRef]
39. Liu, Z.; Niu, L.; Zhang, S.; Dong, G.; Wang, Y.; Wang, G.; Kang, J.; Chen, L.; Zhang, J. Comprehensive Technologies for Iron Ore Sintering with a Bed Height of 1000 mm to Improve Sinter Quality, Enhance Productivity and Reduce Fuel Consumption. *ISIJ Int.* **2020**, *60*, 2400–2407. [CrossRef]

Disclaimer/Publisher's Note: The statements, opinions and data contained in all publications are solely those of the individual author(s) and contributor(s) and not of MDPI and/or the editor(s). MDPI and/or the editor(s) disclaim responsibility for any injury to people or property resulting from any ideas, methods, instructions or products referred to in the content.

materials

MDPI

Article

Reprocessing Possibilities of Poly(3-hydroxybutyrate-co-3-hydroxyvalerate)–Hemp Fiber Composites Regarding the Material and Product Quality

Wiesław Frącz [1], Andrzej Pacana [2,*], Dominika Siwiec [2], Grzegorz Janowski [1] and Łukasz Bąk [1]

1 Department of Materials Forming and Processing, Rzeszow University of Technology, Powstancow Warszawy 8, 35-959 Rzeszow, Poland; wf@prz.edu.pl (W.F.); gjan@prz.edu.pl (G.J.); lbak@prz.edu.pl (Ł.B.)
2 Department of Manufacturing Processes and Production Engineering, Rzeszow University of Technology, Powstancow Warszawy 8, 35-959 Rzeszow, Poland; d.siwiec@prz.edu.pl
* Correspondence: app@prz.edu.pl

Abstract: An important issue addressed in research on the assessment of the quality of polymer products is the quality of the polymer material itself and, in accordance with the idea of waste-free management, the impact of its repeated processing on its properties and the quality of the products. In this work, a biocomposite, based on poly(3-hydroxybutyrate-co-3-hydroxyvalerate) (PHBV) with short hemp fibers, was obtained and repeatedly processed, which is a continuation of the research undertaken by the team in the field of this type of biocomposites. After subsequent stages of processing, the selected mechanical, processing and functional properties of the products were assessed. For this purpose, microscopic tests were carried out, mechanical properties were tested in static tensile and impact tests, viscosity curves were determined after subsequent processing cycles and changes in plastic pressure in the mold cavity were determined directly during processing. The results of the presented research confirm only a slight decrease in the mechanical properties of the produced type of biocomposite, even after it has been reprocessed five times, which gives extra weight to arguments for its commercialization as a substitute for petrochemical-based plastics. No significant changes were found in the used parameters and processing properties with the stages of processing, which allows for a predictable and stable manufacturing process using, for example, the injection molding process.

Keywords: PHBV; biocomposites; injection molding; biocomposites reprocessing; recycling

1. Introduction

Waste management is one of the most urgent challenges of modern society. In particular, the problem of plastic waste is becoming more and more urgent. Plastics are present in many areas of our lives and are an integral part of the modern economy. However, their durability and, in most cases, their lack of biodegradability, coupled with the growing production and consumption of plastics, have led to a growing ecological crisis. When these materials enter the environment, they remain there for dozens or even hundreds of years, contributing to soil and water pollution. Some plastics, especially those containing harmful chemicals, can lead to serious human health problems. For instance, phthalates and bisphenol A (BPA) can disrupt the body's hormonal balance and cause various types of diseases. The lack of effective management of plastic waste results in the waste of mineral resources. Despite progress in the field of plastic recycling, many challenges remain. The lack of a uniform waste segregation and collection system, low recycling rates in some countries and the lack of appropriate processing technologies are serious problems that have prompted a search for new solutions [1–4].

One of the most important steps in solving the problem of plastic waste management is reducing the consumption of these materials. The introduction of innovative packaging, the promotion of a circular economy and consumer education are key in this context. Creating innovative recycling methods, such as the chemical decomposition of plastics into primary components, may be an important step forward [5–7]. Furthermore, the development of alternative biodegradable and compostable materials may contribute to solving the problem and finding a solution. Promoting sustainable materials, such as bioplastics, can lead to a reduced burden on the environment [8,9]. Among a fairly wide range of bioplastics, special attention should be paid to double green polymers, because they are of natural origin and are fully biodegradable. It is worth noting that their possibilities of use may be multiplied due to the continuous circulation of this type of materials and their derivatives in nature [10,11].

The circular economy is an economic model that aims to minimize the amount of waste by maximizing the use and reprocessing of resources. Unlike the traditional linear model, in which raw materials are consumed and discarded, the circular economy focuses on closing product life cycles, reducing the impact on the environment and using resources more efficiently [12]. With the growing problem of plastic pollution and the growing environmental awareness of society, the need to find sustainable alternatives to traditional plastics is becoming an increasingly pressing challenge. Polyhydroxyalkanoates (PHAs) are emerging as a revolutionary category of biodegradable polymers, potentially transforming the way we think about the production, consumption and disposal of materials [13].

Polyhydroxyalkanoates (PHAs) are a type of biodegradable polymer obtained from natural sources, such as bacteria, which use them as energy storage substances. In recent years, interest in these polymers has increased, due to their potential to replace traditional, non-biodegradable plastics. They are distinguished not only by their ability to degrade in the environment, but also by a variety of properties, which means they can be used in many areas, from packaging to medicine [14]. The structure of PHAs may vary depending on the source of the microorganisms which they are obtained from. Their structural diversity depends mainly on the type of monomers from which they are composed [15]. There are several key types, such as poly(3-hydroxybutyric acid) (PHB), poly(3-hydroxyvaleric acid) (PHV), and copolymers such as poly(3-hydroxybutyric acid-co-3-hydroxyvaleric acid) (PHBV). The PHA production process typically involves the use of bacteria. However, there is also a growing interest in the production of PHAs using plant organisms, which opens up new opportunities for the sustainable production of biodegradable plastics [16,17]. The PHA production process uses microorganisms with a high storage capacity and diverse biochemical processes, in order to increase the number of cycles of the "growth" and "starvation" phases [18,19]. It is also worth noting that the sterilization of reactors for the PHA production process is not necessary, and bacterial cultures are able to adapt to various additional waste raw materials [20]. The main advantage of PHA production is the ability to use real fermented waste as raw materials, such as agricultural or food industry by-products, which reduces the costs of substrate use [21,22]. PHAs exhibit a variety of properties that make them attractive for a variety of applications. Their biodegradability is a key element that enables their decomposition in natural conditions. Their biocompatibility makes them useful in the field of medicine, and their thermoplasticity allows them to be formed many times [23]. Despite promising prospects, the implementation of PHA materials is a challenge. Production costs and ethical issues related to the genetic modification of organisms require further research. Nevertheless, technological advances and society's commitment towards sustainable alternatives will contribute to the development of this group of polymers. The prospect of lower production costs and growing ecological awareness opens the door for PHAs to become a key element in the global movement to counteract petrochemical plastic pollution [24,25].

Polyhydroxybutyrate-co-valerate (PHBV) is a type of biodegradable polymer that is a copolymer of poly(hydroxybutyrate) (PHB) and poly(hydroxyvalerate) (PHV). This combination of two different monomers introduces flexibility into the polymer structure, which makes PHBV more elastic than PHB alone [26]. Therefore, this material has a number of applications in various fields, such as the production of packaging, disposable products and biodegradable films. Broader commercial uses of this biopolymer are still difficult due to high production costs and a small difference between the melting point and the degradation temperature of this polymer, as well as low flexibility and quite high brittleness [27–30]. For this reason, further research plans of scientists include improving the mechanical properties and processing window of this biopolymer, as well as the possibility of producing composites based on PHBV [31,32]. It should be noted, however, that despite these advantages, PHBV production still encounters challenges related to, for example, production scaling. However, the development of this technology continues progressing, with the hope of finding more effective and economical production methods. The work by Guo, Stuckey and Murphy [33] presented the possibility of developing a PHBV production system without the use of fossil fuels. PHBV polymers have a slightly lower energy consumption during production per kg of polymer than petrochemical polymers. The current production processes and scale of PHBV production are still largely undeveloped, compared with the well-developed production of petrochemical polymers. It is forecast that further optimization of PHBV production technology and the expansion of its production scale may result in an improvement in the environmental condition. Additionally, the results of the work show that the use of renewable sources, instead of fossil electricity and heat resources required for the production of PHBV, will ensure the effective optimization of the process. These results confirm the view that, due to the expansion in the development of bioplastics production and changes in the sources of generation for the electricity and heat consumed, the bioplastics production industry can be independent from fossil fuel products.

One of the possible ways of commercializing ecological composite materials, especially PHBV, may be the use of natural fillers, e.g., fibrous fillers in biopolymer matrices. It is expected that their use will improve the mechanical properties of the manufactured composites while maintaining complete biodegradability and low production costs, compared with pure biopolymers [34]. Fibers of plant origin are cheaper than synthetic fibers, such as glass fiber. Of course, it should be noted that the prices of natural fibers also depend on geographical location, a very important aspect in terms of the availability of suitable natural fibers [35,36]. In Europe, the main emphasis is placed on the production of flax fibers and, to a lesser extent, hemp fibers, while in Asia, hemp, jute and kenaf fibers are more popular. Kenaf is commercially grown in the United States, while sisal is widely cultivated in tropical African countries, the West Indies and the Far East. The largest global production of plant fibers is bamboo and sugar cane stalks [36]. Many car construction and equipment elements are manufactured on the basis of composites based on thermoplastics and natural fibers. Door panels, seat backrests, trunk elements and upholstery are made of this type of composites [37]. The use of natural fiber composites is mainly due to their lower production costs, weight reduction, recyclability and marketing incentives in the era of environmental protection. Natural fibers, such as linen, hemp, cotton and jute, have long been used in the production of textiles and other products, but now their role is expanding to the area of plastics [38]. Adding natural fibers to polymers can provide a number of benefits. Firstly, there is the potential to significantly reduce the amount of plastic used, which directly translates into reduced waste. Moreover, natural fibers are biodegradable, which means that products containing them will decompose more easily after the end of the product's life [39]. Natural fibers as fillers can also improve the mechanical properties of plastics. For example, the addition of hemp fibers to a polymer can increase its tensile strength and fracture resistance [40]. In addition, the variation in structure in natural fibers adds an aesthetic appearance to products, which is particularly attractive to consumers looking for more ecological options without sacrificing attractive design. However, introducing natural fibers into plastics is not without its challenges. The mixing and forming processes must be adjusted to obtain optimal mechanical properties. Moreover, quality control and the standardization of production processes become crucial to maintain product consistency [41].

An important issue is the possibility of developing modern biodegradable polymers of natural origin, such as PHBV bioplastic and hemp fiber filler. Currently the studies conducted [42–45] indicate an improvement in the properties of the obtained composites, compared with pure PHBV, after the addition of this filler. Another very important issue, that may significantly affect the possibilities of commercialization of this type of composites, is the assessment of the possibility of their multiple reprocessing (recycling) and the examination of the impact of subsequent reprocessing cycles on the functional, mechanical and processing properties of the obtained materials and a quality of products.

2. Materials and Methods

2.1. Research Materials

A copolymer of poly(3-hydroxybutyric acid) and poly(3-hydroxyvaleric acid) PHBV was used to produce the biocomposite, with the trade name ENMAT Y1000P NaturePlast (Mondeville, France), a specific weight of 1250 kg/m^3 and a softening temperature in the range of 165 °C to 175 °C [46]. ENMAT Y1000P belongs to the group of polyhydroxyalkanoates (PHA). The share of PHV in the biopolymer used was 8%.

The fillers used were hemp fibers with a length (L) of approximately 1 mm, produced by EKOTEX company (Kowalowice, Poland) and surface-modified with a 10% sodium hydroxide solution (the fibers were etched to improve adhesion to the polymer matrix). Fibers with an approximate length to diameter ratio (L/d) of 10 were used. The fibers were used to reduce production costs and improve some mechanical, processing and functional properties, compared with pure biopolymers, while maintaining the ability to biodegrade [42–45]. These fibers are characterized by a cellulose content of approximately 68%, a hemicellulose content of approximately 15%, and a lignin and other ingredients content of approximately 10% [37,47].

The produced biocomposite contained 30% of the mass share of hemp fibers and 70% of the mass share of the polymer matrix. The fiber type and the mass fraction of the filler were chosen based on the results of previously conducted research, i.e., to improve the mechanical, processing and functional properties, compared with pure PHBV [42–45].

Due to the large number of produced biocomposite moldings and reprocessed series, markings were introduced for the series of tests, which are summarized in Table 1.

Table 1. List of produced and reprocessed series of biocomposites.

Designation	Biocomposite/Multiple Processing
0x	starting/primary material
1x	reprocessed
2x	twice processed
3x	processed three times
4x	processed four times
5x	processed five times

2.2. Production of Composite and Test Samples

The process of manufacturing the PHBV–hemp fiber biocomposite consisted of several stages. After mixing the hemp fiber with PHBV, the resulting mixture was dried in a Chemland DZ-2BC (Szczecin Stargard, Poland) laboratory dryer equipped with a vacuum pump. The drying process was carried out for 1 h at 90 °C.

The biocomposite of PHBV–hemp fiber was produced using an extrusion technological line, consisting of a single-screw extruder from ZAMAK EHP-25E (produced by ZAMAK Mercator company, Skawina, Poland) [48], a cooling bath and a granulator.

The PHBV–hemp fiber biocomposite was extruded at the extruder processing temperature profile shown in Table 2. The extrusion was carried out at a screw rotation speed of 100 rpm.

Table 2. Temperatures of the extruder heating zones.

Set Temperature [°C]			
Head	**Zone 3**	**Zone 2**	**Zone 1**
175	170	160	150

The last stage of production was the granulation of the produced material and, as a result, the obtaining of biocomposite granules. A Zamak granulator (produced by ZAMAK Mercator company, Skawina, Poland) was used, equipped with a cutting tool (mill), which allowed for the collection of granules with cylindrical geometry. The obtained granulate, before the next stage of the process, was dried in a laboratory dryer for 1 h at 90 °C.

During the injection molding process for the test samples, a BOY55E injection molding machine (produced by BOY Maschines Inc., Exton, PA, USA) with a Priamus system was used, allowing for the control and monitoring of the injection molding machine.

The injection mold was equipped with temperature and pressure sensors (Figure 1) as components of the Priamus system.

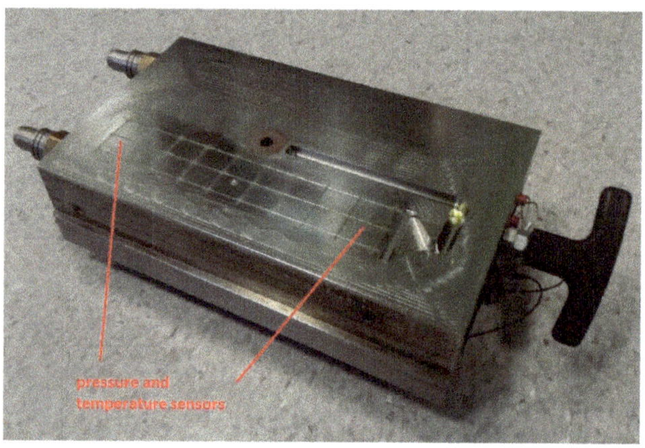

Figure 1. Injection mold with temperature and pressure sensors.

During the injection of the molded parts, the pressure in the mold cavity was measured using Priamus 6002B piezoelectric sensors [49] and the Priased 5080A (produced by Priamus System Technologies AG, Schaffhausen, Szwajcaria) four-channel amplifier integrated with them. The Fill Control software (version 1.0) used allowed for the recording of data from measurement channels for individual zones [50].

The temperature was measured using N-type thermoelectric sensors, which were mounted in the flow path at the same distance as the pressure sensors. The obtained measurements allowed the determination of the rheological characteristics of the composite and its subsequent processed batches.

In the first stage of the research, the samples with dog-bone geometry were produced from granules of the PHBV–hemp fiber biocomposite. The adjustable processing parameters of this material are presented in Table 3. During subsequent injection cycles, the viscosity values of the biocomposite, as a function of the shear rate and the profile of pressure changes, were recorded.

Table 3. Adjustable processing parameters for injection molding of the biocomposite.

Parameter	Value
Mold temperature [°C]	85
Melt temperature [°C]	185
Cooling time [s]	25
Packing time [s]	25
Packing pressure [MPa]	30
Flow rate [cm^3/s]	35

After mechanical property tests, the manufactured molded pieces were ground using a Wanner C17 (Wertheim, Germany) plastics mill. The ground pieces were dried for 1 h at 90 °C, and then the samples with dog-bone geometry were injected again. Again, changes in cavity pressure and viscosity were measured as a function of shear rate. The procedure for subsequent processing of the tested biocomposite was carried out five times. During the processing and testing of subsequent series of biocomposite, the same adjustable parameters were used in the injection molding process. Correction of these parameters was not required, due to the fact that the molded pieces obtained in subsequent series were of good quality, in terms of their shape and dimensions (Figure 2). There were also no organoleptically visible effects of degradation of the biocomposite in the form of flashes, burns or underflows after subsequent series of reprocessing, which may confirm that the quality of the processed biocomposite after repeated processing does not significantly deteriorate, in terms of the possibility of another processing cycle.

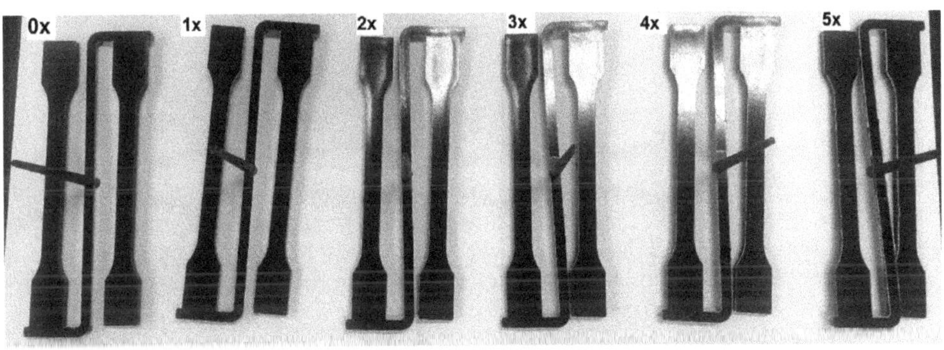

Figure 2. Samples from the original composite (0x) and subsequent reprocessing cycles (1x–5x).

2.3. Research Methods

Uniaxial tensile test.

The Zwick Z030 (produced by Zwick Roell, Ulm, Germany) testing machine was used to test the strength of the obtained composites. The uniaxial tensile test was carried out in accordance with the EN ISO 527-1 standard [51] for molded pieces with dog-bone geometry. Each series of samples consisted of seven molded pieces. Based on the obtained test results, the following were analyzed: the Young's modulus (E), tensile strength (σ_M) and the relative elongation at maximum tensile stress (ε_M). The results were analyzed statistically; the following were determined: the arithmetic mean ($\overline{x}$), standard deviation (s) and coefficient of variation (V).

Brinell hardness

The hardness assessment of the biocomposite was carried out using the Brinell method with the EN ISO 2039-1 standard [52] in two areas of the sample (Figure 3), i.e., in the measurement zone (zone A) and in the gripping part (zone B). A Zwick 3106 (produced by Zwick Roell, Ulm, Germany) hardness tester was used for this purpose. Each series of samples consisted of seven pieces.

Figure 3. Areas of the sample for hardness testing.

Impact tensile test

In order to determine the impact tensile strength of biocomposites, tests were carried out in accordance with the EN ISO 8256 standard [53] using a CEAST 9050 pendulum hammer (produced by Instron Inc. Europe, Buckinghamshire, UK). The samples were cut from the ones for uniaxial tensile testing, in accordance with the requirements of the standard. The notch was made for entire sample packages. Each series included seven samples.

Microstructure studies

To visually assess the sample surfaces and fiber geometry, a Nikon MM-800 workshop (produced by Nikon Inc., Tokyo, Japan) microscope with E-MAX software was used. The dimensions measured were min. fifty fibers on the top layer of the molded piece, for the first and subsequent processed series. The measurements were performed for each sample in the same area of the molded piece.

Microstructure tests were carried out using a HITACHI S-3400 scanning electron microscope (SEM) (produced by Hitachi Inc., Tokyo, Japan), based on specimens from the uniaxial tensile test.

Shrinkage assessment

The shrinkage of the molded parts with dog-bone geometry was tested partly based on the EN ISO 294-4 standard [54] The primary shrinkage was tested after approx. 3 h, and the secondary shrinkage was tested approx. 14 days after the molded pieces were manufactured by means of the injection molding process. The tests were performed for a series of seven samples.

Rheological test

The determination of the viscosity curves for the repeatedly injected biocomposite was done by recording the pressure and temperature, using appropriate sensors mounted in the injection mold cavity. The melted plastic is rheologically described by the Newtonian model. Its viscosity, in this case, can be calculated as [55]:

$$\eta = \frac{\tau}{\gamma} \tag{1}$$

where:

η—viscosity,

τ—shear stress,

and γ—shear rate.

Calculation of the shear stress and shear rate is possible thanks to the knowledge of the geometry of the injection mold cavity and the pressure values in two different places along the flow path of the polymer material. For a rectangular mold cavity cross-section:

$$\gamma = \frac{3 * \dot{Q}}{4 * k^2 * z} \tag{2}$$

$$\tau = \frac{\Delta p * k}{W} \tag{3}$$

where:

$\dot{Q}$—flow rate,

k—cavity height,

z—cavity width,

Δp—pressure difference between pressure sensors,

and W—distance between pressure sensors.

Knowing the relationships for shear rate (2) and shear stress (3), Equation (1), used to determine the viscosity, can be written as:

$$\eta = \frac{\tau}{\gamma} = \frac{\frac{\Delta p * k}{W}}{\frac{3 * \dot{Q}}{4 * k^2 * z}} = \frac{4 * k^3 * z * \Delta p}{3 * \dot{Q} * W} \tag{4}$$

Viscosity curves were determined for newly produced (0x) and multi-reprocessed (1x, 2x, 3x, 4x, 5x) PHBV–hemp fiber biocomposites. Viscosity measurements were made for various temperatures of the melt (180 °C, 185 °C and 190 °C), at injection rates ranging from 10 to 70 cm^3/s.

3. Results and Discussion

The results of the uniaxial tensile test were analyzed, taking into account the Young's modulus, tensile strength and elongation of the samples. When analyzing the values of the Young's modulus (Figure 4), there was no visible trend of a decrease or an increase in its value. However, it was observed that, with the next reprocessing cycle, the dispersion of the results increased, which may indicate a deteriorating homogeneity of the biocomposite. In turn, in the case of tensile strength (Figure 5), there was a visible trend associated with a decrease in its value after subsequent reprocessing cycles, by up to 18% (for the 5x biocomposite), compared with the originally produced biocomposite (0x). There was also a noticeable decrease in the maximum elongation of the sample, by approximately 13% for the 5x biocomposite, compared with the 0x biocomposite. When analyzing these three properties, a significant impact of reprocessing on the mechanical properties of the products can be noted.

When adding natural fibers to the polymer matrix, one can increase the tensile strength of the composite. These fibers act as reinforcement, improving the load-bearing capacity of the composite along the tensile direction [44,56]. After repeated cycles of processing, the fibers become mechanically shortened, and carry loads in their longitudinal direction to a lesser extent, which reduces their tensile strength. The reduction in sample elongation (Figure 6) may be related to the thermal load history of the biocomposite, which becomes less and less flexible after each reprocessing. Additionally, multiple reprocessing can affect the molecular structure of the polymer, which in turn affects its mechanical properties. This may lead to a loss of elasticity and is manifested by a reduced ability to deform before fracture [57,58].

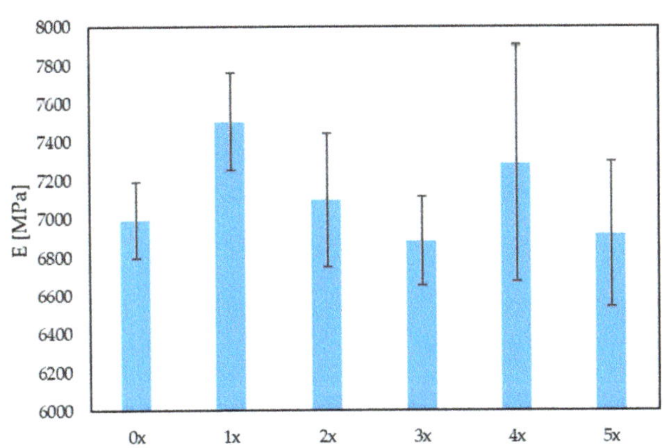

Figure 4. Young's modulus for multi-reprocessed biocomposites.

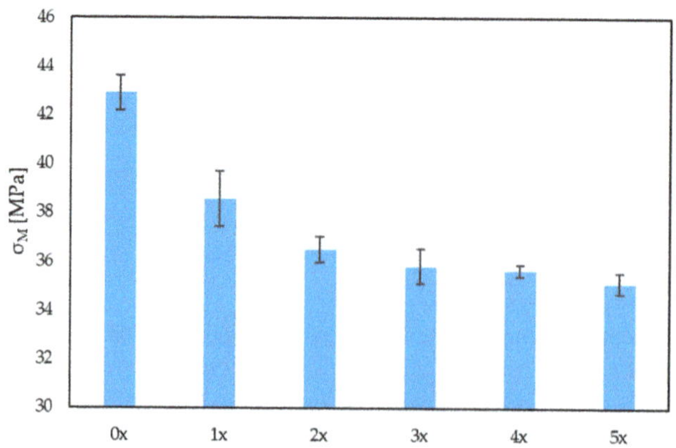

Figure 5. Tensile strength for multi-reprocessed biocomposites.

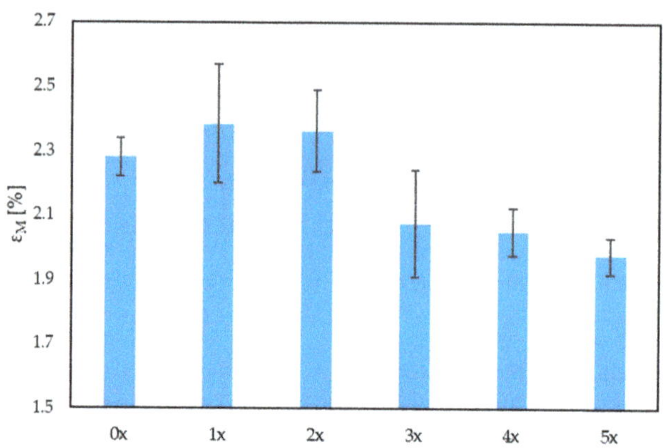

Figure 6. Relative elongation at the maximum tensile stress for multi-reprocessed biocomposites.

The results of the Brinell hardness tests indicate a slight decrease after repeated processing of the biocomposite in area A of the sample, i.e., in the narrowing (Figure 7a). The decrease in value is approximately 6% for the 5-fold recycled biocomposite (5x), compared with the originally produced one (0x), but this result is within the standard deviation. In the case of hardness tests in area B (Figure 7b), there is a noticeable increase in hardness by approximately 15% for the 5x biocomposite, compared with 0x. Basically, it can be noted that, for the initially produced biocomposite (0x), there is a significant difference in hardness in the area A (by approx. 28%) compared with area B. A significant influence of the geometry and fiber distribution in the polymer matrix is visible here. The fibers in the area of the measuring part of the sample are oriented along the length of the sample and are parallel to each other, which results from the constant geometry of the mold cavity. This results, among others, in an even distribution of fibers in layers throughout the thickness of the molded part, including its surface layer, and a significant improvement in its mechanical properties. In turn, in area B, i.e., the gripping part of the sample, the fibers are distributed more chaotically, due to the change in the geometry of the forming cavity. A similar trend is confirmed in publication [44], where the influence of adding hemp, wood and linen filler (15% by mass) on the improvement in hardness in relation to pure PHBV was examined—a significant improvement in hardness was found after adding each of the fibrous fillers in areas A and B. A similar trend was also noted here, i.e., in area A the hardness was significantly higher than in area B. From the tests carried out for the composite reprocessed five times, it can be noted that the hardness values in both areas A and B are similar to each other. This fact may be due to the shortening of the fiber length by the repeated grinding of the samples. As a result of the shortening of their length, the fibers are characterized by a lower aspect ratio (L/d) and lose their mechanical properties as a typical reinforcement of the polymer matrix.

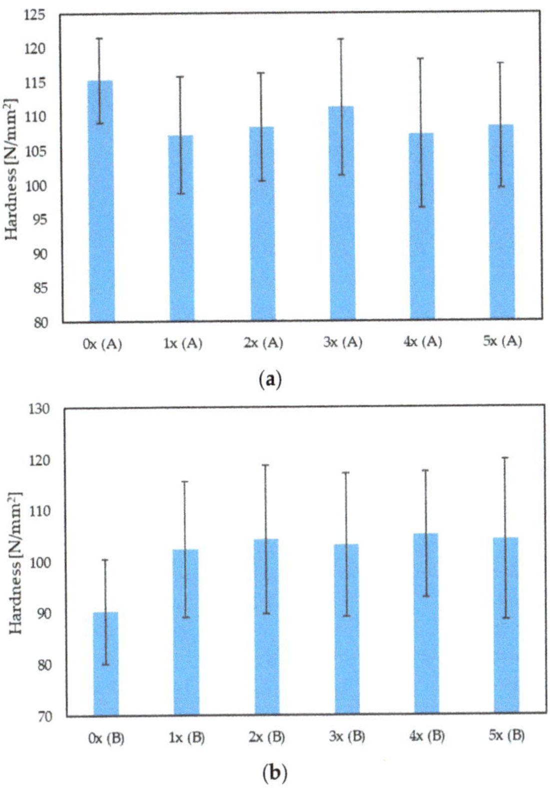

Figure 7. Hardness for multi-reprocessed biocomposites, in areas A (**a**) and B (**b**) of the molding.

Testing the hardness of plastics using the Brinell method is an important tool in the field of materials engineering, allowing for the assessment and control of the quality of materials at various stages of the production process and during operation. In the context of polymer biocomposites, this test is important because it allows one to determine how this material behaves under load, how easily it is deformed and what its mechanical properties are [59]. Hardness test results can be used to evaluate the quality, strength and applications of a given plastic. In the context of the quality of materials evaluation, hardness is an important parameter when assessing the quality of plastic products. Materials with higher hardness are usually more durable and resistant to abrasion, which is crucial for many applications, such as machine components, tools or structural elements [60,61]. Based on the results obtained (Figure 7), in this context, changes in hardness after repeated reprocessing cycles are small, and repeated reprocessing of the biocomposite itself may result in uniform hardness in various areas of the molded product.

When analyzing the results of the impact tensile strength (Figure 8), a significant decrease in this parameter can be noted after repeated cycles of reprocessing. The impact tensile strength of the 5-times processed biocomposite (5x) is approximately 29% lower than the original biocomposite (0x). It is also possible to notice a significant increase in the dispersion of test results after subsequent reprocessing attempts, which proves a greater diversity of sample properties within the tested group. The fibers embedded in the polymer matrix of the originally processed composite had a relatively constant length of approximately 1 mm. After multiple processing cycles, some fibers were shortened as a result of the grinding of the moldings, which varied the length of the fibers and, consequently, influenced the differences in the impact tensile strength of the reprocessed samples.

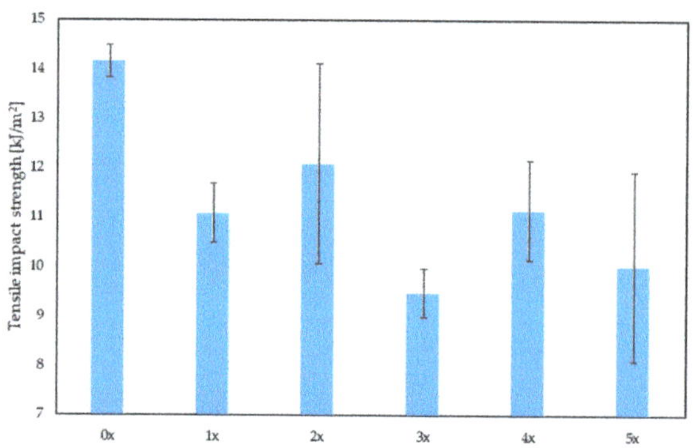

Figure 8. Impact tensile strength for biocomposites depending on the multiplicity of processing.

The influence of the impact strength of biocomposites with fiber filler on the quality of products may be significant, especially in the context of products that are exposed to dynamic loads, such as impacts or vibrations. Fiber fillers such as plant-derived fibers can improve the impact strength of biocomposites [44]. However, based on the impact tensile results (Figure 8), it can be concluded that the mere presence of fibers does not automatically guarantee an increase in impact strength. The impact strength of a biocomposite may be influenced by many factors, such as the type of fibers, their length, their orientation, their quantity and the repeated processing of the biocomposite [62].

Fiber fillers have a significant impact on the longitudinal shrinkage of materials. The fibers reinforce the structure, which may reduce the tendency of longitudinal shrinkage. Fibers of appropriate length and distribution can improve the elasticity of the material, which translates into a reduced risk of deformation under the influence of longitudinal stresses. Fibrous fillers also play a key role in controlling transverse shrinkage. Fibers, dispersed evenly in the material, can prevent excessive transverse shrinkage, which is especially important in the case of materials subject to changing environmental conditions. It is also worth raising the issue of the impact of natural fibers on shrinkage through the thickness of the product. Properly selected fibers can increase the density of the material, which translates into greater thickness. However, there is a subtle balance between the number of fibers and the preservation of mechanical properties. Excess fibers may lead to the compaction of the material, which may consequently reduce the thickness of the products [63–65].

When analyzing the values of longitudinal shrinkage (Figure 9a), a significant increase in the value of this parameter can be found with subsequent reprocessing cycles. The fibers shorten with each subsequent grinding of the moldings, which reduces the importance of this filler as reinforcement in the longitudinal direction. The value of secondary longitudinal shrinkage increased by almost 82% for the 5x biocomposite, compared with the original 0x. In turn, the value of secondary transverse shrinkage (Figure 9b) was reduced by approximately 28% for the 5x biocomposite, compared with 0x. There was also a decrease in the secondary shrinkage in thickness (Figure 9c) by approximately 14% for the 5-times reprocessed biocomposite, compared with the original one. The reduction in shrinkage in these directions may be caused by the fragmentation of the hemp fibers after reprocessing, which are distributed throughout the entire volume of the melt, and not mainly in the longitudinal direction, as was the case with the originally processed biocomposite.

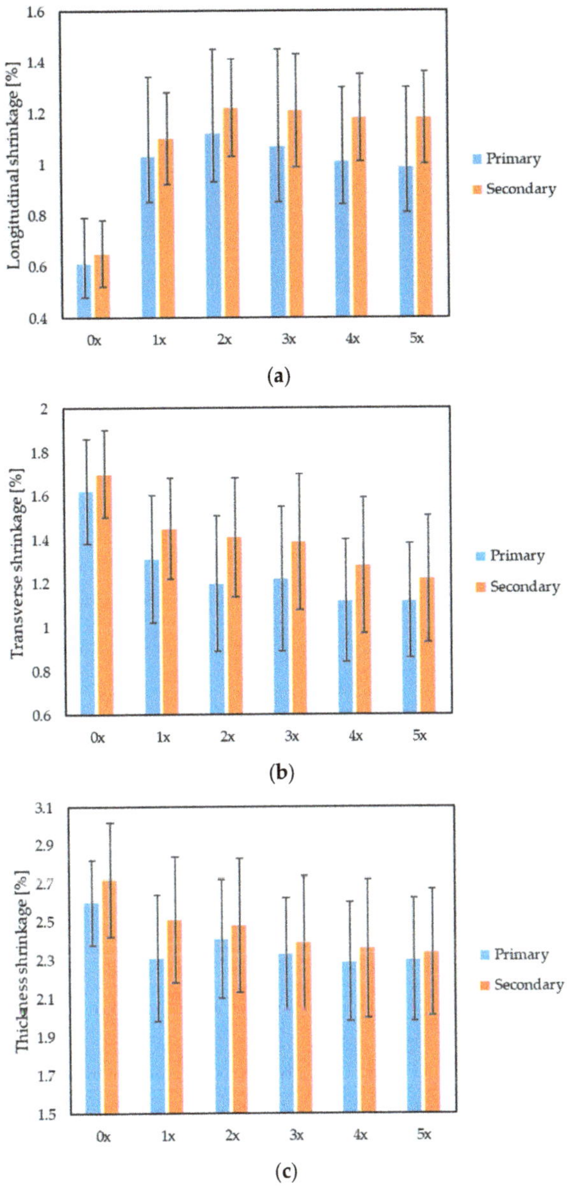

Figure 9. Primary and secondary shrinkage of injection molded parts: (**a**) longitudinal, (**b**) transverse and (**c**) in thickness, for multi-reprocessed biocomposites and the primary biocomposite.

Importantly, variable parameters characterizing the fibers after subsequent series of reprocessing were the length (L) and diameter (d) of the fibers, and, consequently, the so-called aspect ratio: L/d. This coefficient is important in the processing of composites filled with short fibers because, after exceeding the critical length (above the critical L/d value), the fibers do not function as reinforcement. They no longer improve the properties of composites, especially in the direction of loads acting along the fiber axis. A too-small contact surface of the fiber with the polymer matrix makes them crack in the middle of their length under the influence of stress [66–68].

When analyzing the lengths of the fibers presented in Table 4 and the photograph (Figure 10), it can be noticed that, for each processed series of the biocomposite, the fibers are shortened, and, after the final fifth processing cycle (5x), the hemp fibers were approximately 41% shorter in length than the fibers in the originally manufactured moldings (0x). There was no significant change in fiber diameter. As a consequence, the fiber aspect ratio decreased by approximately 42%. This reduction in the value of the L/d parameter results in, among others, the deterioration of the mechanical properties and the increased shrinkage of the molded parts, as evidenced by the results obtained. In turn, referring to the results regarding the rheological characteristics of the material, obtained experimentally during injection, the shortening of the length of the fibers reduces the viscosity of the flowing material (due to lower flow resistance, as the fibers are shorter after each injection cycle) and the occurrence of lower pressure values in the mold cavity. It is noticeable that, with each subsequent reprocessing cycle, the fibers are less and less oriented in a specific direction. In addition, an interesting phenomenon noted during fiber length measurements is the fact that the fibers are less and less visible on the surface layer of the molded part, which may indicate that they tend to stick together in the core of the molded piece.

The fracture structure of the samples after the uniaxial tensile test is shown in Figure 11. SEM photographs were taken for samples originally produced (0x) and samples processed five times (5x). Photographs taken for the extreme conditions show increasing fiber fragmentation with the number of reprocessing cycles. The fibers also thin, and the polymer matrix shows signs of heat stress. The photos also indicate a good, uniform distribution of the fibers in the matrix. It can also be observed that the homogenization of fiber distribution in the matrix increases with the number of processing times.

The processing properties of the PHBV–hemp fiber biocomposites were determined for subsequent processing cycles (0x, 1x, 2x, 3x, 4x, 5x), during their injection into a special injection mold, by determining the viscosity curve. For each biocomposite, viscosity measurements were performed at various temperature of the melt plastic (180 °C, 185 °C and 190 °C) and at various injection speeds, ranging from 10 to 70 cm^3/s. The experimentally determined changes in the biocomposite viscosity in the injection mold cavity are shown, for example, in Figure 12. It was noted that, with increasing temperatures and reprocessing cycles, lower viscosity values were observed for the same shear rates. This relationship can be observed in particular when comparing the determined viscosity curve for the 0x and 5x biocomposites (Figure 13).

Table 4. Average length L, diameter d and L/d aspect ratio for 50 fibers in the biocomposite after subsequent processing cycles.

Average Values from the Measurements of 50 Fibers			
Biocomposite	**L [mm]**	**d [mm]**	**Aspect Ratio L/d**
0x	0.991	0.113	8.770
1x	0.782	0.110	7.109
2x	0.714	0.113	6.319
3x	0.604	0.105	5.752
4x	0.493	0.108	4.563
5x	0.403	0.109	3.696

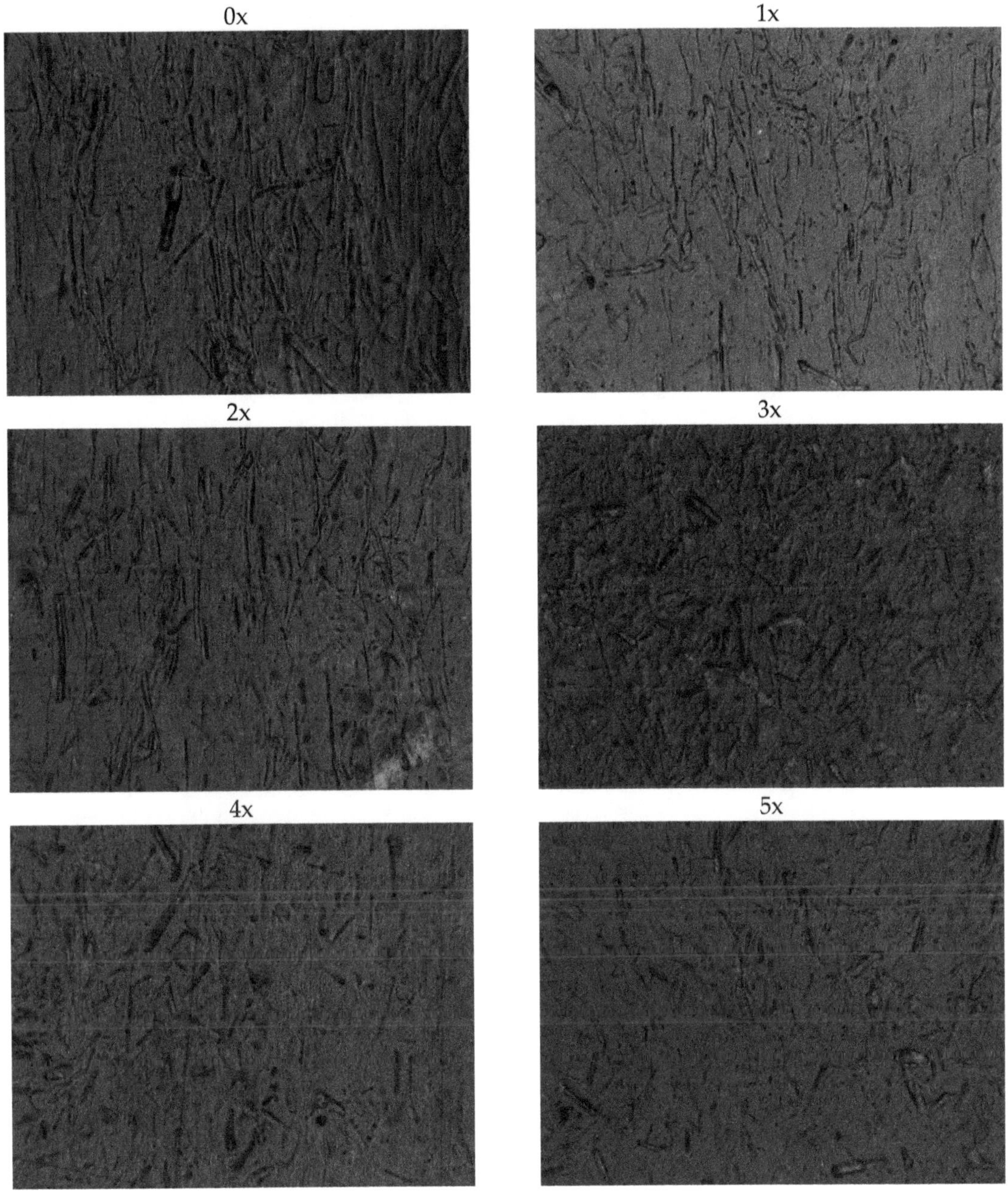

Figure 10. Photographs of fibers on the surface top layer of the molded piece for subsequent biocomposite reprocessing cycles (50× magnification).

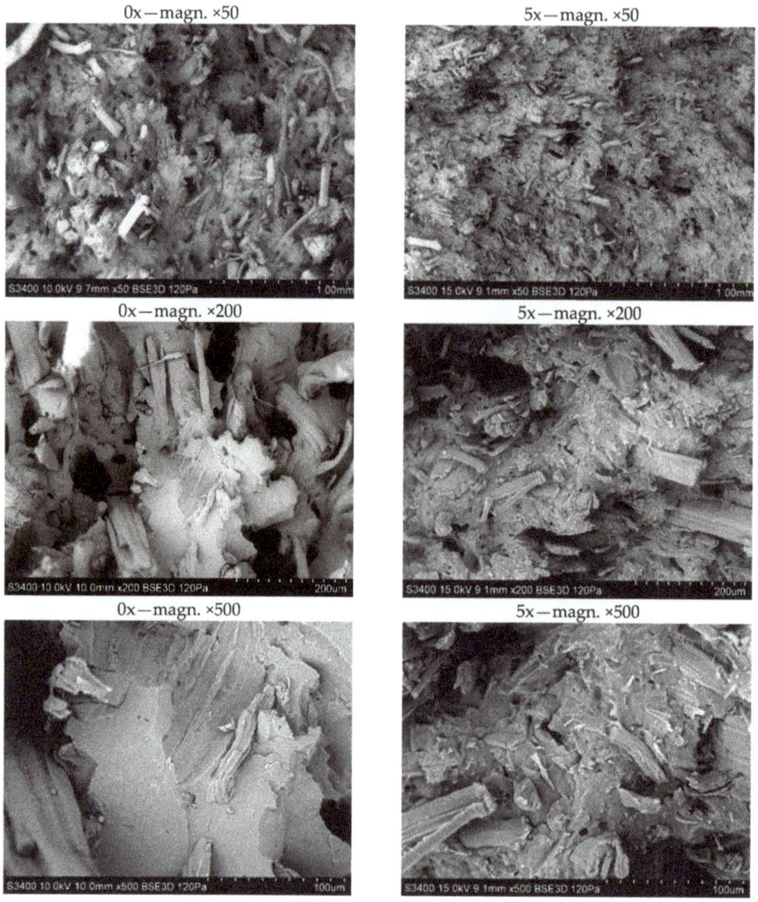

Figure 11. SEM photographs of the fracture surfaces of biocomposite samples for 0x and 5x biocomposites.

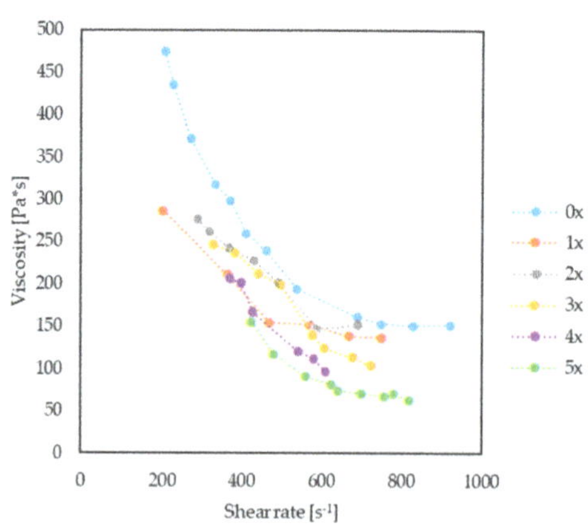

Figure 12. Examples of viscosity curves for biocomposites: 0x, 1x, 2x, 3x, 4x and 5x, at a temperature of 185 °C, obtained based on injection molding tests.

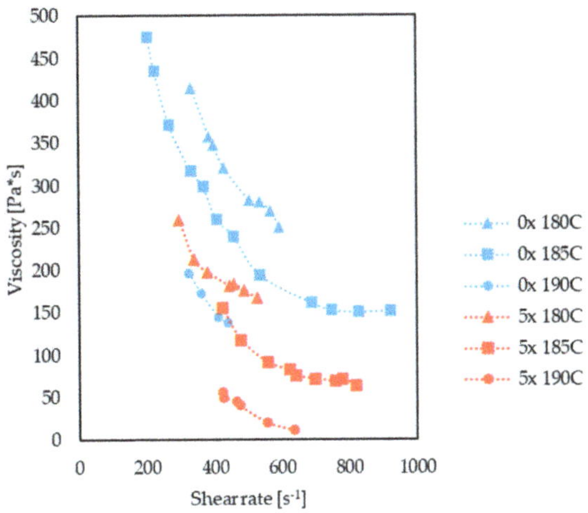

Figure 13. Comparison of viscosity curves for 0x and 5x biocomposites at temperatures of 180 °C, 185 °C and 190 °C.

It should be noted that viscosity curves were experimentally obtained in narrow shear rate ranges. Very often, in practice and in scientific work, special rheological models are used to extrapolate viscosity values to shear rate areas for which no experimental tests have been carried out [55,69,70]. The experimentally obtained viscosity curves were extrapolated using the Cross–WLF (Cross-Williams–Landel–Ferry) model [71,72]:

$$\eta(\gamma, T, p) = \frac{\eta_0(T, p)}{1 + \left(\frac{\eta_0 * \gamma}{\tau^*}\right)^{1-n_c}} \quad (5)$$

where:

η—viscosity,

T—temperature,

p—pressure,

n_c i τ^*—Cross–WLF model constants,

and n_c—zero viscosity.

The change in zero viscosity as a function of temperature can be written as:

$$\eta_0(\mathrm{T}, \mathrm{p}) = D_1 * exp\left[-\frac{A_1 * (T - T_g)}{A_2 + (T - T_g)}\right] \quad (6)$$

where:

$$T_g(p) = D_2 + D_3 * p \quad (7)$$

$$A_2 = A_3 + D_3 * p \quad (8)$$

where:

T_g—glass transition temperature,

and $D_1, D_2, D_3, A_1, A_2, A_3$—Cross—WLF model constants.

For the original and subsequently processed biocomposites, based on experimentally determined data, the values of the Cross–WLF model parameters necessary to determine viscosity curves in the shear rate range of 10^1–10^6 s^{-1} were determined by means of the Data Fit 9.0 software (Table 5). Viscosity curves, calculated using the Autodesk Moldflow Insight 2021 software (based on experimental data), are presented in Figure 14.

Table 5. Parameters of the Cross–WLF model of multi-reprocessed biocomposites.

	Biocomposite Processing Multiplicity					
Parameter	**0x**	**2x**	**3x**	**4x**	**5x**	**6x**
n_c [-]	0.1220	0.1291	0.1312	0.1395	0.1425	0.1587
τ^* [Pa]	74,657.9	79,657.9	80,013.3	85,237.0	92,075.0	98,762.0
D_1 [Pa*s]	1.78×10^{11}	1.60×10^{11}	1.40×10^{11}	1.10×10^{11}	9.00×10^{10}	8.00×10^{10}
D_2 [K]	282.25	282.25	282.25	282.25	282.25	282.25
D_3 [K/Pa]	0	0	0	0	0	0
A_1 [-]	19.87	20.54	21.22	22.17	23.33	24.13
A_2 [K]	51.6	51.6	51.6	51.6	51.6	51.6
Determination ceofficient R^2 [-]	0.1220	0.1291	0.1312	0.1395	0.1425	0.1587

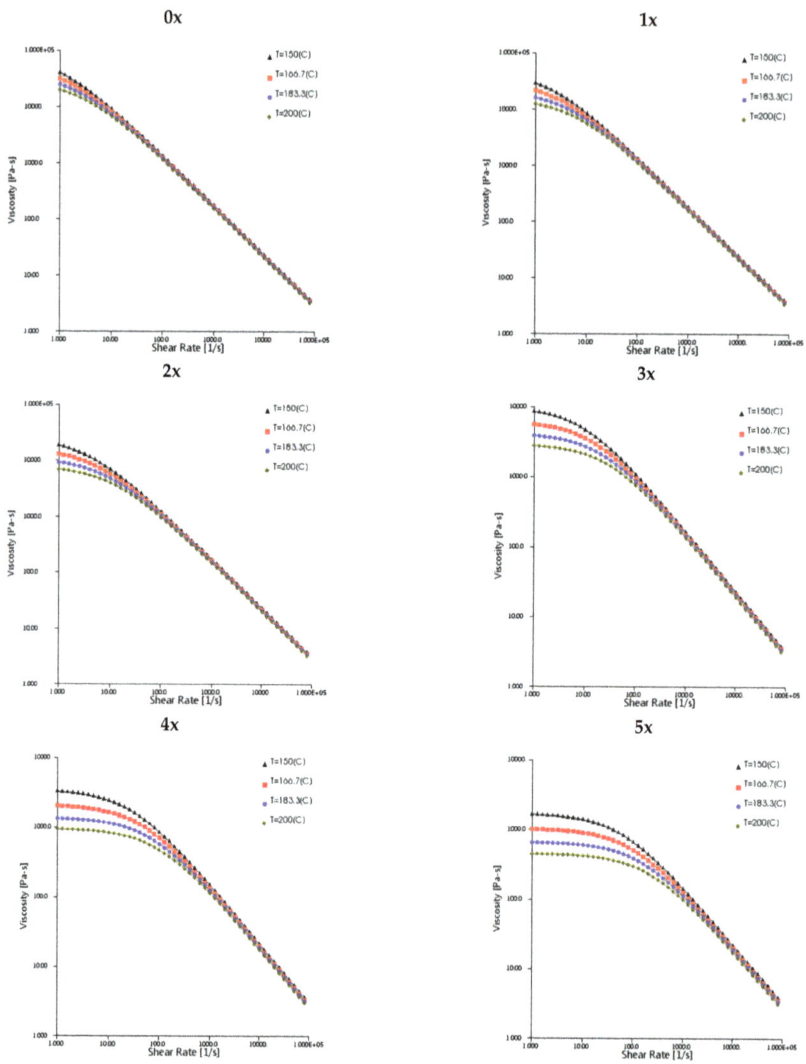

Figure 14. Viscosity curves calculated using Autodesk Moldflow Insight 2021 software for biocomposites processed multiple times (0x, 1x, 2x, 3x, 4x and 5x).

Also in the context of viscosity changes, the impact of decreasing the viscosity of the biocomposite in subsequent processing cycles on the injection process itself was analyzed. Using the PRIAMUS system, the profiles of pressure changes in the molding cavity were recorded for subsequent cycles of the processed biocomposite. It was noted (Figure 15) that, with subsequent cycles, there was a noticeable decrease in the maximum pressure values in the mold cavity at similar shear rate values. A flowing stream of plastic with increasingly lower viscosity causes lower flow resistance, which leads to lower pressure values. This is important when controlling the quality of the injection molding process. Obtaining lower pressures in the mold cavity may result in the stabilization of the injection process and less exploitation of the injection molding machine and injection mold [73,74].

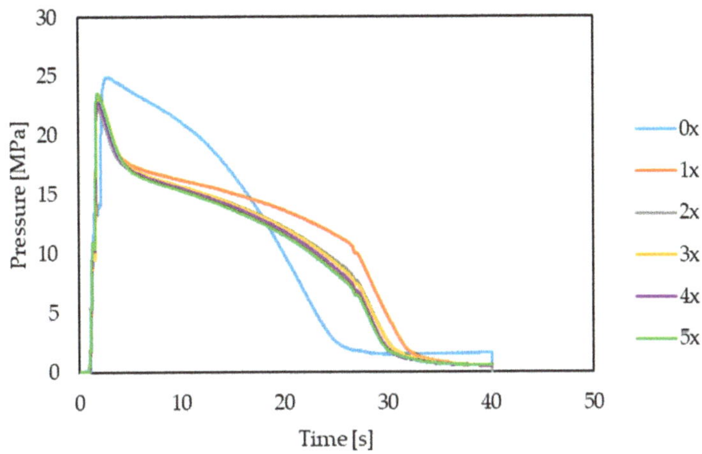

Figure 15. Representative pressure profiles in the mold cavity for subsequent biocomposite processing cycles.

4. Summary and Conclusions

The impact of multi-reprocessing of biocomposites with hemp fibers on the quality of the injection products can be quite complex and depends on many factors. Repeated processing may lead to the gradual degradation of the fibers in the biocomposite. This, in turn, may affect the mechanical properties of the composite. Long-term processing may result in fibers breaking and shortening. The results obtained in the uniaxial tensile test indicate a deterioration of the mechanical properties in terms of tensile strength by up to approx. 18% and in terms of the elongation of the sample by approx. 13%, compared with the originally produced biocomposite. A similar trend was also noted in the context of impact tensile strength, where a decrease of approximately 29% in the value of this parameter was noted after the fifth reprocessing. In turn, when analyzing the hardness results in the two tested measurement areas, it is possible to note a decrease in hardness in the narrowing part of the sample by approximately 6%, where this decrease is within the standard deviation. In turn, in the case of measurements of the gripping part of the sample, an increase in hardness of approximately 15% was noted, which may be due to the fact that the shortened fibers are distributed more evenly in that part of the mold cavity, where the directionality of the flow of the material changes.

Taking into account the quality of the products in terms of their shape and dimensional accuracy, it was found that, with each reprocessing cycle, the longitudinal shrinkage of the product increased, which was the result of shortening of the fibers during the grinding of the molded pieces. With subsequent reprocessing cycles, the length of the fibers decreased, each time by an average of approx. 9–21%, with the largest decrease recorded for the first cycle (1x). This corresponds to the largest change in the value of longitudinal shrinkage, i.e., as much as 69%, and approx. 15% in the case of transverse shrinkage, also for the first cycle.

For the second cycle, negligible changes in transverse shrinkage (approx. 2%) and slight changes in longitudinal shrinkage (approx. 10%) were recorded. Subsequent processing cycles do not bring significant changes, both in terms of fiber length and shrinkage value. Much shorter, fragmented fibers only have a small impact on the shrinkage of the polymer matrix. The observed decrease in the value of transverse and thickness shrinkage is most likely the result of the orientation of the shortened fibers not only in the longitudinal direction (as was the case with the originally processed biocomposite), but in the entire volume of the polymer matrix. The described phenomena, regarding changes in shrinkage, may be confirmed by tests of the composite structure on the surface layer of the molded part, which show that the fibers have actually shortened.

A very important issue in the context of multi-reprocessing of the PHBV–hemp fiber biocomposite are changes in the viscosity of the material, which require adjustments to the adjustable parameters of the processing process. The Priamus system was used in the research. The results indicate a general trend that the viscosity of this material decreases with the number of times the PHBV–hemp fiber biocomposite is processed. This is confirmed by the recorded pressure changes in the molding cavity—there are lower maximum pressure values are obtained after repeated processing. The recorded rheological behavior of the biocomposite may be due to the fact that the fibers, after shortening, generate a lower flow resistance while the plastic flows, which reduces the viscosity and also reduces the length of the polymer chains.

The impact of reprocessing biocomposites with natural fibers on the quality of injected products is an important issue in the fields of materials engineering and industry. Biocomposites, which are a combination of natural fillers and biopolymers, are becoming more and more popular due to their environmental friendliness. However, the repeated processing of these materials may generate challenges related to the quality of the final products. The first key aspect is the degradation of natural fibers due to repeated processing. Natural fibers, such as hemp fibers, may be shortened as a result of mechanical and thermal processes. As a result, their tensile strength and elasticity (as a result of thermal degradation) may decrease, which is an important factor affecting the mechanical properties of the biocomposite.

Another quality feature of injection molded products is dimensional stability. The impact of repeated processing on the structure of the tested biocomposite may lead to undesirable changes in the dimensions of the injected products. Therefore, it is necessary to monitor these changes and adjust the injection process to minimize the impact on dimensional accuracy. The surface quality of products injected from tested biocomposites subject to repeated processing also deserves attention. Possible changes in the structure of the tested material may affect the appearance and smoothness of the surface, which is especially important in the case of products with high aesthetic requirements.

Ultimately, the question of the recyclability of biocomposites is becoming more and more important. Reprocessing must be consistent with the principles of sustainable development, and possible changes in the structure of materials should be taken into account in recycling processes too. The reprocessing of biocomposites may lead to changes in their rheological properties. An increase in the number of reprocessing cycles can affect the viscosity, the ability of the fibers to disperse in the polymer matrix and the ability of the biocomposite to flow during injection. A good understanding of the impact of repeated processing on the processing capabilities and rheology of biocomposites is essential to improve production processes, minimize material losses and achieve the optimal quality of molded products. Research on the impact of the reprocessing of biocomposites with natural fibers on the quality of injection molded products is of fundamental importance for the further development of this promising field of materials engineering.

It is necessary to constantly improve manufacturing processes, monitor changes in the properties of processing materials and search for innovative solutions aimed at minimizing the negative impact on the quality of final products. In this way, biocomposites with natural fibers can become an even more competitive alternative to traditional plastics, combining ecological benefits with high quality products.

Author Contributions: Conceptualization, A.P., W.F. and G.J.; methodology, A.P., W.F. and G.J.; software, D.S.; validation, G.J.; formal analysis, D.S.; investigation, Ł.B. and G.J.; resources, D.S., G.J. and Ł.B.; data curation, W.F.; writing—original draft preparation, G.J.; writing—review and editing, A.P. and W.F.; visualization, A.P.; supervision, A.P. and W.F.; project administration, A.P. and W.F.; funding acquisition, A.P. and W.F. All authors have read and agreed to the published version of the manuscript.

Funding: This research received no external funding.

Institutional Review Board Statement: Not applicable.

Informed Consent Statement: Not applicable.

Data Availability Statement: The data presented in this study are available on request from the corresponding author.

Conflicts of Interest: The authors declare no conflict of interest.

References

1. Bobulski, J.; Kubanek, M. Deep learning for plastic waste classification system. *Appl. Comput. Intell. Soft Comput.* **2021**, *2021*, 6626948. [CrossRef]
2. Chow, C.F.; So, W.M.W.; Cheung, T.Y.; Yeung, S.K.D. Plastic waste problem and education for plastic waste management. In *Emerging Practices in Scholarship of Learning and Teaching in a Digital Era*; Springer: Berlin/Heidelberg, Germany, 2017; pp. 125–140.
3. Kosior, E.; Crescenzi, I. Solutions to the plastic waste problem on land and in the oceans. In *Plastic Waste and Recycling*; Academic Press: Cambridge, MA, USA, 2020; pp. 415–446.
4. Pacana, A.; Radon-Cholewa, A.; Pacana, J.; Wozny, A. The study of stickiness of packaging film by Shainin method. *Przemysł Chem.* **2015**, *94*, 1334–1336.
5. Alaerts, L.; Augustinus, M.; Van Acker, K. Impact of Bio-Based Plastics on Current Recycling of Plastics. *Sustainability* **2018**, *10*, 1487. [CrossRef]
6. Pacana, A.; Bednarova, L.; Liberko, I.; Wozny, A. Effect of selected production factors of the stretch film on its extensibility. *Przemysł Chem.* **2014**, *93*, 1139–1140.
7. Jnr, A.K.L.; Yunana, D.; Kamsouloum, P.; Webster, M.; Wilson, D.C.; Cheeseman, C. Recycling waste plastics in developing countries: Use of low-density polyethylene water sachets to form plastic bonded sand blocks. *Waste Manag.* **2018**, *80*, 112–118.
8. Narancic, T.; O'Connor, K.E. Plastic waste as a global challenge: Are biodegradable plastics the answer to the plastic waste problem? *Microbiology* **2019**, *165*, 129–137. [CrossRef]
9. Kale, G.; Kijchavengkul, T.; Auras, R.; Rubino, M.; Selke, S.E.; Singh, S.P. Compostability of bioplastic packaging materials: An overview. *Macromol. Biosci.* **2007**, *7*, 255–277. [CrossRef]
10. Kalantari, K.; Afifi, A.M.; Jahangirian, H.; Webster, T.J. Biomedical applications of chitosan electrospun nanofibers as a green polymer–Review. *Carbohydr. Polym.* **2019**, *207*, 588–600. [CrossRef]
11. Siwiec, D.; Pacana, A. Model Supporting Development Decisions by Considering Qualitative-Environmental Aspects. *Sustainability* **2021**, *13*, 9067. [CrossRef]
12. Corvellec, H.; Stowell, A.F.; Johansson, N. Critiques of the circular economy. *J. Ind. Ecol.* **2021**, *26*, 421–432. [CrossRef]
13. Wang, Y.; Yin, J.; Chen, G. Q. Polyhydroxyalkanoates, challenges and opportunities. *Curr. Opin. Biotechnol.* **2014**, *30*, 59–65. [CrossRef] [PubMed]
14. Reddy, C.S.K.; Ghai, R.; Rashmi; Kalia, V.C. Polyhydroxyalkanoates: An overview. *Bioresour. Technol.* **2003**, *87*, 137–146. [CrossRef] [PubMed]
15. Kalia, V.C.; Singh Patel, S.K.; Shanmugam, R.; Lee, J.-K. Polyhydroxyalkanoates: Trends and advances toward biotechnological applications. *Bioresour. Technol.* **2021**, *326*, 124737. [CrossRef] [PubMed]
16. Kumar, P.; Ray, S.; Kalia, V.C. Production of co-polymers of polyhydroxyalkanoates by regulating the hydrolysis of biowastes. *Bioresour. Technol.* **2016**, *200*, 413–419. [CrossRef] [PubMed]
17. Venkateswar Reddy, M.; Venkata Mohan, S. Influence of aerobic and anoxic microenvironments on polyhydroxyalkanoates (PHA) production from food waste and acidogenic effluents using aerobic consortia. *Bioresour. Technol.* **2012**, *103*, 313–321. [CrossRef]
18. Gedde, U.W. Crystalline Polymers. In *Polymer Physics*; Springer: Dordrecht, The Netherlands, 1999. [CrossRef]
19. Beun, J.J.; Dircks, K.; Van Loosdrecht, M.C.M.; Heijnen, J.J. Poly-β-hydroxybutyrate metabolism in dynamically fed mixed microbial cultures. *Water Res.* **2002**, *36*, 1167–1180. [CrossRef]
20. Serafim, L.S.; Lemos, P.C.; Albuquerque, M.G.; Reis, M.A. Strategies for PHA production by mixed cultures and renewable waste materials. *Appl. Microbiol. Biotechnol.* **2008**, *81*, 615–628. [CrossRef]
21. Beccari, M.; Bertin, L.; Dionisi, D.; Fava, F.; Lampis, S.; Majone, M.; Valentino, F.; Vallini, G.; Villano, M. Exploiting olive oil mill effluents as a renewable resource for production of biodegradable polymers through a combined anaerobic–aerobic process. *J. Chem. Technol. Biotechnol.* **2009**, *84*, 901–908. [CrossRef]

22. Albuquerque, M.G.E.; Martino, V.; Pollet, E.; Avérous, L.; Reis, M.A.M. Mixed culture polyhydroxyalkanoate (PHA) production from volatile fatty acid (VFA)-rich streams: Effect of substrate composition and feeding regime on PHA productivity, composition and properties. *J. Biotechnol.* **2011**, *151*, 66–76. [CrossRef]
23. Pandey, A.; Adama, N.; Adjallé, K.; Blais, J.-F. Sustainable applications of polyhydroxyalkanoates in various fields: A critical review. *Int. J. Biol. Macromol.* **2022**, *221*, 1184–1201. [CrossRef]
24. Kumari, S.V.; Pakshirajan, K.; Pugazhenthi, G. Recent advances and future prospects of cellulose, starch, chitosan, polylactic acid and polyhydroxyalkanoates for sustainable food packaging applications. *Int. J. Biol. Macromol.* **2022**, *221*, 163–182. [CrossRef] [PubMed]
25. Alves, A.A.; Siqueira, E.C.; Barros, M.P.; Silva, P.E.; Houllou, L.M. Polyhydroxyalkanoates: A review of microbial production and technology application. *Int. J. Environ. Sci. Technol.* **2022**, *20*, 3409–3420. [CrossRef]
26. Policastro, G.; Panico, A.; Fabbricino, M. Improving biological production of poly(3-Hydroxybutyrate-co-3-hydroxyvalerate) (PHBV) co-polymer: A critical review. *Rev. Environ. Sci. Bio/Technol.* **2021**, *20*, 479–513. [CrossRef]
27. Vogel, R.; Tändler, B.; Voigt, D.; Jehnichen, D.; Häußler, L.; Peitzsch, L.; Brünig, H. Melt spinning of bacterial aliphatic polyester using reactive extrusion for improvement of crystallization. *Macromol. Biosci.* **2007**, *7*, 820–828. [CrossRef] [PubMed]
28. Arakawa, K.; Yokohara, T.; Yamaguchi, M. Enhancement of melt elasticity for Poly(3-hydroxybutyrate-Co-3-hydroxyvalerate) by addition of weak gel. *J. Appl. Polym. Sci.* **2007**, *107*, 1320–1324. [CrossRef]
29. Blackburn, R. *Biodegradable and Sustainable Fibres*; Taylor & Francis: London, UK, 2005.
30. Vogel, R.; Tändler, B.; Häussler, L.; Jehnichen, D.; Brünig, H. Melt spinning of poly(3-hydroxybutyrate) fibers for tissue engineering using α-cyclodextrin/polymer inclusion complexes as the nucleation agent. *Macromol. Biosci.* **2006**, *6*, 730–736. [CrossRef] [PubMed]
31. Bledzki, A.K.; Jaszkiewicz, A. Mechanical performance of biocomposites based on PLA and PHBV reinforced with Natural Fibres—A comparative study to PP. *Compos. Sci. Technol.* **2010**, *70*, 1687–1696. [CrossRef]
32. Chen, G.X.; Hao, G.J.; Guo, T.Y.; Song, M.D.; Zhang, B.H. Structure and mechanical properties of poly (3-hydroxybutyrate-co-3-hydroxyvalerate)(PHBV)/clay nanocomposites. *J. Mater. Sci. Lett.* **2002**, *21*, 1587–1589. [CrossRef]
33. Guo, M.; Stuckey, D.C.; Murphy, R.J. Is it possible to develop biopolymer production systems independent of fossil fuels? case study in energy profiling of polyhydroxybutyrate-Valerate (PHBV). *Green Chem.* **2013**, *15*, 706. [CrossRef]
34. Meereboer, K.W.; Pal, A.K.; Cisneros-López, E.O.; Misra, M.; Mohanty, A.K. The effect of natural fillers on the marine biodegradation behaviour of poly(3-hydroxybutyrate-co-3-hydroxyvalerate) (PHBV). *Sci. Rep.* **2021**, *11*, 911. [CrossRef]
35. Dittenber, D.B.; GangaRao, H.V.S. Critical Review of recent publications on use of natural composites in infrastructure. *Compos. Part A Appl. Sci. Manuf.* **2012**, *43*, 1419–1429. [CrossRef]
36. Faruk, O.; Bledzki, A.K.; Fink, H.-P.; Sain, M. Biocomposites reinforced with natural fibers: 2000–2010. *Prog. Polym. Sci.* **2012**, *37*, 1552–1596. [CrossRef]
37. Ferreira, F.; Pinheiro, I.; de Souza, S.; Mei, L.; Lona, L. Polymer composites reinforced with natural fibers and nanocellulose in the automotive industry: A short review. *J. Compos. Sci.* **2019**, *3*, 51. [CrossRef]
38. Brebu, M. Environmental degradation of plastic composites with natural fillers—A review. *Polymers* **2020**, *12*, 166. [CrossRef] [PubMed]
39. Saba, N.; Tahir, P.; Jawaid, M. A review on potentiality of Nano Filler/natural fiber filled polymer hybrid composites. *Polymers* **2014**, *6*, 2247–2273. [CrossRef]
40. Pickering, K.L.; Sawpan, M.A.; Jayaraman, J.; Fernyhough, A. Influence of loading rate, alkali fibre treatment and crystallinity on fracture toughness of random short hemp fibre reinforced polylactide bio-composites. *Compos. Part A Appl. Sci. Manuf.* **2011**, *42*, 1148–1156. [CrossRef]
41. Pacana, A.; Siwiec, D. Model to Predict Quality of Photovoltaic Panels Considering Customers' Expectations. *Energies* **2022**, *15*, 1101. [CrossRef]
42. Janowski, G.; Frącz, W.; Bąk, Ł.; Trzepieciński, T. The Effect of the Extrusion Method on Processing and Selected Properties of Poly(3-hydroxybutyric-co-3-hydroxyvaleric Acid)-Based Biocomposites with Flax and Hemp Fibers. *Polymers* **2022**, *14*, 5370. [CrossRef]
43. Janowski, G.; Frącz, W.; Bąk, Ł. The Mechanical Properties Prediction of Poly [(3-hydroxybutyrate)-co-(3-hydroxyvalerate)] (PHBV) Biocomposites on a Chosen Example. *Materials* **2022**, *15*, 7531. [CrossRef]
44. Frącz, W.; Janowski, G.; Smusz, R.; Szumski, M. The Influence of Chosen Plant Fillers in PHBV Composites on the Processing Conditions, Mechanical Properties and Quality of Molded Pieces. *Polymers* **2021**, *13*, 3934. [CrossRef]
45. Frącz, W.; Janowski, G.; Bąk, Ł. Influence of the Alkali Treatment of Flax and Hemp Fibers on the Properties of PHBV Based Biocomposites. *Polymers* **2021**, *13*, 1965. [CrossRef] [PubMed]
46. TianAn Biologic Materials Co. *Safety Data Sheet Enmat, Y1000P*; TianAn Biologic Materials Co.: Ningbo, China, 2009.
47. Kim, J.K.; Pal, K. *Recent Advances in the Processing of Wood-Plastic Composites*; Springer: Berlin/Heidelberg, Germany, 2010.
48. ZAMAK Mercator. *Technical and Start-Up Documentation of the Extruder ZAMAK EHP-25E*; ZAMAK Mercator: Skawina, Poland, 2011.
49. Priamus System Technologies. *Technical Documentation of Sensors Priamus 6002B*; Priamus System Technologies: Schaffhausen, Switzerland, 2016.

50. Priamus System Technologies. *Technical Documentation of the System Priamus*; Priamus System Technologies: Schaffhausen, Switzerland, 2019.
51. *EN ISO 527-1*; Plastics—Determination of Tensile Properties. In Part 1: General Principles. International Organization for Standardization: Geneva, Switzerland, 2019.
52. *EN ISO 2039-1*; Plastics—Determination of Hardness Part 1: Ball Indentation Method. International Organization for Standardization: Geneva, Switzerland, 2001.
53. *EN ISO 8256*; Plastics—Determination of Tensile-Impact Strength. International Organization for Standardization: Geneva, Switzerland, 2023.
54. *EN ISO 294-4*; Plastics—Injection Moulding of Test Specimens of Thermoplastic Materials-Part 4: Determination of Moulding Shrinkage. International Organization for Standardization: Geneva, Switzerland, 2018.
55. Frącz, W.; Janowski, G. Determination of viscosity curve and PVT properties for wood-polymer composite. *Wood Res.* **2018**, *2*, 321–334.
56. Jagadeesh, P.; Puttegowda, M.; Thyavihalli Girijappa, Y.G.; Rangappa, S.M.; Siengchin, S. Effect of natural filler materials on fiber reinforced hybrid polymer composites: An overview. *J. Nat. Fibers* **2020**, *19*, 4132–4147. [CrossRef]
57. Singh, M.K.; Mohanty, A.K.; Misra, M. Upcycling of waste polyolefins in natural fiber and sustainable filler-based biocomposites: A study on recent developments and future perspectives. *Compos. Part B Eng.* **2023**, *263*, 110852. [CrossRef]
58. Augustia, V.A.; Chafidz, A. Production and recycling of Biocomposites: Present Trends and Future Perspectives. *Recycl. Polym. Blends Compos.* **2023**, 389–403.
59. Kaymakci, A.; Ayrilmis, N. Investigation of correlation between Brinell hardness and tensile strength of wood plastic composites. *Compos. Part B Eng.* **2014**, *58*, 582–585. [CrossRef]
60. Yusuf, S.S.; Islam, N.; Ali, H.; Akram, W.; Siddique, A. Optimum process parameters selection for Brinell hardness number of natural fiber reinforced composites using Taguchi method. *Saudi J. Eng. Technol.* **2019**, *4*, 422–427. [CrossRef]
61. Sydor, M.; Pinkowski, G.; Jasińska, A. The Brinell method for determining hardness of wood flooring materials. *Forests* **2020**, *11*, 878. [CrossRef]
62. Kuan, H.T.; Tan, M.Y.; Shen, Y.; Yahya, M.Y. Mechanical properties of particulate organic natural filler-reinforced polymer composite: A Review. *Compos. Adv. Mater.* **2021**, *30*, 263498332110075. [CrossRef]
63. Yeo, J.-S.; Lee, J.-H.; Hwang, S.-H. Effects of lignin on the volume shrinkage and mechanical properties of a styrene/unsaturated polyester/lignin ternary composite system. *Compos. Part B Eng.* **2017**, *130*, 167–173. [CrossRef]
64. Tsujimoto, A.; Barkmeier, W.W.; Takamizawa, T.; Latta, M.A.; Miyazaki, M. Mechanical properties, volumetric shrinkage and depth of cure of short fiber-reinforced resin composite. *Dent. Mater. J.* **2016**, *35*, 418–424. [CrossRef] [PubMed]
65. Santos, J.D.; Fajardo, J.I.; Cuji, A.R.; García, J.A.; Garzón, L.E.; López, L.M. Experimental evaluation and simulation of volumetric shrinkage and warpage on polymeric composite reinforced with short natural fibers. *Front. Mech. Eng.* **2015**, *10*, 287–293. [CrossRef]
66. Fu, S. Effects of fiber length and fiber orientation distributions on the tensile strength of short-fiber-reinforced polymers. *Compos. Sci. Technol.* **1996**, *56*, 1179–1190. [CrossRef]
67. Vas, L.M.; Ronkay, F.; Czigány, T. Active fiber length distribution and its application to determine the critical fiber length. *Polym. Test.* **2009**, *28*, 752–759. [CrossRef]
68. Heim, D.; Hartmann, M.; Neumayer, J.; Klotz, C.; Ahmet-Tsaous, Ö.; Zaremba, S.; Drechsler, K. Novel method for determination of critical fiber length in short fiber carbon/carbon composites by double lap joint. *Compos. Part B Eng.* **2013**, *54*, 365–370. [CrossRef]
69. Peydró, M.A.; Parres, F.; Crespo, J.E.; Juárez, D. Study of rheological behavior during the recovery process of high impact polystyrene using cross-WLF model. *J. Appl. Polym. Sci.* **2010**, *120*, 2400–2410. [CrossRef]
70. Olley, P.; Mulvaney-Johnson, L.; Coates, P.D. Simulation of the gas-assisted injection moulding process using a viscoelastic extension to the cross-wlf viscosity model. *Proc. Inst. Mech. Eng. Part E J. Process Mech. Eng.* **2011**, *225*, 239–254. [CrossRef]
71. Shi, X.Z.; Huang, M.; Zhao, Z.F.; Shen, C.Y. Nonlinear fitting technology of 7-parameter cross-WLF viscosity model. *Adv. Mater. Res.* **2011**, *189–193*, 2103–2106. [CrossRef]
72. Baum, M.; Anders, D.; Reinicke, T. Approaches for numerical modeling and simulation of the filling phase in injection molding: A Review. *Polymers* **2023**, *15*, 4220. [CrossRef]
73. Kurt, M.; Saban Kamber, O.; Kaynak, Y.; Atakok, G.; Girit, O. Experimental investigation of plastic injection molding: Assessment of the effects of cavity pressure and mold temperature on the quality of the final products. *Mater. Des.* **2009**, *30*, 3217–3224. [CrossRef]
74. Araújo, C.; Pereira, D.; Dias, D.; Marques, R.; Cruz, S. In-cavity pressure measurements for failure diagnosis in the injection moulding process and correlation with numerical simulation. *Int. J. Adv. Manuf. Technol.* **2023**, *126*, 291–300. [CrossRef]

Disclaimer/Publisher's Note: The statements, opinions and data contained in all publications are solely those of the individual author(s) and contributor(s) and not of MDPI and/or the editor(s). MDPI and/or the editor(s) disclaim responsibility for any injury to people or property resulting from any ideas, methods, instructions or products referred to in the content.

materials

Article

Polypropylene Color Masterbatches Containing Layered Double Hydroxide Modified with Quinacridone and Phthalocyanine Pigments—Rheological, Thermal and Application Properties

Magdalena Kozłowska, Magdalena Lipińska *, Michał Okraska and Joanna Pietrasik

Institute of Polymer and Dye Technology, Lodz University of Technology, 90-924 Łódź, Poland; magdalena.stefaniak@dokt.p.lodz.pl (M.K.); michal.okraska@p.lodz.pl (M.O.); joanna.pietrasik@p.lodz.pl (J.P.)
* Correspondence: magdalena.lipinska@p.lodz.pl

Abstract: Polypropylene color masterbatches containing modified layered double hydroxides, LDHs, were created. The simple, industry-acceptable method of LDH surface modification with quinacridone and phthalocyanine pigments using the pulverization method in ball mills was applied. It was reported that the modification parameters such as time and rotational speed affected the tendency to create the aggregates for modified fillers. TGA analysis of the modified LDH showed that modification with phthalocyanine pigment shifted the temperature at which 5%, $T_{5\%}$, and 10% of mass loss, $T_{10\%}$, occurred compared with that for unmodified LDH. The viscoelastic properties of prepared masterbatches were investigated. The incorporation of the modified fillers instead of neat pigments led to an increase in the loss shear modulus, G'', indicating a stronger influence on the dissipation of energy by the melted masterbatch. The similar values of tan, δ, were determined for melted masterbatches containing phthalocyanine pigment and green modified LDH filler. The incorporation of both LDHs modified by phthalocyanine and quinacridone pigment fillers slightly increased the zero-shear viscosity, η_0, compared with that of the masterbatches based on the neat pigments. The Cole–Cole plots and the analysis of the Maxwell and continuous relaxation models showed that modified colored LDH fillers facilitated the relaxation of the melted masterbatch, and shorter relaxation times were observed. The phthalocyanine-modified LDH filler improved the thermal stability of the masterbatches. Additionally, the impact of pigments and modified, colored LDH on the crystallization of polypropylene was investigated.

Keywords: polypropylene masterbatches; layered double hydroxide; quinacridone pigments; phthalocyanine pigments; viscoelastic properties

1. Introduction

Polypropylene is one of the most important commodity plastics used in variety of applications, among them fibers [1], composite materials [2,3], and food packaging products [4]. Commonly, additives such as stabilizers and plasticizers [4], fillers [5,6], flame retardants [1] or dyes and pigments [7,8] are incorporated into polypropylene to adjust the mechanical and processing properties and to improve appearance.

Linear trans-quinacridone and phthalocyanine organic pigments have received attention for their mass coloration of the polypropylene fiber [1,9].

Quinacridone, 5,12-dihydroquinolino(2,3-b)acridine-7-14-dione is one of the most important red–violet-shade pigments [10], and due to its photovoltaic activity, efficient emission, effective carrier mobility, it is used in organic thin-film transistors [11]. Quinacridone particles form strong a particle–particle network via intermolecular N-H·····O hydrogen bonding and π–π stacking [12] that results in the insolubility of the quinacridone pigment in common organic solvents and strong difficulty in controlling particle size as well as the aggregation of quinacridone particles after synthesis.

Phthalocyanines are highly conjugated, aromatic, planar macrocyclic organic compounds with eighteen delocalized π-electrons [13,14]. Phthalocyanines exist as both free-metal and metal complexes, and they have been used in applications in catalysis, sensors, light-harvesting dyes in solar cells (organic photovoltaics), and absorbers in non-linear optic and textile dyes [13–15]. Phthalocyanine dissolved in s suitable solvent and complexed with heavy metal salts forms organic complexes with bright colors; thus, the commonly used blue and green organic pigments are phthalocyanine pigments [16,17]. Organic phthalocyanine pigments face the same problems as do quinacridone pigments in the process of practical application, including poor dispersion, uneven particle size and easy agglomeration properties leading to poor color quality of the final product.

Polyolefin products, including polypropylene, to be attractive for clients, need to have a desired color. Coloring substances are usually introduced into polymer melts, during the phase of obtaining a granulate (mixing extrusion), before the forming process (injection and extrusion) [18]. Most often, color concentrates, otherwise known as masterbatches, are used for dyeing polymers [19]. White color masterbatches based on titanium dioxide pigments are often used to produce color injection-molded products such as spoons, forks, knives [19,20]. The masterbatches reinforced with nanoclays were designed to produce food trays with barrier properties [21]. Various organic pigments, among them quinacridone pigments, ultramarine blue, diketopyrrolopyrole, are used to prepare polypropylene masterbatches by commercial companies [22]. A color masterbatch is a highly concentrated pigment or a mixture of pigments/dyes enclosed in the form of granules. They may additionally contain additional components, such as antioxidants, antistatic agents, plasticizers, etc. The production process is multi-stage. It starts (in the plasticizing system of the extruder) with the dosing of the ingredients, their melting and mixing with the provision of thermal energy, homogenization and the subsequent cooling of the material ribbon coming out of the head. The last stage is cutting the web into homogeneous granules using a granulator [18].

The obtained color concentrate offers plastic processors a number of advantages resulting from their use (Scheme 1). First is a reduction in material costs compared to those of ready-made mixtures dyed in mass. Another advantage is the ability to obtain the desired properties of the finished product. An important aspect is a reduction in or the complete elimination of the inhalation of toxic dust, which occurs when powders are used for dyeing.

The biggest disadvantage of using masterbatches is the possibility of the incomplete mixing of the ingredients of the composition. This process shows some non-uniformity in homogenization. The random nature of mixing polymer components to a certain extent, with temporary or permanent effects of the segregation of mixed components, causes significant disturbances in the uniformity of mixed components, and consequently an uneven coloring of the produced product [18]. Therefore, the prevention of pigment aggregation is crucial. The surface free energy and polarity of organic pigments differ from those of polypropylene, which can generate interfacial tension in polar/non-polar systems [23]. To prevent the aggregation of dyes, various strategies were applied, such as the addition of dispersants [24], the grafting of polymers on pigments [25], and the absorption of the pigment on polysaccharide materials via the presence of hydrogen bonding [26]. One of the promising methods is the modification of layered double hydroxides with organic dyes [27].

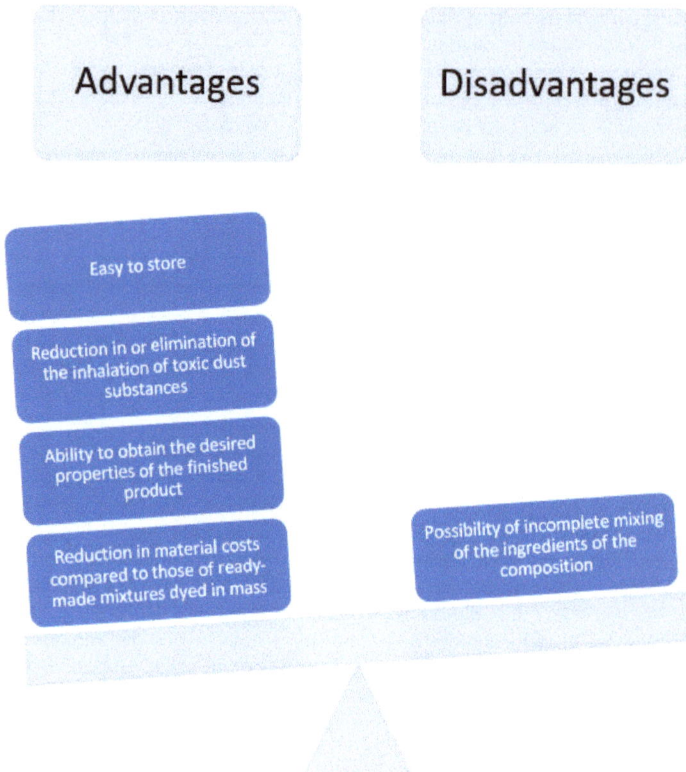

Scheme 1. Masterbatch or raw pigment?

Layered double hydroxides, LDHs, are composed of positively charged brucite-like layers containing anion and water molecules in the interlayer gallery, with the general formula $[M^{2+}{}_{1-x}M^{3+}x(OH^-)_2]^{x+}[(A^{n-})_{x/n}{\cdot}mH_2O]^{x-}$, where M^{2+} and M^{3+} are cations that occupy octahedral positions in hydroxide layers, A^{n-} [28]. Modified LDHs are able to improve the flame retardancy, thermal stability and combustion behavior of polyolefine material [29]. The Zn–Al LDH was used as inorganic host for the encapsulation of negatively charged anionic dyes; the hybrid organic–inorganic pigment was synthetized using the co-precipitation method [30]. Maragoni R. et al. [31] used layered hydroxide salts intercalated and adsorbed with anionic blue and orange dyes as coloring substances for poly(vinyl alcohol), PVA. Various blue dye molecules were incorporated via the co-precipitation method within the galleries of layered double hydroxide, and blue colored hybrid organic–inorganic pigments were applied as coloring additives to polystyrene PS [27]. The new color-tunable hybrid dark-red to violet pigments based on layered double hydroxides and 1,2-dihydroxyantraquinone dye were obtained via the modification of a filler surface in a water/alcohol dye solution [32]. Coiai S. et al. [33] prepared co-intercalated fluorescent LDH particles via both anion exchange and calcination–rehydration methods using, as modifying substances, fluorescein and alkyl sulfate anions; modification was an effective strategy for preventing the aggregation of the dye. Polypropylene nanocomposites with better tensile strength and enhanced durability under accelerated aging were prepared via the incorporation of layered double hydroxide modified with a UV-absorbing azo-dye, 3(4-anilinophenylazo) benzenesulfonic acid [34].

It can be concluded that it is possible to use layered double hydroxide as a carrier for coloring substances and to produce colored polymer nanocomposites. Thus, in our work colored LDH fillers modified with quinacridone and phthalocyanine pigments were used as the additive to the masterbatch. The industry-acceptable method of modification via the pulverization and application of ball mills was applied to obtain modified fillers. The rheological properties of polypropylene masterbatches containing modified colored LDH fillers were investigated. The impact of chosen color systems on the relaxation behavior of melted polypropylene was reported. It was expected that via the incorporation of the synergistic system LDH/pigment, the thermal properties of the masterbatch would be improved. Additionally, the effect of pigments and modified colored LDH on the crystallization of polypropylene was predicted.

2. Materials and Methods

2.1. Materials

The layered double hydroxide, LDH, hydrotalcite Pural MG70, product no. 595070, produced by Sasol Germany GmbH (Hamburg, Germany), further in the text denoted as HT, was used to prepare a modified hybrid colored filler. The properties of used hydrotalcite were as follows: a $MgO:Al_2O_3$ ratio of 70:30, and a surface area mg^2/g, measured using BET methods after calcination for 3 h at 550 °C, of 196 m^2/g.

The XRD patterns of used hydrotalcite are illustrated in Figure S1 (Supporting Information). The observed peaks were similar to those of the crystallographic pattern described in the literature for Mg–Al layered double hydroxides [35,36]. The existence of an ordered layered structure of the layered double hydroxide was confirmed via the presence of the main sharp and symmetrical peak at $2\theta = 11.40°$ (determined via the 003 reflection in XRD analysis). The interlayer spacing calculated based on the main peak according to Bragg equation [37] was 0.78 nm and did not changed after modification. Similar interlayer spacing was determined by other authors [35] for the unmodified layered double hydroxide containing carbonate anions in the interlayer space. The small characteristic peak was found at $2\theta = 5.6°$, indicating expanded interlayer spacing (d = 1.58 nm) due to presence of water in the interlayer space [36].

Figure 1 shows the morphology of the HT particles used to prepare the hybrid colored filler.

Figure 1. The SEM pictures of the layered double hydroxide, hydrotalcite, HT, used to prepare the hybrid colored filler (SEM microscope, LEO 1530 Gemini, producent Zeiss/Leo, Oberkochen, Germany).

Two various pigments were used to prepare the modified hybrid colored HT pigments: (1) pigment Green 7, phthalocyanine green $C_{32}C_{16}CuN_8$, no. CAS: 1328-53-6, producent Sigma-Aldrich (Poznań, Poland), further denoted as PG7; (2) pigment Red 122, 2,9-dimethylquinacridone,((2,9-dimethyl-5,12-dihydroquinolino [2,3-b]acridine-7,14-dione), $C_{22}H_{16}N_2O_2$, no. CAS 980-26-7, producent Sigma-Aldrich (Poznań, Poland), further denoted as PR122. The chemical structures of the used pigments are shown in Figure 2.

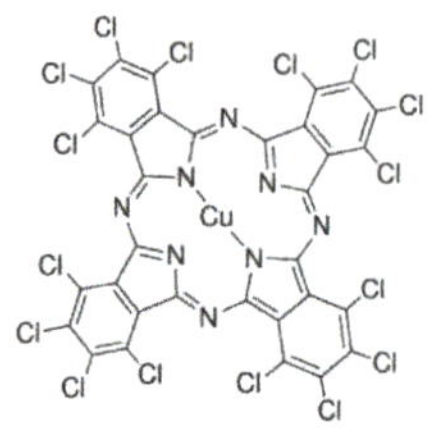

(a)

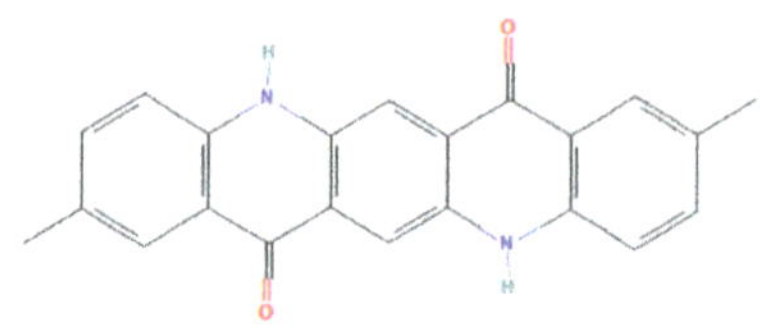

(b)

Figure 2. The chemical structures of used pigments: (**a**) pigment Green 7, phthalocyanine green, PG7; (**b**) pigment Red 122, 2,9-dimethylquinacridone, PR122.

2.2. Preparation of Modified Colored Filler

The modification was carried out by using the PM 200 planetary-ball mill (Pulverisette 5, Fritsch-GmBH, Idar-Oberstein, Germany) according to the Scheme 2. First, placing 10 g of pigment and 100 g of filler was placed in a steel vessel with grinding balls. A rotational speed of 50 rpm for 15 min was set, or 60 rpm for 10 min in order to find more favorable mixing parameters in terms of energy. In this case, the following sample designations were adopted: HT-Green 50 rpm, 60 rpm and HT-Red 50 rpm, 60 rpm for hybrid colored pigments modified, respectively by phthalocyanine green and 2,9-dimethylquinacridone using procedure described above.

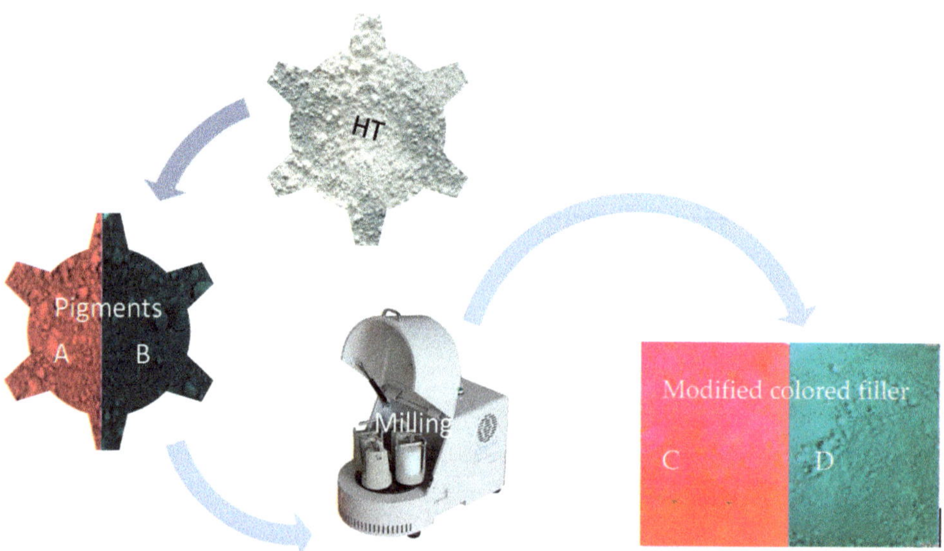

Scheme 2. Preparation of modified colored filler. A—pigment Red 122, 2,9-dimethylquinacridone,((2,9-dimethyl-5,12-dihydroquinolino[2,3-b]acridine-7,14-dione), PR122; B—pure pigment Green 7, phthalocyanine green, PG7; C—modified colored filler HT-Red 50 rpm; D—modified colored filler HT—Green 50 rpm.

The ball milling process is a green technology regarded as a "top-down" approach in the production of fine particles. The ball milling process can be used to increase the particle surface area and further to optimize the dispersion state of particles in biocomposites [38]. The mechanical energy created during milling allows the physical breakdown of coarse particles into finer ones; thus, it is used to produce fine drug particles [39]. Ball milling technology can be used for the preparation and chemical functionalization of polymers, e.g., nanocellulose derivatives [40]. Similarly, various milling techniques are applied for the simultaneous attainment of particle dispersion and the surface modification of solids such as metal oxides [41] or layered fillers [42]. The simplicity and low cost of filler modification using ball mills compared with other methods of filler surface modification, e.g., the solvent method, are great advantages for the industry. Among other surface modification methods, it allows an avoidance of the use of organic solvents and thus is regarded as being an environmentally friendly technique [40]. Here, it is also important that different types of ball milling equipment are available at low prices. Thus, we selected this method as an easy-to-use, fast, profitable and ecological method to modify the surface of the layered double hydroxide using selected pigments.

2.3. Preparation of Masterbatches

The carrier of the coloring substance was polypropylene, PP. As a polymeric carrier of the masterbatch, the polypropylene Sabic PP520P (Sabic Europe, Geelen, The Netherlands), particularly designed for the extrusion of cast films, and typically used for food, hygiene and textile packaging and lamination films, was used. The parameters of the PP were a melt flow rate, MFR, of 10.5 $g{\cdot}min^{-1}$ at 230 °C and 2.16 kg (ASTM D 1238 [43]), and a density of 905 $kg{\cdot}m^{-3}$ (ASTM D792 [44]).

The modified colored pigments HT-Green and HT-Red were added to the masterbatch as a replacement for the neat pigment in the masterbatch composition. The amount of incorporated modified pigments and neat pigments was 0.25 g for 100 g of matrix polypropylene for every masterbatch produced. The prepared samples were further denoted as follows: PP—neat polypropylene sample; PP/HT—polypropylene sample containing unmodified layered double hydroxide; PP/PG7—masterbatch containing neat pigment Green 7, phthalocyanine green; PP/PR122—masterbatch containing neat pigment Red 122, 2,9-dimethylquinacridone; PP/HT-Green—masterbatch containing layered double hydroxide modified with phthalocyanine green; PP/HT-Red—masterbatch containing layered double hydroxide modified with 2,9-dimethylquinacridone.

The production of color concentrates was achieved using a single-screw extruder at 220 °C. The method of their preparation is presented in the diagram below (Scheme 3).

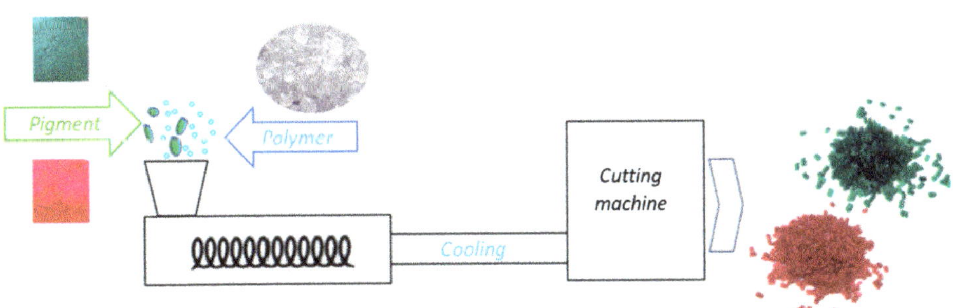

Scheme 3. Production of polypropylene color concentrate containing modified HT. C—modified colored filler HT-Red 50 rpm; D—modified colored filler HT—Green 50 rpm; E—masterbatch PP/HT-Red 50 rpm.; F—masterbatch PP/HT-Green 50 rpm.

Additionally, the test samples for viscoelastic studies in the form of plates with a thickness of 2 mm were made by using a single-screw extruder and an injection molding machine.

2.4. The Characterization of Modified Colored LDH Fillers

To estimate the tendency toward aggregation after modification, aggregate size analysis of the prepared modified fillers was performed by using the dynamic light scattering, DLS, technique. The size of the unmodified layered double hydroxide HT, neat pigments PG7 and PR122 and modified colored fillers HT-Green and HT-Red was determined using Zetasizer Nano Serie S90 (Malvern Panalytical Ltd., Malvern, UK). The size of the particles was measured for the water dispersions, and the concentration of the dispersion was 5 g of modified filler per 100 mL of dispersing medium. To estimate the tendency of modified colored fillers in a non-polar medium, the size of the agglomerates was additionally measured in paraffin oil (model of polypropylene); paraffin oil produced by PHU Olmax S.J., Lodz, Poland was used. Before the measurements the dispersions were stabilized via ultrasonic treatment for 10 min (Ultrasonic bath, Bandelin Sonorex DT 255, Bandelin GmbH, Berlin, Germany).

The optical microscope images were obtained using the Opta-Tech Lab40 microscope (Opta-Tech, Warsaw, Poland) connected with digital camera Mi6 with 6 megapixel sensor-IMX178, Resolution -3072 × 2048 (Opta-Tech, Warsaw, Poland), and computer program Capture 2.3 (Opta-Tech, Warsaw, Poland). The results of the studies are compiled in the Supporting Information, Figures S2–S4.

The surface energy of modified colored fillers was calculated using the Owens–Wendt–Rabel–Kaelble, work, method [23,45]. The Owens–Wendt–Rabel–Kaelble method is the standard procedure for calculating the surface free energy of a solid. The tested solid is wetted with several measuring liquids, which enables the division of the surface energy into two components: polar and dispersive. The OWRK method enables the optimization and observation of polarization changes between two surfaces. The test was performed at ambient temperature using the K100 MKII tensiometer (KRÜSS GmbH, Hamburg, Germany). Polar (water and chloroform) and non-polar (1,4-dioxane) liquids were used. At least three measurements were taken for each testing liquid. In the addition, the polarity was calculated based on the simple equation polarity = γ^p/γ, where γ^p is the polar component (dynes·cm^{-1}) and γ is the total free energy (dynes·cm^{-1}).

2.5. Viscoelastic Properties at Processing Temperature 200 °C

The dynamic viscoelastic properties of PP masterbatch mixtures were studied at 200 °C. A similar temperature for the rheology tests was used by other authors [5] for polypropylene composites containing layered double hydroxides. The oscillation rheometer Ares G2 (TA Instruments, New Castle, DE, USA) equipped with a plate–plate geometry (diameter: 25 mm) was used during the tests.

To perform oscillation testing in the plate–plate geometry, the sample was loaded between plates and it oscillated back and forth at a given stress or strain amplitude and frequency. The applied motion can be represented as a sinusoidal wave with the stress or strain amplitude. The ratio of the applied stress (or strain) to the measured strain (or stress) is a quantitative measure of material stiffness, and it gives the complex modulus, G* (Equation (1)) [46].

$$G^* = \frac{\sigma_{max}}{\gamma_{max}} \tag{1}$$

For an elastic material (stress is proportional to strain), the maximum stress occurs at the maximum strain and both stress and strain are said to be in-phase. For viscous materials, stress and strain are out-of-phase by 90° or $\pi/2$ radians. For viscoelastic materials, the phase difference (the phase angle δ) between stress and strain is between two extremes. This phase difference allows the viscous (loss modulus, G″) and elastic component (storage modulus, G′) ratio to the total material stiffness (G*) to be determined in accordance with the equation (Equation (2)):

$$G^* = \sqrt{\left(G'^2 + G''^2\right)};\ G' = G^*\cos\delta;\ G'' = G^*\sin\delta \tag{2}$$

The relationship can also be presented in the form of Equation (3):

$$G^* = G' + iG'' \tag{3}$$

where i is an imaginary number equal to $\sqrt{1}$.

Complex viscosity, η^*, is a measure of the total resistance to flow as a function of angular frequency (ω) and it is calculated in accordance with Equation (4):

$$\eta^* = \frac{G^*}{\omega} \tag{4}$$

It can be divided into two component parts, that include dynamic viscosity, η', and the out-of-phase component of complex viscosity, η''. Both represent the real and the imaginary parts of η^*, respectively, in the following form (Equation (5)):

$$\eta^* = \eta' + i\eta'' \tag{5}$$

The amplitude sweep tests at 200 °C, at a constant value of angular frequency, 10 rads^{-1}, were performed; based on the tests the averages values of the storage shear modulus, G', loss shear modulus, G'', and loss factor tan, δ, for the linear viscoelastic region were calculated. The frequency sweep tests at 200 °C, at a constant value of the oscillation amplitude, 0.5%, for a varied frequency range, 0.1–628 rad·s^{-1}, were performed.

The analysis of the relaxation of the melted masterbatches containing neat hydrotalcite, HT, particles and pigment-modified HT was performed based on the frequency sweep test using various relaxation models.

The Maxwell model [47] was applied. The linear viscoelastic data were recalculated by using Equations (6) and (7). The discrete relaxation spectrum was obtained, and the relaxation times, λ_i, and values of the relaxation modulus, G_i, were calculated.

$$G'(\omega) = \sum_{i=1}^{N} G_i \frac{(\omega\lambda_i)^2}{1 + (\omega\lambda_i)^2} \tag{6}$$

$$G''(\omega) = \sum_{i=1}^{N} G_i \frac{(\omega\lambda_i)}{1 + (\omega\lambda_i)^2} \tag{7}$$

The six Maxwell elements were sufficient for the recalculation of the values of the storage, G', and loss shear modulus, G'', for the frequency sweep experimental data. The level of correlation was R^2 = 0.999.

The following model equations, Equations (8) and (9) [48–50], with n terms were used to extract the continuous relaxation spectrum by fitting the oscillation data.

$$G'(\omega) = \int_{-\omega}^{+\omega} H(\ln\tau) \frac{\omega^2\tau^2}{1 + \omega^2\tau^2} d\ln\tau \tag{8}$$

$$G''(\omega) = \int_{-\omega}^{+\omega} H(\ln\tau) \frac{\omega\tau}{1 + \omega^2\tau^2} d\ln\tau \tag{9}$$

The spectrum, H(lnτ), was discretized in the order of 100 steps. The spectrum represents all the pairs of fitted {Hi,τi} parameters.

The numerical fitting of the storage shear modulus, G', and loss shear modulus, G'', to the relaxation models was conducted using the TRIOS® Software (TRIOS v3 1.5.3696) provided by TA Instruments (New Castle, DE, USA).

Additionally, the zero-shear viscosity, η_0, and characteristic mean relaxation times, τ_m, were calculated from viscosity Cole–Cole plots (plots of η'' vs. η', where η' is dynamic viscosity and η'' is the out-of-phase component of complex viscosity, η^*) as proposed in the literature [51–54].

2.6. The DSC and TGA Analysis

DSC analysis was performed using a DSC1 apparatus (Mettler Toledo, Ithaca, NY, USA). All tests were performed in a nitrogen atmosphere. The samples were subjected to three heating/cooling/heating steps from −150 °C to 200 °C with a heating rate of 10 °C·min^{-1}. The first heating step was performed to eliminate the thermal history of the sample. The degree of crystallinity, χ_c, was calculated from both the first and second cycle of heating in accordance with Equation (6) [55–57].

$$\chi_c = \frac{\Delta H_m}{\Delta H_m^0} \cdot \frac{100}{w} \tag{10}$$

where ΔH_m is the experimental melting enthalpy, ΔH_m^0 is the the enthalpy of melting of 100% crystalline PP (207.1 Jg^{-1} [55–57]), and w is the weight fraction of polypropylene.

TGA analysis was performed using a TGA/DSC1 (Mettler Toledo, USA) analyzer. Samples were heated from 25 °C to 600 °C in argon, and from 600 °C to 900 °C in air, with a heating rate of 10 °C·min^{-1}.

2.7. CIELab Measurements

In order to determine the influence of the hybrid pigment on the color profile of the masterbatch, the CIELab model was used, where the lightness of color is L with a scale from 0 (black) to 100 (white). Two coordinates, a,b, can take both positive and negative values. Positive values of the *a* coordinate determine the share of red, and negative value determine the share of green green. Positive values of the *b* coordinate refer to the share of yellow, and negative values refer to the share of blue. The difference between two colors in the space is calculated on the basis of a mathematical formula (Equation (11)):

$$\Delta E = \sqrt{(\Delta L)^2 + (\Delta a)^2 + (\Delta b)^2} \tag{11}$$

The measurements were taken using a Konica Minolta CM-36dG apparatus (Konica Minolta Inc. Japan, Chyoda-Tokio, Japan).

3. Results

3.1. Surface Energy, the Tendency toward Aggregation and the Thermal Stability of Colored Layered Double Hydroxide Fillers

Layered double hydroxides demonstrate a strong tendency to agglomerate due to electrostatic interactions and the possibility to form hydrogen bonds via hydroxyl groups present on the filler surface. The particle size and the distribution of the filler aggregates in the polymer matrix can be important factors influencing the rheological behavior of filled polymer materials. Uneven dispersion and the occurrence of large agglomerates can be disadvantages from the processing point of view. They can generate problems during the further mixing of a masterbatch with the polymer as well as affect the color stability of the masterbatch. The modification of the surface can both reduce the tendency toward agglomeration or enhance the formation of aggregates. Therefore, the aggregate size in water and paraffin oil for modified colored fillers was studied to estimate the influence of modification on the tendency to agglomerate in polar and non-polar mediums.

The modification of layered double hydroxide with pigments influenced the size of the formed aggregates in the water medium (Table 1). Hydrotalcite modification with both PG7 and PR122 pigments resulted in an increase in the range of the aggregate size in the polar medium. A much higher tendency of HT-Green toward aggregation was noticed than that of the composition of HT-Red.

Table 1. Range of the aggregate size formed in the polar water medium for the modified, colored fillers.

Sample	Range of Aggregates Size (nm) *	Size of the Main Fraction (nm) *	Percentage as Number of Main Fraction (%) *
Unmodified HT	940–2118	1332	38.0
HT-Green 50 rpm	1484–7456	3091	27.0
HT-Green 60 rpm	1718–7456	4145	26.7
HT-Red 50 rpm	1246–1589	1471	24.1
HT-Red 60 rpm	1351–1893	1763	23.4

* Measurements were taken in the polar water medium.

After the modification of the layered double hydroxide surface with PG7 pigment, the range of the formed aggregates in the non-polar paraffin oil medium increased, but still the formed aggregates were not bigger than 3000 nm (Table 2). Oppositely, for HT-Red fillers modified with PR122 pigment a lower tendency toward agglomeration was indicated. It should be noted that PG7 is a polar compound and due to presence of chloride atoms in the structure its surface free energy strongly differs from that of the non-polar medium resulting in worse stability of the dispersion. The DLS plots of aggregate sizes as a function of the percentage by volume are shown in Figures S5 and S6 in the Supporting Information.

Table 2. Range of the aggregate size formed in the non-polar paraffin oil medium for the modified, colored fillers.

Sample	Range of Aggregates Size (nm) *	Size of the Main Fraction (nm) *	Percentage by Number of Main Fraction (%) *
Unmodified HT	995–1545	1270	35.6
HT-Green 50 rpm	1152–2780	2376	27.8
HT-Green 60 rpm	2073–2780	2586	54.9
HT-Red 50 rpm	1152–1790	1438	34.3
HT-Red 60 rpm	1152–2073	1561	33.1

* Measurements were taken in the non-polar paraffin oil medium.

The optical microscope images were taken for pure hydrotalcite, HT, and neat pigments before milling (Figure S2, Supporting Information) as well as for HT–pigment powders after milling at various milling speeds (Figures S3 and S4, Supporting Information). The optical microscopy studies confirmed that the hydrotalcite before milling formed grains, with diameters larger than 80 μm. Both pigments formed smaller aggregated grains, but still the agglomerates with diameters larger than 80 μm were present. The optical microscopy analysis confirmed that after the modification, phthalocyanine green PG7 pigment was present mostly on the surface of the hydrotalcite grains, and did not form separated grains. Milling at a 50 rpm speed for 15 min reduced the diameter of the largest aggregates. Oppositely, for HT-Green 60 rpm, larger aggregates were observed. This is in agreement with the findings of DLS studies. For HT-Green 60 rpm, larger aggregates were also determined to be present in water as well as in the non-polar paraffine oil medium compared with those with HT-Green 50 rpm. Similarly, for the modified HT-Red fillers, pigment was mostly present on the surface of HT grains (Figure S4, Supporting Information). A reduction in the size of HT-Red grains was observed independently of the applied speed of milling.

The surface free energies of tested samples were In the range of 10.7–27.8 dynes·cm^{-1} (Table 3). The calculated polarity index was from 0.05 up to 0.44. The polar component for PG7 and PR122 was due to their chemical structure. All tested modified hydrotalcite fillers showed a more non-polar character. It is worth noting the decrease in the polar composition of HT-Green 50 rpm and HT-Red 50 rpm with respect to unmodified HT and pure pigments PG7 and PR122, which is beneficial from the point of view of mixing them with non-polar polypropylene, PP.

Table 3. Calculated values of surface free energy, γ, dispersive, γ^d, and polar, γ^p, components of surface free energy and the polarity index, γ^p/γ.

Sample	γ^p dynes·cm^{-1}	γ^d dynes·cm^{-1}	γ dynes·cm^{-1}	γ^p/γ
Unmodified HT	3.0	16.1	19.2	0.16
PG7	5.8	17.2	23.0	0.25
HT-Green 50 rpm	2.7	21.8	24.5	0.11
HT-Green 60 rpm	8.1	10.5	18.6	0.44
PR122	1.0	11.9	12.8	0.08
HT-Red 50 rpm	0.5	10.2	10.7	0.05
HT-Red 60 rpm	8.9	18.9	27.8	0.32

The thermal stability of the obtained colored fillers was analyzed using the TGA method (Figures 3 and 4). The DTGA plots for neat PG7 and PR122 confirmed the high thermal stability of both pigments. The degradation and weight loss for the PG7 pigment was reported in the temperature range of 520–780 °C and for that for the PR122 pigment was reported in the temperature range of 450–820 °C (Figure 3). A two-step mechanism of mass loss was observed for the PR122 pigment, identified as two peaks in the DTGA plots, with the maximum occurring at 601 °C and 733 °C. A single degradation peak was observed for the PG7 pigment with a maximum at 697 °C.

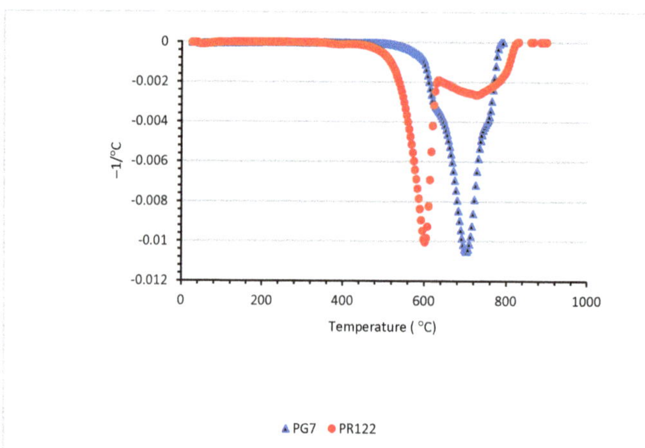

Figure 3. DTGA plots of neat PG7 and PR122 pigments.

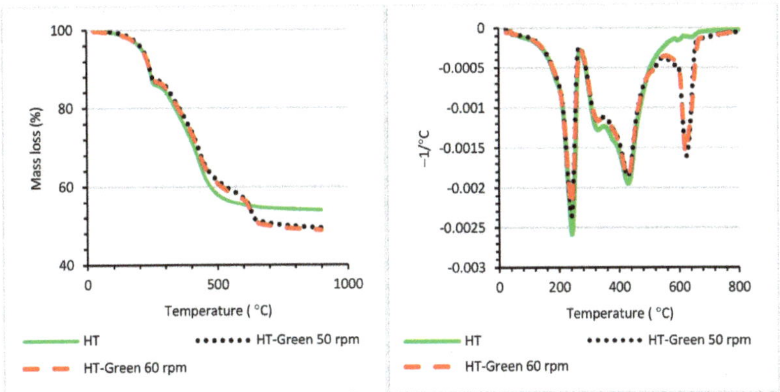

Figure 4. TGA and DTGA plots of HT-Green fillers.

The temperatures at 5, 10, 20% of weight loss, and the temperatures at the maximum of the DTA peaks for the studied pigments and colored fillers are compiled in Table 4. The thermogravimetric analyses (TGA) of unmodified hydrotalcite, HT, and modified, colored fillers are presented in Figures 4 and 5.

Table 4. The temperatures at 5, 10 and 20% of weight loss, $T_{5\%}$, $T_{10\%}$, and $T_{20\%}$, and the temperatures of the maximum of the DTGA peaks, T_{max}.

	$T_{5\%}$ (°C)	$T_{10\%}$ (°C)	$T_{20\%}$ (°C)	T_{max} (°C)
PG7	598	622	649	697
PR122	532	559	583	601; 733
HT	211	238	337	241; 430
HT−Green 50 rpm	217	244	355	241; 431; 625
HT−Green 60 rpm	214	241	352	241; 430; 622
HT−Red 50 rpm	181	223	355	226; 430; 622
HT−Red 60 rpm	181	223	352	229; 433; 625

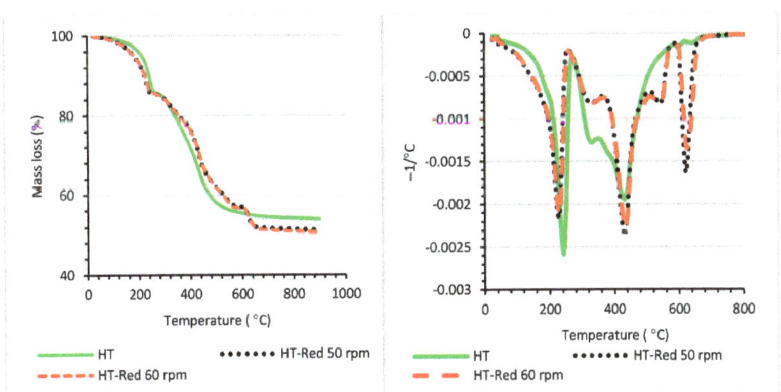

Figure 5. TGA and DTGA plots of HT-Red fillers.

Two loss regions at the temperature range of 30–270 °C and of 300–600 °C characteristic of hydrotalcite can be seen for the neat and modified fillers. The first mass loss region is attributed to physically and chemically bonded water and the second to the interlayer anions and interlayer water associated with the anions. Similar behavior, and a two-step mechanism of mass loss was observed by other authors for various layered double hydroxides [58]. An additional loss mass region at a temperature higher than 600 °C was observed for HT-Green and HT-Red fillers, and it was attributed to the thermal decomposition of adsorbed pigments on the filler surface. The layered double hydroxides HT-Green 50 rpm and HT-Green 60 rpm modified with PG7 pigment started to lose water below 95 °C and the observed mass loss was lower than that for the unmodified layered double hydroxide. Further, also, the mass loss observed for the second region, attributed to the removal of the interlayer water, was lower.

The layered double hydroxides HT-Red 50 rpm and HT-Red 60 rpm modified with PR122 pigment started to lose water at around 100 °C, but the dehydration and the removal of physically associated water occurred faster and the mass loss in first region was stronger than that for unmodified layered double hydroxide. The modification of the layered double hydroxide with pigment PR122 strongly influenced the thermal behavior of the HT-Red fillers in the second region, and this was attributed to the removal of the interlayer water associated with anions present in the interlayer space. The removal of the interlayer water started at a higher temperature. Probably, the PR122 pigment was not only adsorbed on the outer layer of the filler via interactions with hydroxyl groups present on the layered double hydroxide surface, but was also able to form associations with the interlayer water.

3.2. The Viscoelastic Properties and Relaxation Behavior of the Melted Masterbatches at a Processing Temperature of 200 °C

Dynamic rheological tests are a very sensitive method that allow an estimation of changes in the rheological behavior of a melted polymer after the incorporation of additives. The viscoelastic properties of the masterbatches at a processing temperature of 200 °C were investigated; the storage shear modulus, G', and the loss shear modulus, G'', were determined as functions of angular frequency and are compiled in Figures S7 and S8 in the Supporting Information.

The incorporation into the masterbatch of the additives, layered double hydroxide HT, pigment PG7 or PR122 and modified HT-Green and HT-Red pigments did not strongly influence the viscoelastic properties of melted the polypropylene, PP. The values of the storage shear modulus, G', and loss modulus, G'', for the PP compositions containing layered double hydroxide HT or the modified pigment based on the HT filler slightly decreased compared with those of neat PP. Various factors should be considered here; first, the process of masterbatch production via the extrusion of the composition could have influenced the changes in the viscosity of the base polymeric material, polypropylene. As we show in Figure S9 (Supporting Information), for samples studied at 220 °C the extrusion and granulation of the pure palettes of polypropylene under similar conditions as those used in masterbatch production slightly decreased the viscosity of the melted material. Second, the additives, especially pigments, could have influenced the processes of degradation but here more in-depth studies are needed to estimate the effect of pigments on the possible thermo-mechanical or thermo-oxidative degradation of polypropylene. As we analyzed further, the changes in viscosity occurred to a higher extent for the masterbatches containing pure pigments. Another factor influencing the values of the storage modulus, G', might have been the enhanced mobility (relaxation) of confined polymeric chains at the interface of the PP–LDH layer. This effect was reported by other authors [5] and attributed to the reduced values of the storage shear modulus, G', for polypropylene layered double hydroxide composites under low layered double hydroxide loading.

From the industrial point of view, what is more important is that the color additives or changes in the composition do not significantly affect the rheological behavior in the melt state of the produced masterbatch.

The average values of the viscoelastic parameters, storage modulus, G' (Pa), loss modulus, G'' (Pa), and loss factor, tan δ (−), measured at 200 °C at 10 rad·s are compiled in Table 5.

Table 5. The viscoelastic parameters of masterbatches at 200 °C.

Composition	G'_{LVR} (Pa)	G''_{LVR} (Pa)	Tan δ (−)
PP	4614 ± 49	8371 ± 57	1.81 ± 0.01
PP/HT	3292 ± 28	6056 ± 44	1.84 ± 0.01
PP/PG7	3476 ± 85	6478 ± 101	1.86 ± 0.02
PP/HT-Green 50 rpm	4360 ± 55	8125 ± 93	1.86 ± 0.01
PP/HT-Green 60 rpm	4154 ± 59	7724 ± 113	1.86 ± 0.04
PP/PR122	3665 ± 115	7207 ± 159	1.97 ± 0.02
PP/HT-Red 50 rpm	5413 ± 95	9264 ± 84	1.71 ± 0.02
PP/HT-Red 60 rpm	3895 ± 88	6938 ± 92	1.78 ± 0.03

G′—storage shear modulus; G″—loss shear modulus; δ—loss factor tan δ measured at 200 °C at an angular frequency of 10 $rad \cdot s^{-1}$.

Considering the viscoelastic properties, it can be concluded that pigment-modified HT could have been applied as an additive to the masterbatch instead of neat pigments PG7 and PR122. The storage shear modulus, G′, of the masterbatches containing HT-Green and HT-Red fillers were slightly higher compare with those of the masterbatch based on the neat pigments. A stronger impact of the HT-Green and HT-Red prepared by using a 50 rpm speed of milling on the values of the loss shear modulus, G″, was observed; this was because the dissipation of energy could have been affected by the presence of solid particles. Further, the state of the dispersion of the filler in the melt and its tendency to agglomerate could have been a factor influencing the dissipation of energy by the melted masterbatch. The application of various milling speeds during the preparation of modified HT fillers affected the tendency of the modified HT to form aggregates, as was analyzed in the previous paragraph. It affected the viscoelastic behavior of the melted composition, leading to the higher values of the storage shear modulus, G′, and loss modulus, G″, observed for the compositions based on HT-Green 50 rpm and HT-Red 50 rpm.

The similar values of tan, δ (Table 5), were determined for compositions containing PG7 pigment and HT-Green. Differently, higher values of tan, δ, were reported for the masterbatch containing pure PR122 pigment compared with those of compositions containing modified HT-Red 50 rpm. The higher mixing speed during the modification of HT could have been the factor leading to the reduced size of the formed aggregates. The viscoelastic properties, storage modulus, G′, and loss modulus, G″, of the masterbatch, containing the modified HT-Red filler prepared using a higher speed of mixing were similar to those of the composition containing pure pigment PR122. This confirmed that the optimization of the process of the modification of the filler during the preparation of colored HT-Red fillers was necessary, leading to a reduction in the tendency of aggregation, and further influenced the viscoelastic behavior of the masterbatch containing the modified HT-Red 60 rpm filler.

The viscosity of the material in the melt state was an important processing parameter. The influence of the modified colored HT on the complex viscosity, η*, dynamic viscosity, η′, and out-of-phase component of complex viscosity, η″, was investigated. The values of complex viscosity, η*, measured as a function of angular frequency at 200 °C are shown in Figure 6.

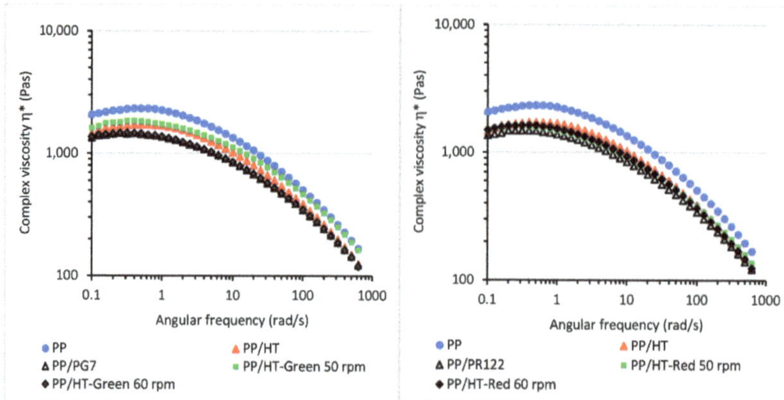

Figure 6. Complex viscosity, η* (Pas), at 200 °C, and at an applied oscillation strain of 0.5%.

The plots of complex viscosity, η*, showed a plateau, which is typical for thermoplastic homopolymers at low values of frequency, and shear tinning behavior as the frequency was increased. The values of the complex viscosity, η*, at 200 °C for all compositions containing HT or modified colored HT were lower compared with those of neat PP. The lower viscosity of the studied compositions resulted from the reduction in the viscosity of the polypropylene during masterbatch preparation. The preparation of the masterbatch compositions was conducted in two steps. First, the components of the masterbatch recipe were mixed with PP using an extruder. The obtained material was granulated. Then, the colored pallets were used as an additive to the pure PP thermoplastic material during injection molding. The process of masterbatch preparation could have influenced the viscoelastic properties, leading to the changes in the viscosity of the thermoplastic PP base. During extrusion, the thermo-mechanical and thermo-oxidative aging of the material could have occurred due to the presence of oxygen. The applied shear rate could have resulted in the chain scission of the material and the changes in the dispersity of the material and its molecular weight leading to changes in viscosity. The presence of additives, especially pigments, can promote the thermo-oxidative aging of a material, leading to stronger changes in viscosity during the processing of a masterbatch. The values of complex viscosity, η*, measured for the plateau region (frequency range 0.1–1 rads^{-1}) were as follows: PP η* = 2252 ± 68 Pa·s; PP/HT η* = 1657 ± 63 Pa·s; PP/PG7 η* = 1421 ± 32 Pa·s; PP/HT-Green 50 rpm η* = 1778 ± 47 Pa·s; PP/HT-Green 60 rpm η* = 1432 ± 28 Pa·s; PP/PR122 η* = 1449 ± 35 Pa·s; PP/HT-Red 50 rpm η* = 1560 ± 35 Pa·s; PP/HT-Red 60 rpm η* = 1587 ± 33 Pa·s.

The influence of the various additives on the relaxation behavior of the melted polymer and its viscoelastic properties can be estimated based on the viscosity Cole–Cole plots. The viscosity Cole–Cole plots are the plots of out of phase component of complex viscosity η* versus dynamic viscosity η″. For the homopolymers the Cole–Cole plots usually form the semicircular arcs and the rheological parameters such as the zero-shear viscosity η_0 and characteristic relaxation time can be derived [50–53]. The Cole–Cole plots are depicted in Figure 7.

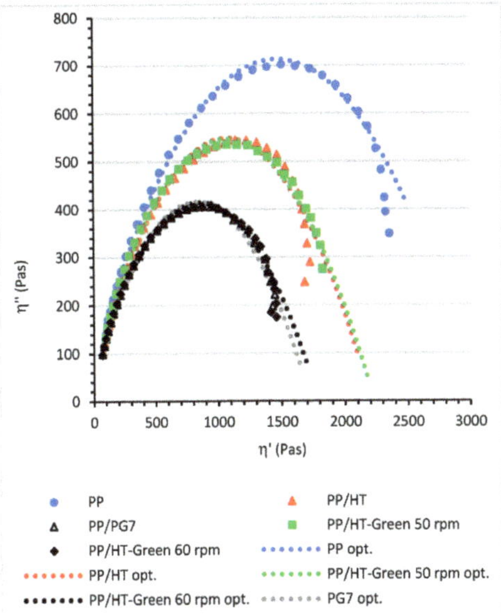

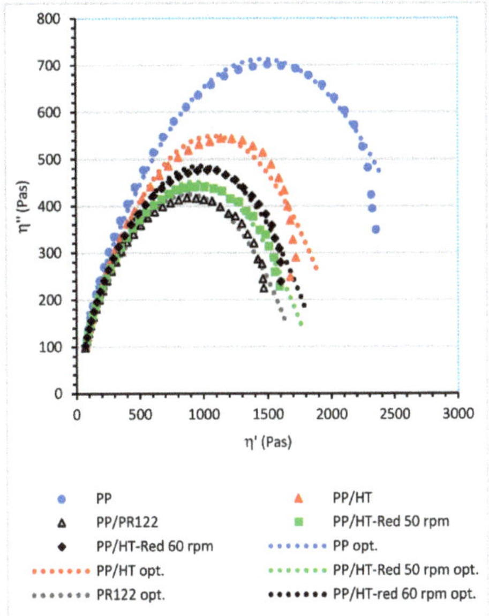

Figure 7. The Cole–Cole plots at 200 °C; η′—dynamic viscosity (Pas), η″—out of phase component of complex viscosity η* (Pas).

All compositions showed Cole–Cole plots of a semicircular shape, which is typical of homopolymers. The incorporation of layered double hydroxide HT shifted the maximum of the arc compared with that of neat PP. The relaxation of the melted polypropylene was affected by the presence of layered double hydroxide particles. The maximum of the arc was presented at lower values of dynamic viscosity, η′, and at a higher value of angular frequency (a shorter relaxation time). The incorporation of HT particles facilitated the relaxation of melted PP. Further, both pigments, PG7 and PR122, strongly facilitated the relaxation of the masterbatch. The maximum of the arc was shifted to a meaningfully lower value of dynamic viscosity, η′, after the addition of the PG7 pigment. The modification of the layered double hydroxide surface by pigments caused a further facilitation of the relaxation of the masterbatch compared with that of unmodified HT. The selected additives reduced the values of dynamic viscosity, η′. The values of the characteristic relaxation time, τ_m, and the zero-shear viscosity, η_0, calculated based on the Cole–Cole plot are compiled in Table 6.

Table 6. The characteristic relaxation time, τ_m (s), and the zero-shear viscosity, η_0 (Pas).

Composition	τ_m (s)	η_0 (Pas)
PP	1.123	3041
PP/HT	0.892	2120
PP/PG7	1.001	1728
PP/HT-Green 50 rpm	0.708	2225
PP/HT-Green 60 rpm	0.795	1789
PP/PR122	1.013	1829
PP/HT-Red 50 rpm	0.885	1942
PP/HT-Red 60 rpm	0.889	2014

The Maxwell model [47] was applied to calculate the discrete relaxation spectra of the studied masterbatches. The calculated values of the relaxation modulus, G_i, and relaxation times, λ_I, are compiled in Tables S1 and S2 (Supporting Information) and depicted in Figure 8.

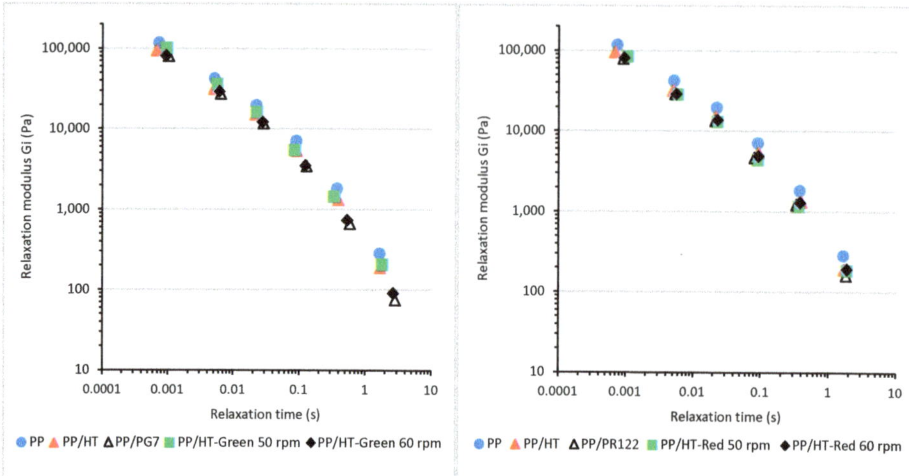

Figure 8. The discrete relaxation spectra calculated using the Maxwell model.

A more significant influence of the additives on the relaxation spectra was observed for the masterbatches containing the modified HT-Red filler, and lower values of the relaxation modulus, G_i, and shorter relaxation times, λ_I, confirmed the facilitated relaxation of the melted PP in the presence of both HT-Red 50 rpm and HT-Red 60 rpm. A similar effect was observed when the modified HT-Green filler prepared using a higher mixing speed (60 rpm) was incorporated into the masterbatch. The incorporation of the unmodified layered double hydroxide slightly increased the values of the relaxation modulus, G_i. Both PG7 and PR122 pigments slightly facilitated the relaxation of the masterbatch. The effect on the relaxation times was less evident than that for the alternative modified coloring fillers HT-Green and HT-Red.

The influence of the modified colored fillers on the relaxation of the melted masterbatches was confirmed via the calculation, based on the frequency sweep tests, of the continuous relaxation spectra, which are shown in Figure 9.

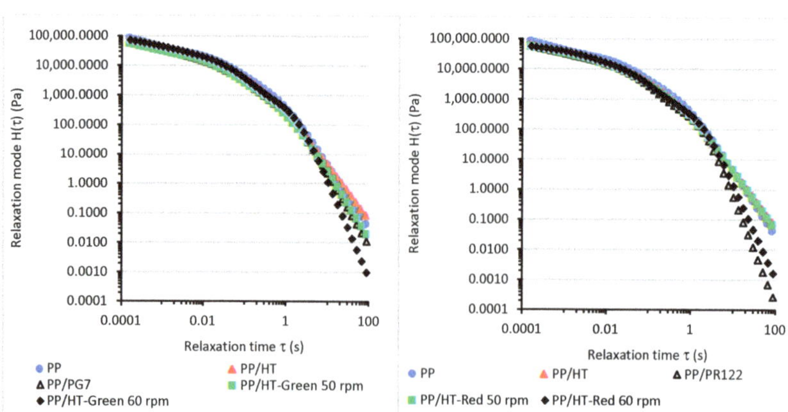

Figure 9. Continuous relaxation spectra for the masterbatches melted at 200 °C.

The incorporation of the both HT-Green and HT-Red instead of the neat pigments PG7 and PR122 facilitated the relaxation of the melted masterbatches, and the reduction in the values of the relaxation modulus, H(τ), occurred faster. This is a processing advantage, and the faster relaxation of the material together with the lower values of the storage shear modulus in a low frequency range were the crucial factors influencing the processing of the material during extrusion, reducing the die-swell of the masterbatch.

3.3. *Thermal Properties and Crystallization of Masterbatches*

The influence of the colored fillers on the thermal properties of the masterbatches was analyzed using the TGA method (Figures 10 and 11).

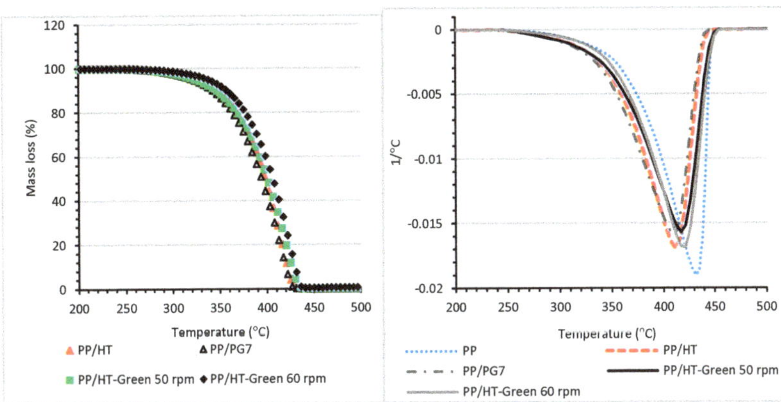

Figure 10. TGA and DTGA plots of masterbatches containing PG7 pigment and modified HT-Green fillers.

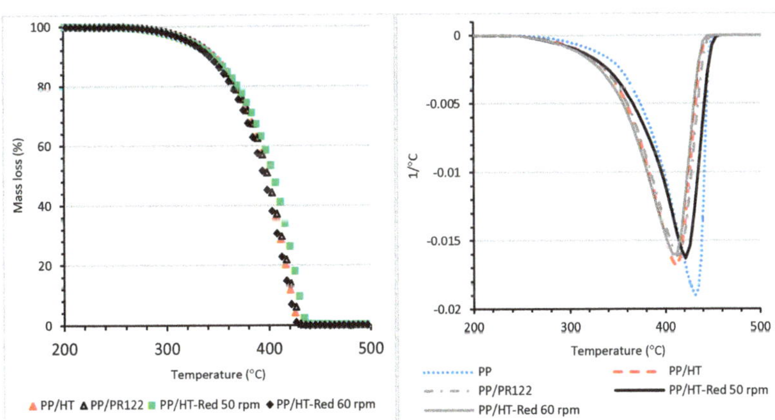

Figure 11. TGA and DTGA plots of masterbatches containing PR122 pigment and modified HT-Red fillers.

It can be seen from the TGA and DTGA plots (Figure 10) that the incorporation of both HT-Green 50 rpm and HT-Green 60 rpm instead of neat pigment PG7 improved the thermal stability of the masterbatch. This effect was stronger for the HT-Green filler prepared by using a higher speed of milling during the modification of layered double hydroxide. A reduced tendency of agglomeration for the HT-Green filler at 60 rpm and better dispersion in the masterbatch were responsible for the enhancement in the thermal stability of the masterbatch. Similarly, the incorporation of the colored HT-Red filler instead of the PR122 pigment improved the thermal stability of the masterbatch (Figure 11). Here, the shift in the temperature at which the maximum weight loss occurred toward higher values of temperature was stronger after the incorporation of the HT-Red filler at 50 rpm. The temperatures at which 2, 5, 10, 20% ($T_{2\%}$, $T_{5\%}$, $T_{10\%}$, and $T_{20\%}$) and the maximum weight loss, T_{max}, occurred are compiled in Table 7.

Table 7. The temperatures at which 2, 5, 10 and 20% of weight loss, $T_{2\%}$, $T_{5\%}$, $T_{10\%}$, and $T_{20\%}$, occurred and the temperatures of the maximum of the DTGA peaks, T_{max}.

	$T_{2\%}$ °C	$T_{5\%}$ °C	$T_{10\%}$ °C	$T_{20\%}$ °C	T_{max} °C
PP	316	337	361	385	433
PP/HT	305	328	347	370	427
PP/PG7	291	319	342	361	428
PP/HT-Green 50 rpm	296	324	347	370	431
PP/HT-Green 60 rpm	310	338	356	379	431
PP/PR122	300	323	342	366	429
PP/HT-Red 50 rpm	300	324	347	370	437
PP/HT-Red 60 rpm	300	324	342	366	427

The temperature at which the maximum weight loss, T_{max}, of the masterbatches occurred was lower compared with that of PP pallets used during the preparation of the masterbatches. This resulted from the previously described thermo-mechanical degradation of the polypropylene base during the extrusion and the preparation of the masterbatch. The incorporation of the modified, colored filler HT-Green instead of neat pigment PG7 strongly shifted the temperatures of 2, 5, 10, and 20% of weight loss toward higher values of temperature. The shift in the temperatures of 5% of weight loss, $T_{5\%}$, for the masterbatch after the replacement of neat PG7 pigment with the HT-Green filler at 50 rpm was 5 °C and after the incorporation of the HT-Green filler at 60 rpm it was 19 °C. The enhancement of thermal stability after incorporation into polymeric material layered double hydroxide was reported by other authors [28], but here, for the masterbatch containing HT-Green at 60 rpm it also resulted from the synergistic effect of the filler and pigment. The higher thermal stability of the masterbatch containing HT-Green is a great processing advantage. The incorporation of HT-Red instead of the neat PR122 pigment, to a lesser extent, influenced the temperatures of 2 and 5% of the weight loss, $T_{2\%}$ and $T_{5\%}$. The incorporation of the HT-Red filler at 50 rpm instead of the neat PR122 pigment shifted the temperatures of 10 and 20% of the weight loss. The modified HT-Red hybrid colored filler at 50 rpm had the lowest polarity index, γ^P/γ, as determined by us via the measurement of surface free energy. The enhancement of the non-polar character and the decrease in the polar composition of HT-Red 50 rpm with respect to unmodified HT and PR122 was, here, important from the point of view of mixing them with non-polar polypropylene, PP. This resulted in better homogenization with polypropylene during mixing and influenced the thermal properties of the obtained PP/HT-Red 50 rpm masterbatch causing the strongest changes in T_{max} °C. Thus, the temperature at the maximum weight loss of the masterbatch occurred after the replacement of the neat pigment PR122 by HT-Red 50 rpm shifted toward higher values of temperature of about 8 °C.

The influence of the modified colored fillers on the melting, T_m, and crystallization, T_c, temperatures was analyzed. The melting temperatures, T_m, and the calculated degree of crystallinity for the masterbatches produced via extrusion and cooled using air are compiled in Table 8. Additionally, the melting temperature, T_m, of the polypropylene processed in a similar way was added. The melting temperatures of all produced masterbatches did not vary significantly. Further, the difference in melting of the produced masterbatches and polypropylene was not observed. This is an advantage when considering the application of the masterbatches as the coloring system for the polypropylene products. The influence of the additives on the degree of the crystallinity, χ_c, was observed. The addition of the neat pigments PG7 and PR122 enhanced the degree of crystallinity, χ_c, of the masterbatch compared with that of the neat polypropylene or material containing layered double hydroxide. Similarly, the degree of crystallinity, χ_c, increased after the addition of the colored fillers compared with that of the neat polypropylene or material containing unmodified layered double hydroxide. Here, it should be noted that the produced masterbatches were cooled using air, and thus the calculated values of degree of crystallinity, χ_c, showed the percentage of the crystalline phase after the process of masterbatch production.

Table 8. The melting temperatures, T_m, the values of the enthalpy of melting, ΔH_m, and the degree of crystallinity, χ_c, of the produced masterbatches.

	T_m (°C) *	ΔH_m ($J\cdot g^{-1}$) *	χ_c (%) *
PP	170	86.70	41.9
PP/HT	169	84.58	40.9
PP/PG7	169	92.86	44.9
PP/HT-Green 50 rpm	168	100.79	48.8
PP/HT-Green 60 rpm	168	95.32	46.1
PP/PR122	167	101.68	49.2
PP/HT-Red 50 rpm	169	94.16	45.6
PP/HT-Red 60 rpm	168	102.92	49.8

* The parameters were determined via DSC for the produced masterbatches.

To estimate the influence of the additives on the crystallization temperature, T_c, and melting temperature, T_m, the samples were additionally analyzed after removing the thermal history of samples and after crystallization at a defined speed of cooling. In Table 9, the crystallization, T_c, and melting, T_m, temperatures determined via DSC are compiled, together with the calculated values of the degree of crystallinity, χ_c, for samples cooled and further heated at a speed of 10 °C·min^{-1}. All DSC plots for the studied materials are compiled in the Supporting Information (Figures S10–S16).

Table 9. The crystallization, T_c, and melting, T_m, temperatures, the values of crystallization, ΔH_c, and the melting ΔH_m enthalpy, and the degree of crystallinity, χ_c.

	T_c (°C) *	H_c ($J\cdot g^{-1}$) *	T_m (°C) *	ΔHm ($J\cdot g^{-1}$) *	χ_c (%) *
PP/HT					
PP/PG7	119.7	96.47	166.5	91.95	44.5
PP/HT-Green 50 rpm	127.2	92.95	170.8	81.7	39.6
PP/HT-Green 60 rpm	128.0	102.14	169.1	89.24	43.2
PP/PR122	127.8	95.24	170.5	84.26	40.8
PP/HT-Red 50 rpm	129.6	100.94	170.3	87.74	42.5
PP/HT-Red 60 rpm	129.1	98.95	172.2	83.45	40.4
	129.1	98.63	169.7	88.99	43.1

* The parameters were determined via DSC for the produced masterbatches after removing, during the measurement, the thermal history of the sample. An applied speed of cooling and heating of 10 °C·min^{-1} was used.

Figure 12 shows the DSC cooling step for the studied materials. The addition of both pigments, PG7 and PR122, or the colored modified HT-Green and HT-Red fillers strongly influenced the crystallization temperature, T_c, of the masterbatches compared with that of the sample containing unmodified layered double hydroxide. Here, higher temperatures of crystallization, T_c, were detected. The shift in crystallization temperatures between samples PP/HT and PP/PG7 was about 7.5 °C, and between samples PP/HT and PP/PR122 it was about 9.9 °C. Similarly, the addition of the colored modified HT-Green and HT-Red fillers shifted the crystallization temperatures towards higher value of temperature. The impact of the color additives on the melting temperatures, T_m, was not so evident. It is well known that cooling speed is a factor influencing the formed crystalline phase [9]. The determined the degree of crystallinity, χ_c, after the removal of the thermal history of the sample was higher for the sample containing unmodified layered double hydroxide. Both fillers and pigments could act as nucleating agents of polypropylene enhancing the formation of the crystalline phase [7,8]. The presence of the layered double hydroxide was the main factor influencing the crystallization of the masterbatch. However, more in-depth studies are needed to determine the impact of the surface covering of the layered double hydroxide HT by chosen pigments on the crystallization of the masterbatches.

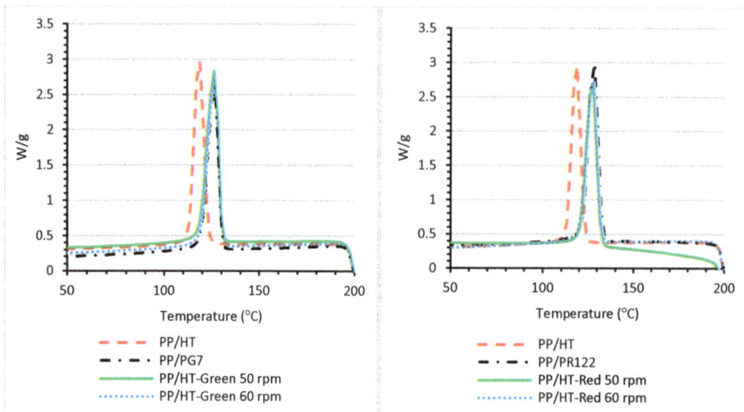

Figure 12. DSC plots of the cooling step for masterbatches containing the PG7 and PR122 pigments and modified fillers HT-Green and HT-Red. A speed of cooling of 10 °C·min^{-1} was used.

3.4. The Influence of Hybrid Pigment on the Color Profile of Masterbatch

Figure 13 shows 1-neat PG7, 2-HT-Green 50 rpm and 3-HT-Green 60 rpm, respectively. Based on the coordinates, a significant change in the color of the modified pigments in relation to the hydrotalcite can be seen, which was obvious and visible to the naked eye. Comparing the color change of HT-Green to that of the unmodified pigment, an increase in brightness could be seen via the increasing L parameter and shifting X coordinate toward a red color.

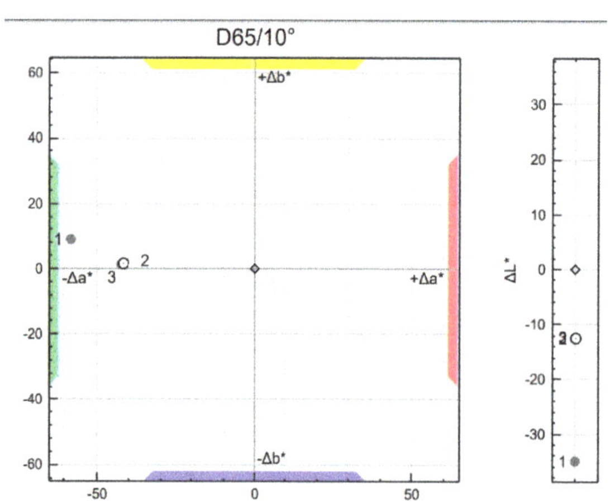

Figure 13. Color coordinates for modified PG7.

Figure 14 shows 1-neat PR122, 2-HT-Red 50 rpm and 3-HT-Red 60 rpm. Comparing the color change in HT-Red to that of the unmodified pigment, an increase in brightness could be seen via increasing the L parameter and shifting X coordinate toward a green color and that of the a coordinate toward a bluer color.

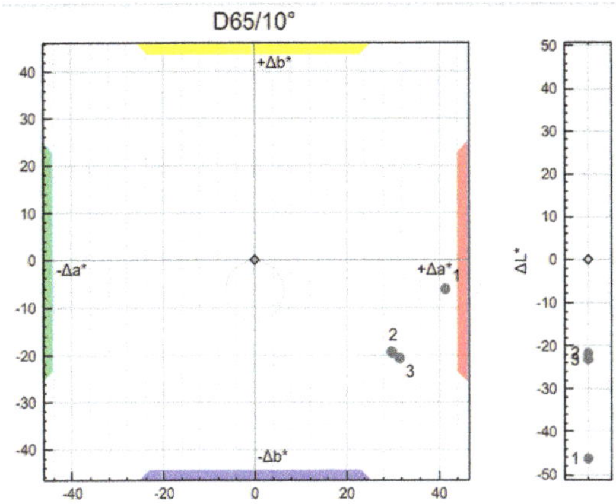

Figure 14. Color coordinates for modified PR122.

4. Conclusions

The surface of the layered double hydroxide was modified using quinacridone and phthalocyanine green pigments using the optimized milling process. The applied method influenced the size of the aggregates formed by layered double hydroxide HT.

The tensiometer measurement showed a decrease in the polar component of surface free energy, γ^P, after the modification of the hydrotalcite with both pigments when a 50 rpm speed of milling was used. This is advantageous from the point of view of mixing modified LDH with non-polar polypropylene.

It is worth noting that hydrotalcite modified using phthalocyanine green, HT-Green, are characterized by higher thermal stability compared with that of pure hydrotalcite, which has been proven using the thermogravimetric technique. After analyzing the results obtained via the DTGA measurement, the temperature at which 5, 10 and 20% mass loss occurred was determined to be shifted to a higher value of temperature. It turns out that the PC7 pigment has a higher thermal resistance than PR122 does, and therefore, the modified HT-Green color filler also has a higher thermal resistance compared to that of HT-Red.

The modification of hydrotalcite with pigments slightly influenced the rheological properties of the masterbatches. It was noticed that this additive slightly increased the rheological parameters such as the storage shear modulus, G', and loss shear modulus, G''. This is very beneficial from an industrial point of view and from that of the future application of a new modified colored filler instead of the previously used pure pigments in masterbatch formulation. Further analyzing the rheological properties of the tested systems, it can be seen that the storage modulus, loss modulus and loss factor tan, δ, value of the PP/PR122 system have higher values than those in the case of the PP/PG7 system. Taking into account the system containing hydrotalcite modified with phthalocyanine green, it was noticed that PP/HT-Green had a stronger influence on the values of the storage modulus, loss modulus and loss factor tan, δ, where HT–pigment mixtures were prepared for 15 min and at a mixing speed of 50 rpm.

The addition of HT-pigments facilitated the relaxation of melted PP/HT–pigment system compared with that of pure polypropylene or PP/HT composites. Hydrotalcite modified by using phthalocyanine green, HT-Green 60 rpm, had the strongest effect on the relaxation of the melted masterbatch. The lower values of characteristic relaxation times calculated based on the viscosity Cole–Cole plots were reported. Similarly, also, other models applied to calculate the relaxation times (the Maxwell model and continuous relaxation model) showed a shortening of the relaxation times.

The positive effect of modification on the thermal stability of the polypropylene masterbatches was noticed, and higher temperatures of the maximum weight loss, T_{max}, were determined for PP/HT–pigment masterbatches compared to those of masterbatches with PP/HT. The maximum weight loss temperature, T_{max}, of the masterbatch after the replacement of the red quinacridone pigment PR122 via modified HT-Red 50 rpm shifted toward higher values of temperature of about 8 °C.

Supplementary Materials: The following supporting information can be downloaded at https://www.mdpi.com/article/10.3390/ma16186243/s1. Figure S1: XRD diffraction patterns of layered double hydroxide Pural MG70; Figure S2: The optical microscope images of hydrotalcite, HT, pigment Green 7, phthalocyanine green, PG7 (c), and pigment Red 122, 2,9-dimethylquinacridone,((2,9-dimethyl-5,12-dihydroquinolino[2,3-b]acridine-7,14-dione), PR122 (d), at a magnification of 1000×.; Figure S3: The optical microscope images of hydrotalcite, HT, modified with pigment Green 7, phthalocyanine green, PG7, after milling at various speeds: 50 rpm (a,b); 60 rpm (c,d). The images were taken at a magnification of 500× (a–c) and 1000× (b–d); Figure S4: The optical microscope images of hydrotalcite, HT, modified with pigment Red 122, 2,9-dimethylquinacridone,((2,9-dimethyl-5,12-dihydroquinolino[2,3-b]acridine-7,14-dione), PR122, after milling at various speeds: 50 rpm (a,b); 60 rpm (c,d). The images were taken at a magnification of 500× (a) and 1000x (b–d); Figure S5: The DLS plots of aggregate sizes as a function of percentage by volume for hydrotalcite and modified HT-Green fillers; Figure S6: The DLS plots of aggregate sizes as a function of percentage by volume for hydrotalcite and modified HT-Red fillers; Figure S7: The viscoelastic properties, the storage shear modulus, G' (Pa), and loss shear modulus, G'' (Pa), for PP masterbatches based on the PG7 pigment and HT-Green; Figure S8: The viscoelastic properties, storage shear modulus, G' (Pa), and loss shear modulus, G'' (Pa), for PP masterbatches based on the PR122 pigment and HT-Red; Figure S9: The values of complex viscosity as a function of frequency at 220 °C for the palettes of polypropylene and the masterbatches extruded and processed under similar conditions as those of the polypropylene masterbatches; Table S1: Values of relaxation modulus, G_i (Pa), and relaxation times, λ_i (s), calculated using Maxwell models; Table S2: Values of relaxation modulus, G_i (Pa), and relaxation times, λ_i (s), calculated using Maxwell models; Figure S10: DSC plots for sample PP/HT; Figure S11: DSC plots for sample PP/PG7; Figure S12: DSC plots for sample PP/PR122; Figure S13: DSC plots for sample PP/HT-Green 50 rpm; Figure S14: DSC plots for sample PP/HT-Green 60 rpm; Figure S15: DSC plots for sample PP/HT-Red 50 rpm; Figure S16: DSC plots for sample PP/HT-Red 60 rpm.

Author Contributions: Conceptualization, M.K. and M.L.; methodology, M.K., M.L. and M.O.; validation, M.L. and J.P.; formal analysis, M.K. and M.L.; investigation, M.K. and M.L; data curation, M.L. and M.K; writing—original draft preparation, M.L. and M.K; writing—review and editing, M.L. and J.P.; visualization, M.L.; supervision, M.L. and J.P. All authors have read and agreed to the published version of the manuscript.

Funding: This research received no external funding.

Institutional Review Board Statement: Not applicable.

Informed Consent Statement: Not applicable.

Data Availability Statement: Data available on request.

Acknowledgments: This work was completed while the first author was a doctoral Candidate in the Interdisciplinary Doctoral School at the Lodz University of Technology, Poland.

Conflicts of Interest: The authors declare no conflict of interest. The funders had no role in the design of the study; in the collection, analyses, or interpretation of data; in the writing of the manuscript; or in the decision to publish the results.

References

1. Broda, J.; Gawlowski, A.; Slusarczyk, C.; Wlochowicz, A.; Fabia, J. The influence of additives on the structure of polypropylene fibres. *Dye. Pigment.* **2007**, *74*, 508–511. [CrossRef]
2. Paleo, A.J.; Krause, B.; Mendes, A.R.; Tavares, C.J.; Cerqueira, M.F.; Muñoz, E.; Pötschke, P. Comparative thermoelectric properties of polypropylene composites melt-processed using Pytograf® III carbon nanofibers. *J. Compos. Sci.* **2023**, *7*, 173. [CrossRef]
3. Harekrushna, S.; Mishra, B.; Senapti, P.; Murmu, R.; Sahu, D. Mechanical, thermal, and morphological properties of graphene nanoplatelet-reinforced polypropylene nanocomposites: Effects of nanofiller thickness. *J. Compos. Sci.* **2021**, *5*, 24. [CrossRef]
4. Tsagdi, A.; Drossos, I.; Georgiou, D.; Exarhopoulos, S.; Karasiotas, G.; Kallitsis, J.K.; Kalogianni, E.P. Injection molded PP foams using food ingredients for food packaging applications. *Polymers* **2021**, *13*, 288. [CrossRef]
5. Wang, Q.; Zhang, X.; Wang, C.J.; Zhu, J.; Guo, Z.; O'Hare, D. Polypropylene/layered double hydroxide nanocomposites. *J. Mater. Chem.* **2012**, *22*, 19113. [CrossRef]
6. Patra, S.C.; Swain, S.; Senapti, P.; Sahu, H.; Murmu, R.; Sutar, H. Polypropylene and graphene nanocomposites: Effects of selected 2D-Nanofiller's Plate Sizes on Fundamental Physicochemical Properties. *Inventions* **2023**, *8*, 8. [CrossRef]
7. Janostik, V.; Senkerik, V.; Manas, L.; Stanek, M.; Cvek, M. Injection-molded isotactic polypropylene colored with green transparent and opaque pigments. *Int. J. Mol. Sci.* **2023**, *24*, 9924. [CrossRef] [PubMed]
8. Broda, J. Structure of polypropylene fibres coloured with a mixture of pigments with different nucleating ability. *Polymer* **2003**, *44*, 6943–6949. [CrossRef]
9. Broda, J.; Włochowicz, A. Influence of pigments on supermolecular structure of polypropylene fibers. *Eur. Polym. J.* **2000**, *36*, 1283–1297. [CrossRef]
10. Paulus, E.F.; Leusen, F.J.J.; Schmidt, M.U. Crystal structures of quinacridones. *Cryst. Eng. Comm.* **2007**, *9*, 131–143. [CrossRef]
11. Kanbur, Y.; Coskun, H.; Głowacki, E.D.; Irimia-Vladu, M.; Sariciftci, N.S.; Yumusak, C. High temperature-stability of organic thin-film transitors based on quinacridone pigments. *Org. Electron.* **2019**, *66*, 53–57. [CrossRef]
12. Chen, P.; Liu, G.-J.; Wang, Y.; Zhang, S.X.-A. A stable aggregated system of silyl ether substituted quinacridone and its aggregation-state changes induced by fluoride-ions: Inspiration for a dual guaranteed strategy for probe design. *RSC Adv.* **2016**, *6*, 25986. [CrossRef]
13. Schmidt, A.M.; Calvete, M.J.F. Phthalocyanines: An Old Dog Can Still Have New (Photo)Tricks! *Molecules* **2023**, *26*, 2823. [CrossRef] [PubMed]
14. Yahya, M.; Nural, Y.; Seferoğlu, Z. Recent advances in the nonlinear optical (NLO) properties of phthalocyanines: A review. *Dye. Pigment.* **2022**, *196*, 109960. [CrossRef]
15. Zeinidenov, A.K.; Aimukanov, A.K.; Kambar, D.S.; Ilyassov, B.R.; Zavgorodniy, A.V. Effects of phthalocyanine nanostructure on photovoltaic performance of its polymer composite thin films. *Mat. Chem. Phys.* **2021**, *267*, 124680. [CrossRef]
16. Pająk, A.; Rybiński, P.; Janowska, G.; Kucharska-Jastrząbek, A. The thermal properties and the flammability of pigmented elastomeric materials. Part I. Phthalocyanine pigments. *Therm. Anal. Calorim.* **2014**, *117*, 789–798. [CrossRef]
17. Tawiah, B.; Asinyo, B.K.; Badoe, W.; Zhang, L. Phthalocyanine green aluminium pigment prepared by inorganic acid radical/radical polymerization for waterborne textile applications. *Int. J. Ind. Chem.* **2017**, *8*, 17–28. [CrossRef]
18. Chorobiński, M.; Skowroński, Ł.; Bieliński, M. Methodology for determining selected characteristics of polyethylene dyeing using CIELab system. *Polimery* **2019**, *64*, 690–696. [CrossRef]
19. Chen, C.; Gu, J.; Weng, Y.; Huang, Z.; Qiu, D.; Shao, S. Optimization of the preparation process of biodegradable masterbatches and characterization of their rheological and application properties. *Polym. Test.* **2018**, *70*, 526–532. [CrossRef]
20. Cabello-Alvarado, C.J.; Quiñones-Jurado, Z.V.; Cruz-Delgado, V.J.; Avila-Orta, C.A. Pigmentation and degradative activity of TiO_2 on polyethylene films using masterbatches fabricated using variable-frequency ultrasound-assisted melt-extrusion. *Materials* **2020**, *13*, 3855. [CrossRef]
21. Olivier-Ortega, H.; Tresserras, J.; Julian, F.; Alcalà, M.; Bala, A.; Espinach, F.X.; Méndez, J.A. Nanocomposites Materials of PLA reinforced with nanoclays using a masterbatch technology: A study of the mechanical performance and its sustainability. *Polymers* **2021**, *13*, 2133. [CrossRef] [PubMed]
22. Ullah, J.; Harkin-Jones, E.; McIlhagger, A.; Magee, C.; Tormey, D.; Dave, F.; Sherlock, R.; Dixon, D. The effect of masterbatch pigments on the crystallization, morphology, and shrinkage behaviour of isotactic polypropylene. *J. Polym. Res.* **2022**, *29*, 183. [CrossRef]
23. Wu, S.; Brzozowski, K.J. Surface free energy and polarity of organic pigments. *J. Coll. Interface Sci.* **1971**, *37*, 686–690. [CrossRef]
24. Agbo, C.; Acheampong, C.; Liping, Z.; Li, M.; Wang, D.; Fu, S. Synthesis and application of novel dispersant using a dichlorotriazine azo moiety and dodecan-1-ol. *Prog. Org. Coat.* **2019**, *127*, 1–7. [CrossRef]
25. Zhou, Y.; Chen, K.; Liu, L.; Wen, S.; Gui, T. The design and preparation of antibacterial polymer brushes with phthalocyanine pigments. *Coatings* **2023**, *13*, 1114. [CrossRef]
26. Saito, Y.; Iwamoto, S.; Tanaka, Y.; Hontama, N.; Endo, T. Suppresing aggregation of quinacridone pigment and improving its color strength by using chitosan nanofibres. *Carbohydr. Polym.* **2021**, *255*, 117365. [CrossRef]
27. Marangoni, R.; Tavlot-Guého, C.; Illaik, A.; Wypych, F.; Leroux, F. Organic inorganic dye filler for polymer: Blue coloured layered double hydroxides into polystyrene. *J. Coll. Interface Sci.* **2008**, *326*, 366–373. [CrossRef]
28. Mochane, M.J.; Magagula, S.I.; Sefadi, J.S.; Sadiku, E.R.; Mokhena, T.C. Morphology, thermal stability, and flammability properties of polymer-layered double hydroxide (LDH) nanocomposites: A Review. *Crystals* **2020**, *10*, 612. [CrossRef]

29. Zhang, T.; Wang, C.; Wang, Y.; Wang, Y.; Han, Z. Effects of modified layered double hydroxides on the thermal degradation and combustion behaviors of intumescent flame retardant polyethylene nanocomposites. *Polymers* **2022**, *14*, 1616. [CrossRef]
30. Hwang, S.-H.; Jung, S.-C.; Yoon, S.-M.; Kim, D.-K. Preparation and characterization of dye-intercalated Zn-Al-layered double hydroxide and its surface modification by silica coating. *J. Phys. Chem Solids* **2008**, *69*, 1061–1065. [CrossRef]
31. Marangoni, R.; Mikowski, A.; Wypych, F. Effect of adsorbed/intercalated anionic dyes into the mechanical properties of PVA: Layered zinc hydroxide nitrate nanocomposites. *J. Coll. Interface Sci.* **2010**, *351*, 384–391. [CrossRef]
32. Marzec, A.; Szadkowski, B.; Rogowski, J.; Maniukiewicz, W.; Kozanecki, M.; Moszyński, D.; Zaborski, M. Characterization and properties of new color-tunable hybrid pigments based on layered double hydroxides (LDH) and 1,2-dihydroxyantraquinone dye. *J. Ind. Eng. Chem.* **2019**, *70*, 427–438. [CrossRef]
33. Coiai, S.; Javarone, S.; Cicogna, F.; Oberhauser, W.; Onor, M.; Pucci, A.; Minei, P.; Iasilli, G.; Passaglia, E. Fluoroscent LDPE and PLA nanocomposites containing fluorescein-modified layered double hydroxides and their ON/OFF responsive behavior towards humidity. *Eur. Polym. J.* **2018**, *99*, 189–201. [CrossRef]
34. Kutlu, B.; Leuteritz, A.; Häuβler, L.; Oertel, U.; Heinrich, G. Stabilization of polypropylene using dye modified layered double hydroxides. *Polym. Degrad. Stab.* **2014**, *102*, 9–14. [CrossRef]
35. Ran, B.; Chen, F.; Li, J.; Li, W.; Yang, F. Adsorption capability for anionic dyes on 2-hydroxyethylammonium acetate-intercalated layered double hydroxide. *Colloids Surfaces A: Physicochem. Eng. Asp.* **2016**, *511*, 312–319. [CrossRef]
36. Marek, A.A.; Verney, V.; Totaro, G.; Sisti, L.; Celli, A.; Cionci, N.B.; Di Gioia, D.; Massacrier, L.; Leroux, F. Organo-modified LDH fillers endowing multi-functionality to bio-based poly (butylene succinate): An extended study from the laboratory to possible market. *Appl. Clay Sci.* **2020**, *188*, 105502. [CrossRef]
37. Bragg, W.H.; Bragg, W.L. The reflection of X-rays by crystals. *Proc. R. Soc. Lond.* **1913**, *88*, 428–438. [CrossRef]
38. Asma, B.; Hammoui, Y.; Adjeroud, N.; Djerrada, N.; Madani, K. Effect of filler load and high-energy ball milling process on properties of plasticized wheat gluten/olive pomace biocomposites. *Adv. Powder Technol.* **2018**, *29*, 1230–1238. [CrossRef]
39. Loh, Z.H.; Samantha, A.K.; Heng, P.W.S. Overview of milling techniques for improving the solubility of poorly water-soluble drugs. *Asian J. Pharmaceut. Sci.* **2015**, *10*, 255274. [CrossRef]
40. Piras, C.C.; Fernández-Prieto, S.; De Borggraeve, W.M. Ball milling: A green technology for the preparation and functionalisation of nanocellulose derivatives. *Nanoscale Adv.* **2019**, *1*, 937. [CrossRef]
41. Zhang, C.; Tominaga, Y.; Soto, K.; Imai, Y. Simultaneous attainment of particle dispersion and surface modification of Al_2O_3 nanoparticles via wet-jet milling. *J. Compos. Mater.* **2020**, *55*, 521–530. [CrossRef]
42. Barczewski, M.; Mysiukiewicz, O.; Hejna, A.; Biskup, R.; Szulc, J.; Michałowski, S.; Piasecki, A.; Kloziński, A. The effect of surface treatment with isocyanate and aromatic carbodiimide of thermally expanded vermiculite used as a functional filler for polylactide-based composites. *Polymers* **2021**, *13*, 890. [CrossRef] [PubMed]
43. *ASTM D 1238*; Standard Test Method for Melt Flow Rates of Thermoplastics by Extrusion Plastometer. ASTM: West Conshohocken, PA, USA, 2013.
44. *ASTM D792*; Standard Test Methods for Density and Specific Gravity (Relative Density) of Plastics by Displacement. ASTM: West Conshohocken, PA, USA, 2020.
45. Kaelble, D.H. Dispersion-polar surface tension properties of organic solids. *J. Adhes.* **1970**, *2*, 66–81. [CrossRef]
46. Malvern Instruments. *A Basic Introduction to Rheology*; Malvern Instruments White Paper; Malvern Instruments worldwide: Malvern, UK, 2016; pp. 1–20. Available online: https://cdn.technologynetworks.com/TN/Resources/PDF/WP160620BasicIntroRheology.pdf (accessed on 11 August 2023).
47. Cuadri, A.A.; Martin-Alfonso, J.E. Thermal, thermo-oxidative and thermomechanical degradation of PLA: A comparative study based on rheological, chemical and thermal properties. *Polym. Degrad. Stab.* **2018**, *150*, 37–45. [CrossRef]
48. Malkin, A.Y. Continuous relaxation spectrum–Its advantages and methods of calculation. *Int. J. Appl. Mech. Eng.* **2006**, *11*, 235–243.
49. Malkin, A.Y.; Vasilyev, G.B.; Adrianov, A.V. On continuous relaxation spectrum. Method of calculation. *Polym. Sci. Ser. A.* **2010**, *52*, 1137–1141. [CrossRef]
50. Bartenev, G.M.; Valishin, A.A.; Panchuk, I.I. Relaxation spectrometry of elastomers. *Vysokomol. Soyed.* **1977**, *19*, 187–193. [CrossRef]
51. López-Barrón, C.R.; Macosko, C.W. Rheology of compatibilized immiscible blends with droplet-matrix and cocontinous morphologies during coarsening. *J. Rheol.* **2014**, *58*, 1935–1953. [CrossRef]
52. Verney, V.; Michael, A. Influence de la polydispersité sur le comportement rhéologique á l'état fondu du polypropylene. *Rheol. Acta* **1985**, *24*, 627–631. [CrossRef]
53. Martin, G.; Labaig, J.J.; Monge, P. Dynamic viscosity of entangled polymers. *Polymer* **1975**, *16*, 223–226. [CrossRef]
54. Garcia-Franco, C.A.; Mead, D.W. Rheological and molecular characterization of linear backbone flexible polymers with the Cole-Cole relaxation spectrum. *Rheol. Acta* **1999**, *38*, 34–47. [CrossRef]
55. Dobrzynska-Mizera, M.; Dutkiewicz, M.; Sterzynski, T.; Di Lorenzo, M.L. Isotactic polypropylene modified with sorbitol-based derivative and siloxane-silsesquioxane resin. *Eur. Polym. J.* **2016**, *85*, 62–71. [CrossRef]
56. Barczewski, M.; Mysiukiewicz, O.; Andrzejewski, J.; Piasecki, A.; Strzemięcka, B.; Adamek, G. The inhibiting effect of basalt powder on crystallization behavior and the structure-property relationship of α-nucleated polypropylene composites. *Polym. Test.* **2021**, *103*, 107372. [CrossRef]

57. Alghyamah, A.A.; Elnour, A.Y.; Shaikh, H.; Haider, S.; Poulose, A.M.; Al-Zahrani, S.M.; Almasry, W.A.; Park, S.Y. Biochar/polypropylene composites: A study on the effect of pyrolysis temperature on crystallization kinetics, crystalline structure, and thermal stability. *J. King Saud Univ.–Sci.* **2021**, *33*, 101409. [CrossRef]
58. Bernard, E.; Zucha, W.J.; Lothenbach, B.; Mäder, U. Stability of hydrotalcite (Mg-Al layered double hydroxide) in presence of different anions. *Cement Concrete Res.* **2022**, *152*, 106674. [CrossRef]

Disclaimer/Publisher's Note: The statements, opinions and data contained in all publications are solely those of the individual author(s) and contributor(s) and not of MDPI and/or the editor(s). MDPI and/or the editor(s) disclaim responsibility for any injury to people or property resulting from any ideas, methods, instructions or products referred to in the content.

materials

MDPI

Article

Effect of an Early-Age Exposure on the Degradation Mechanisms of Cement Paste under External Sulfate Attack

Othman Omikrine Metalssi [1,*], Rim Ragoug [2], Fabien Barberon [3], Jean-Baptiste d'Espinose de Lacaillerie [4], Nicolas Roussel [5], Loïc Divet [6] and Jean-Michel Torrenti [1,*]

1 Laboratoire Matériaux pour une Construction Durable (UMR MCD), University Gustave Eiffel, Cerema, F-77454 Marne-la-Vallée, France

2 Organisation Professionnel Prévention Bâtiment Travaux Public, 25 Avenue General LECLERC, 92600 Boulogne Billancourt, France; rim.touhami@oppbtp.fr

3 EQIOM, 49 Avenue Georges Pompidou, 92300 Levallois-Perret, France; fabien.barberon@eqiom.com

4 Soft Matter Science and Engineering, ESPCI Paris, UMR CNRS 7615, Sorbonne Université, 75005 Paris, France; jean-baptiste.despinose@espci.fr

5 Laboratoire Navier, Université Gustave Eiffel, CPDM, F-77454 Marne-la-Vallée, France; nicolas.roussel@univ-eiffel.fr

6 Civil Engineering Department, Sherbrooke University, 2500 Boulevard de l'Université, Sherbrooke, QC J1K 2R1, Canada; loic.divet@outlook.com

* Correspondence: othman.omikrine-metalssi@univ-eiffel.fr (O.O.M.); jean-michel.torrenti@univ-eiffel.fr (J.-M.T.); Tel.: +33-1-81-66-83-63 (O.O.M.)

Abstract: Among the most significant causes of concrete degradation is ESA (external sulfate attack). The majority of studies are currently conducted on samples that have been saturated and matured. Concrete structures, however, are exposed to the environment once the formwork has been removed. The purpose of this study is to determine what effects early exposure to external sulfates may have on degradation mechanisms. Microstructure, physical, and chemical behavior are monitored using a variety of experimental techniques, including NMR (^{27}Al and ^{29}Si), ICP, XRD, MIP, and SEM. Based on expansion measurements, mature Portland cement paste, unlike the early-age case, degraded rapidly due to the presence of compressed ettringite and gypsum, highlighted by SEM analysis. During ESA, sulfate ions diffuse through the cement matrix and are bound by chemical agents. Chemical analyses indicate that the chemical mechanism varies with the duration of curing. At an early age, external sulfates and aluminates are the most important reagents. For matured cases, these reagents include external sulfates, calcium derived from CH dissolution, and aluminates derived from the total dissolution of AFm.

Keywords: external sulfate attack; physicochemical behavior; early-age effect; sulfate ingress

1. Introduction

Civil engineers continue to struggle with the long-term behavior of concrete structures. In most cases, when the formwork is removed from concrete structures, a variety of physico-chemical aggressions occur. As a consequence of this early-age exposure, the future transport properties of the structural element are influenced, which are important parameters determining the durability of the material. Transport properties affect the material's ability to sustain the degradation mechanisms resulting from the penetration in the cementitious matrix's inner porous microstructure of deleterious ions and aggressive elements (CO_2, sulfates, and chloride ions, etc.) by the external environment [1–7]. Early-age exposure can also influence the mechanisms of internal reactions, such as the well-known alkali–aggregate reaction (AAR) and delayed ettringite formation (DEF) [8–10]. Two phenomena, the alkali's content of the concrete and the temperature reached during the concrete hardening and relative humidity, influence the extent of degradation at an early age. Alkalis are primarily obtained from cement, although some may also be released from aggregates. The presence of a high-alkali content at an early age will cause sulfates to be released, which provide the reagents needed to form ettringite. In terms of temperature, DEF effects are related to the material's thermal history. Most scientists believe that an increase in temperature or duration of heating during hydration at an early age increases the kinetic and ultimate values of expansion. The expansion of the material increases with the increase in humidity, which is higher during the first phase of hydration. Ettringite also contains water as one of its constituents. Since most standards and laboratory tests pertaining to durability are conducted on mortar or concrete samples that have been cured for 28 or 91 days, the effect of early-age exposure is still unclear and requires more laboratory investigations.

External sulfate attack (ESA), considered to be the second leading cause of concrete structure degradation after corrosion, is one of the most hazardous phenomena that can cause concrete structure degradation. ESA is also one of the most cited causes of concrete structure deterioration. In general, this phenomenon can be attributed to the formation of ettringite and gypsum, under some conditions, when sulfate ions and alumina constituents react in cement [11–20]. There is general agreement that the precipitation of these newly formed elements in the cement matrix leads to considerable expansion and cracking in structures. This can cause surface spalling accompanied by an overall loss of structural integrity highlighted by reductions in stiffness and strength [14,21,22].

ESA is a complex multiscale and multiphysics phenomenon that involves physical, chemical, and mechanical interactions [14,20,23–28]. Although these interactions are relatively well understood individually, it remains difficult to predict how they pair together and the overall phenomena remains unclear and highly controversial, even for ordinary Portland cement, for which, unlike other cements, substantial results have been obtained [29]. The complexity of the problem begins with the variety of sulfates that can damage concrete (calcium, sodium, and magnesium sulfate). In the case of sodium sulfate, one school of thought holds that the mechanism of ESA begins with gypsum formation before expansive ettringite formation [30], while another school of thought suggests the opposite process [31]. Furthermore, the mechanism of the ESA depends on the amount of sulfate present in the external solution. Based on Biczok's results [32], in low-concentration sulfate environments, ettringite forms as the predominant product, while in high-concentration environments, gypsum forms as the dominant product.

Although many publications and research studies have been conducted about the ESA mechanisms [33], there remain several unanswered questions and, in particular, the early-age behavior of structures exposed to ESA still requires further study. This early age has a significant impact on the long-term service life of concrete structures, as constituent proportions of concrete and its microstructure undergo continuous and rapid changes during this period. It may be difficult to achieve high-quality concrete when the early-age properties of concrete are not sufficiently taken into account. Material can undergo internal and/or external deformations that can result in cracking and the loss of its mechanical properties. As a result, the material may become weaker and more susceptible to penetration by aggressive agents, such as ESA. Therefore, at an early age, the coupling between the hydration process and mechanical properties is more critical than in mature concrete. It is also possible for poorly cured concrete to seriously compromise its mechanical performance and long-term durability. The proper curing of concrete after placement is essential to maintain satisfactory moisture content and the necessary temperature during this early age of concrete's development in order to achieve its intended properties.

This study examines the kinetics of sulfate ingress and its impact on the microstructure and degradation mechanisms of cement paste at an early age. Both at early ages and after one year of curing in water, the sulfate concentration profiles were determined by ICP (inductively coupled plasma analysis). SEM analysis was used in conjunction with these measurements to investigate the microstructural changes. Furthermore, other investigations were conducted using nuclear magnetic resonance (NMR), ^{27}Al and ^{29}Si, and X-ray diffraction (XRD) to analyze the chemical composition of newly formed elements during

ESA. Lastly, a visual comparison of the degradation of the samples and the expansion measurements for the two different curing conditions were conducted.

2. Materials and Methods

2.1. Formulations and Exposure Conditions

For the purpose of this study, Portland cement (CEM I) with a chemical composition as shown in Table 1 was used for casting cement paste specimens. This composition was obtained by a combination of inductively coupled plasma analysis (ICP) and thermogravimetric analysis (TGA). According to Bogue's approach, the clinker phase contents are as follows: C_3S = 65.2, C_2S = 8.8, C_3A = 7.9, and C_4AF = 8.9. According to these values, there is a significant amount of aluminate in the form of C_3A and C_4AF, which makes it likely that secondary ettringite will be formed after a reaction with a large amount of sulfate ions.

Table 1. Chemical composition of ordinary cement CEM I.

Chemical Composition (%)	CEM I
CaO	62.81
SiO_2	20.22
Al_2O_3	4.85
Fe_2O_3	2.92
CaO (free)	1.58
MgO	0.84
SO_3	2.88
S	0
K_2O	0.77
Na_2O	0.34
Ignition Loss	2.59

For the preparation of the specimens, cylindrical samples with a 10 cm diameter and 15 cm length were cast with cement paste and labeled as the OPC samples. To maximize porosity and enhance sulfate ingress, a high water-to-cement mass ratio (W/C) of 0.6 was selected. As part of the setting procedure, the cement paste was prevented from bleeding with the use of a sample slow-rotation device. This may ensure the homogeneity of the samples (Figure 1a). Once the specimens had been removed from their molds for 24 h, they were cut into 5 cm long cylinders. One third of the samples were then kept directly in sulfate solution (early-age exposure case) and the remaining two thirds were cured in non-renewed water for one year (matured case). One of these two thirds was further exposed to external sulfate solution after a long curing period of one year (matured exposure case), under the same conditions as the early-age case samples. Finally, as a reference case, the last third of the samples was cured in tap water. All cylindrical samples were coated with epoxy resin on their lengths, but not on their bases in order to ensure unidirectional diffusion of sulfate ions (Figure 1b). Hence, there was only one base of the specimen that was in contact with the test solution (semi-immersion) (Figure 2), which was prepared by dissolving 15 g/L of sodium sulfate in deionized water (10 g/L of sodium sulfate). There was a 25 cm^2/L ratio between the surface area of the sample and the volume of the solution. As discussed in more detail below, a pH regulator was used to maintain the pH of the sulfate solution at (8 + 0.1).

(a) (b)

Figure 1. A device used during the cement paste's hardening (**a**), specimens used for the study (**b**).

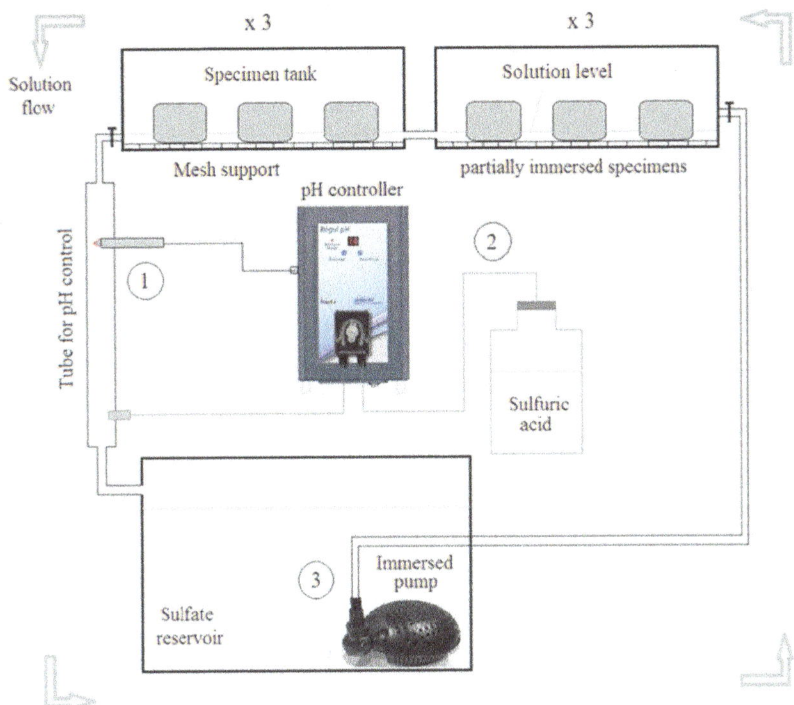

Figure 2. A pH-control scheme for accelerated tests at ESA.

Companion cylindrical specimens with a 10 cm diameter and 15 cm length were used to measure the axial expansion during the degradation process for the early-age and matured cases. After demolding, three specimens for each exposure condition were equipped with stainless steel pins to ensure that the axial dimension was monitored. A digital extensometer and a steel reference length bar were used for these measurements.

2.2. Accelerated Test for ESA

To study the resistance of cementitious materials to ESA, using a constant pH = 8 ± 0.1 is preferable to field conditions (e.g., seawater) [34]. Due to the constant sulfate concentration and pH of our proposed accelerated test method, the experimental conditions are more representative of field conditions. Several studies have used high sulfate amounts in their experiments [35–37]. Figure 2 shows a schematic of the test apparatus.

In this simplified device, sulfate solution with a concentration of 15 g/L is contained in one large tank that serves as a reservoir for approximately 60 L of solution. The sulfate solution was circulated through six smaller tanks containing 10 L each with the aid of an immersion pump with a maximum flow rate of 750 L per hour (Figure 2). Before flowing back into the sulfate reservoir at the output of these tanks, a controller continuously monitors the pH of the solution (process 1). The pH of the sodium sulfate solution was determined through continuous titration with sulfuric acid H_2SO_4 (0.05%) (process 2). At the exit of the tanks containing the specimens, a pH electrode is used to continuously measure the pH of the solution. As part of this system, an actuator injects diluted sulfuric acid into the solution whenever the pH level increases due to leaching, so that it returns to the target pH level. The added sulfuric acid is well mixed since the sulfate solution is constantly flowing between the sulfate reservoir and the specimen tanks (process 3).

In each of the six tanks, semi-immersed samples of similar chemistry were arranged on mesh supports so that the cement paste and sulfate ions were exposed to the ESA to maximize the chance of the cement paste reacting with it. Only 1 cm thickness of each specimen was immersed in the sulfate solution, while the other 4 cm of thickness remained out of solution. Sulfate concentrations were minimized by adjusting the volume ratio between the specimens and sulfate solution. This parameter was kept approximately constant by weekly solution renewals. Sulfate ions were maintained almost constant in solution over time due to the use of sulfuric acid in the titration. The same device was used for expansion measurements in sulfate solutions.

2.3. *Experimental Techniques for Investigation*

The kinetics and effects of sulfate ingress in ordinary cement pastes' porous media were examined using a variety of test methods. All physicochemical investigation methods were applied to a powder ground from the surface of the test samples every 1 mm. As part of this process, Germann Instruments' Profile Grinder tool (Denmark) is used to reduce concrete, mortar, or cement paste into fine powder by precision grinding to small depth increments of 0.5 to 2 mm. As a result, it is possible to determine with high precision the profiles of ionic contents and chemical elements at each depth. In our case, the ground area has a diameter of 73 mm (which justifies the selection of cylinder samples with a diameter of 100 mm). This diameter makes it possible to obtain about 9 g of powder for 1 mm depth. The average fineness of the powder is 315 μm. Between each depth, the powder is carefully picked up and the surface is vacuumed to prevent contamination.

The sulfate content of the samples was determined using inductively coupled plasma atomic emission spectroscopy (ICP). Sulfate concentration is calculated from the total sulfur concentration obtained during this experiment. The exposed surface layer of the samples to a depth of 1.5 cm was ground. The resulting powders had an average size of 300 μm. An acid attack ionized all sulfates in this powder. An ICP-AES measurement was then conducted on the resulting solution. The experimental procedure used in this part of the study which details the procedure for preparing samples to measure the concentration of SO_3 is explained in [23]. Additionally, Mercury Intrusion Porosimetry (MIP) measurements were conducted to examine the effect of ESA on pore size distribution.

Additionally, the spatial distribution of different cement paste phases was also determined by scanning electron microscopy (SEM, Quanta 400 from FEI Company, Hillsboro, OR, USA). Despite the fact that this technique provides only qualitative information, SEM images can be used to distinguish between phases that are well crystallized and do not harm the microstructure, such as compressed ettringite, from phases that are crystallizing in well-distributed free volumes.

Furthermore, a METTLER TOLEDO TGA/DSC1 was used for thermogravimetric analysis (TGA) coupled with differential thermal analysis (DTA). During the ESA, these methods were used to quantitatively characterize cement pastes. Using these techniques, this study could explain the mechanism by which AFt is formed from the consumption of other cement hydrates.

In addition, X-ray diffraction (XRD) was carried out in this research. XRD can be used to identify well-crystallized hydrates in cement paste, such as portlandite or ettringite. As a result, a mineralogical profile is drawn and different types of mineralogical phases are identified according to the type of exposure. The results obtained by XRD were also confirmed by ^{27}Al and ^{29}Si NMR using an Avance III Bruker spectrometer (Martin Dracinsky Group, Czech Republic) ^{27}Al MAS NMR was performed to calculate the amount of different hydrated alumina phases (AFm and AFt) within the cement. ^{29}Si NMR was used to investigate the C-S-H structure changes due to sulfate attack. ^{27}Al MAS NMR one-pulse spectra were obtained by modeling the Czjzeck lineshape using the DMFIT software [38] following the procedure described in [39]. ^{29}Si NMR one-pulse spectra were obtained by using a Gaussian lineshape with the DMFIT software, and the mean chain length (MCL) calculated according to [40].

As a final step, macroscopic observations were conducted on samples exposed to ESA for various periods of time.

3. Results

3.1. Expansion Measurements

Figure 3 shows the average expansions and standard deviations obtained from axial length measurements on three 10 cm × 15 cm cylindrical samples. These curves are for early- and mature-exposure cases. A significant expansion is observed at the end of the test for the matured exposure case (about 0.47%), with the speed divided into two parts: a latent period up to about 30 days in which expansion remains weak, followed by a period in which the expansion speed increases rapidly and remains high until the end of the test (around three months). This change in regime was a result of changes in the microstructure of the samples. Indeed, the acceleration of the expansion is due to the clogging of the porosity by the progressive formation of secondary ettringite and gypsum. It has been observed in contrast that, in the early-age exposure case, the sample exhibited modest axial expansion (up to 0.12%), followed by a gradual slowdown over time. These contrasted results could only be due to variation in the microstructure since the bulk quantities of reagents present in both exposure conditions were identical. These results are in concordance with those of [13]. The details of this will be discussed in the following sections.

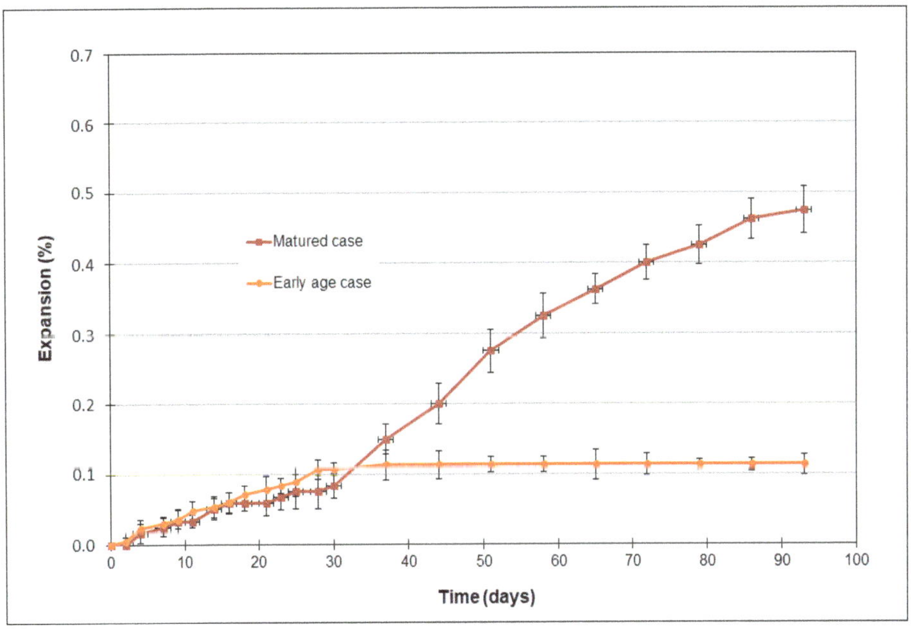

Figure 3. Expansion variation for the early- and mature-exposure cases.

3.2. Sulfates Profiles

Figure 4 illustrates the evolution of sulfate content in OCP samples versus depth under early-age exposure. The curves are divided into two parts: one displaying the ingress of sulfate into the cement, and the second maintaining the average sulfate concentration of the reference cement without being exposed to sulfate solution.

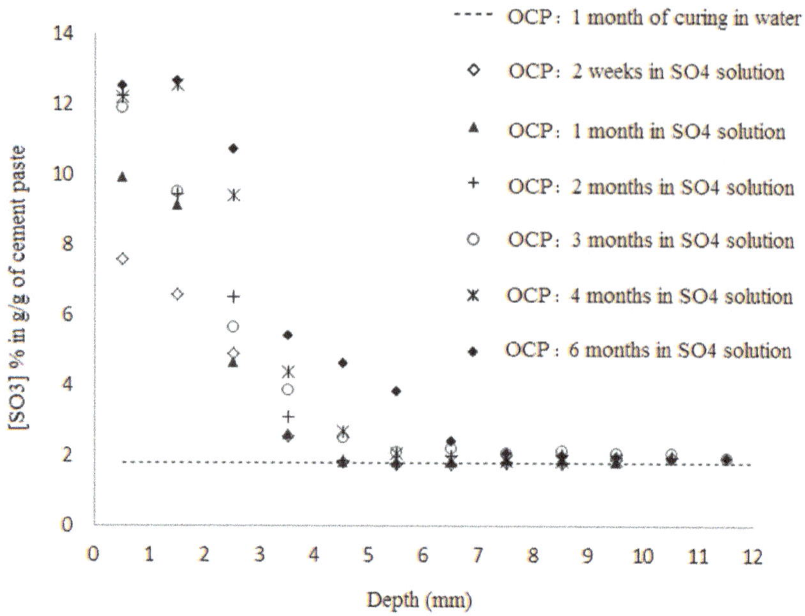

Figure 4. Profiles of sulfate at early age measured by ICP, for different exposure duration to the ESA.

As the exposure time increases, the sulfate content of the cement increases until it stabilizes at approximately 13% g/g near the surface. This value differs considerably from what might be expected from a simple equilibrium between the internal pore solution and the external sulfate solution (about 0.2% g/g of cement). Therefore, the pore solution is dominated by physical or chemically bound sulfates, which explains the decrease in sulfate content with depth. Furthermore, the extremely high ratio of solid or adsorbed sulfates in the cement matrix required to free sulfates in solution, estimated to be of the order of 60, suggests that the fixation process involves an extremely high level of energy, assuming the free sulfates' concentration in the pore solution is of the same order as the external solution sulfate concentration. Thus, sulfate physical adsorption alone is not sufficient to achieve this binding, which is mostly chemical in nature. As stated in [41], the quantity of adsorbed sulfate on C-S-H was estimated to be 1% g/g, which is considerably lower than the average sulfate content reported here of 13% g/g. As the exposure time increases, the penetration depth increases slightly as well. As a result of two weeks' exposure to sulfate solution, it measures approximately 4.5 mm, and at six months, it measures approximately 7.5 mm.

A further illustration of this development is shown in Figure 5. This illustrates the evolution of sulfate content in OCP materials over time for both early and matured exposure modes to sulfate solutions. Despite similar sulfate profiles in both cases after two months of sulfate exposure, there is a difference in the maximum sulfate content attained. Early-age exposure to the first three millimeters results in a higher maximum value. It is a result of the high porosity and permeability of the material at this early stage. In an area near the propagation front, the difference is negligible above this depth. The results of these experiments indicate that although sulfate ions are initially trapped in the porous medium via a diffusion process, they are physically and chemically trapped in the microstructure of the paste at an important kinetic rate. In addition, capillary suction may also contribute to sulfate ingress, especially when samples are exposed at an early age.

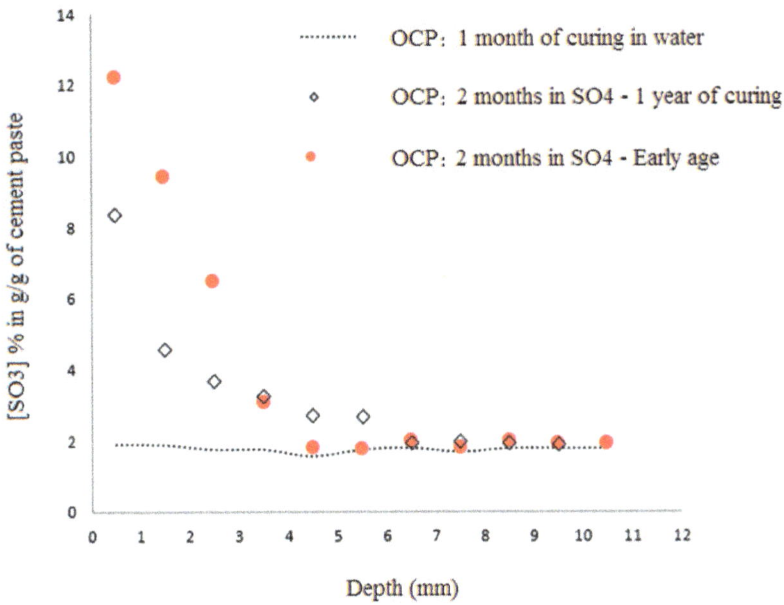

Figure 5. Comparison of the sulfate profiles between the two exposure conditions: early-age and mature cases.

The maximum sulfate penetration depth also increases in matured samples. As a result of the samples' exposed surface having developed some cracks after two months of exposure, this may be the cause of the results. Indeed, after long curing in water, the material becomes rigid and little ettringite formation or other expansive products can cause cracking. The situation is different for materials that are still in the early stages of development, when the matrix is not yet rigid, and expansive products are able to find space to form without causing internal pressures and, consequently, without cracking. Sulfate ions and other deleterious ions may be able to pass through these cracks.

3.3. Chemical Mechanisms

Sulfate ions will certainly affect the chemical equilibrium of the cement paste as they are absorbed by the cement matrix. In cement, sulfate ions form secondary ettringite and gypsum when they react with calcium, aluminum, and water. On the other hand, when cement samples are leached in a sulfate solution, the pH difference between that solution and the pore solution of the samples results in an increase in the calcium content of the pore solution which in turn favors AFt formation [23–27]. This exemplifies how ESA and leaching are strongly coupled and compete with one another. According to Figure 6, the phase assembly in the cement paste underwent a change after being exposed to sulfate solution. A combination of XRD and ^{27}Al NMR analyses indicates that, on the one hand, secondary ettringite was formed as a result of reactions between portlandite and sulfate ions; and on the other hand, that AFm transformed to AFt. In addition, the XRD analysis indicates that there is a considerable amount of gypsum on the exposed surface of the samples, as this area is almost in equilibrium with the pH of the sulfate solution (pH = 8). This pH value results in both ettringite and gypsum precipitation. Moreover, the high rate of anhydrous compounds at an early age allows for the production of more secondary ettringite. Indeed, the aluminates present in C_3A and C_4AF react directly with the sulfate ions to produce ettringite. In the matured case, this mechanism does not exist. All anhydrous compounds have already been consumed during the hydration process.

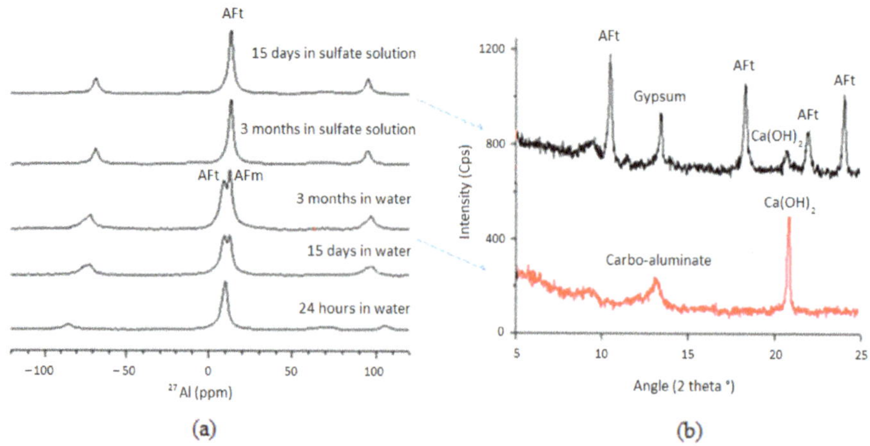

Figure 6. Determination of phases newly formed by ^{27}Al MAS NMR (**a**) and XRD (**b**) after exposure of OCP samples to the sulfate solution.

Figure 7 illustrates the results of TGA. These results are consistent with those obtained by XRD. In fact, portlandite is the first hydrate to be consumed by sulfate ions that penetrate the first millimeters of the surface of the samples (Figure 7a). It is important to note, however, that the consumption does not appear to be the same for both cure conditions. As the material is completely hydrated after one year of curing in water, the amount of portlandite is much greater in the mature case (Figure 7b). In the presence of sulfate ions in the cementitious matrix, they thermodynamically destabilize the hydrates, particularly portlandite, resulting in its dissolution. Calcium from this hydrate reacts with sulfates to form gypsum, especially at neutral pH levels. Accordingly, portlandite dissolves more rapidly when subjected to a sulfate attack than when not exposed. In the process, gypsum replaces calcium hydrates, which results in a decrease in portlandite content.

Figure 7. Quantification of CH content by TGA: (**a**) early-age case, (**b**) comparison early-age/matured cases.

Figure 8a shows the ^{29}Si MAS NMR spectra for both early and mature exposures. The peak of the tetrahedral at the end of the C-S-H chains, Q^1, has decreased while that of the tetrahedral in the middle of the chains, Q^2, has increased. This means that the C-S-H mean chain length (MCL) increased and that the C-S-H structure changed. The MCL of the C-S-H chains increased to about 7.5 in the early-age exposure case and to about 10 in the matured exposure case (Figure 8b). During C-S-H decalcification (C/S ratio decrease), it is well known that the MCL in C-S-H increases [40].

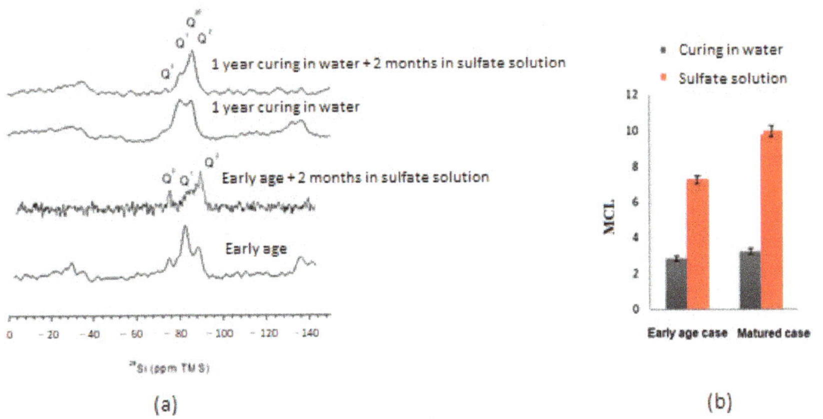

Figure 8. Changes of C-S-H structure investigated by 29Si NMR: (**a**) spectrum, (**b**) mean chain lengths in C-S-H.

Once the calcium has been consumed by dissolving the portlandite, the C-S-H are decalcified, increasing the calcium content in the interstitial solution. The calcium then reacts with excess sulfates to produce gypsum and secondary ettringite. In the mature case, calcium content is greater, suggesting that the development of ESA is more dependent upon the presence of a high-hydrate content.

There is a drastic difference in the AFt/AFm balance between the two exposure cases (Figure 9). In the case of early-age exposure, this ratio is higher. In Figure 9a, the secondary ettringite is partly produced as a result of the progressive dissolution of AFm. The mature-exposure case (Figure 9b) results in the complete consumption of AFm. These results suggest that ESA develops differently depending on the length of time the curing process is carried out prior to exposure. In sum, at an early age, AFt is formed through the reaction of the sulfates with part of the AFm and also through the reaction of the aluminates produced by the anhydrous compounds C_3A and C_4AF. When a mature paste is exposed, this formation of AFt is mainly duc to thc total consumption of AFm and to the contribution of calcium coming from CH dissolution and C-S-H decalcification.

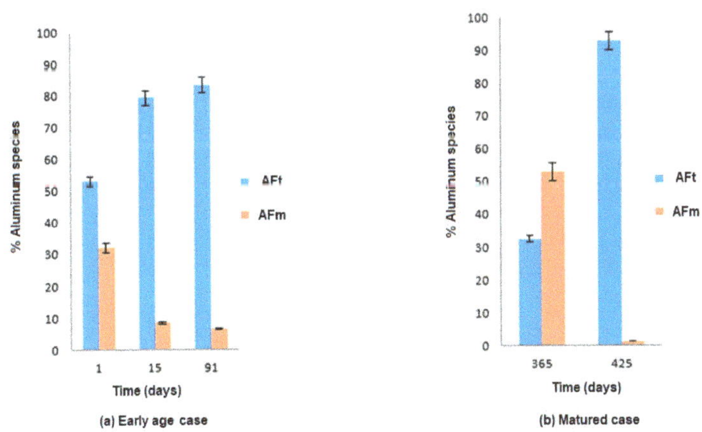

Figure 9. Quantification of Aft and AFm content for the two curing durations.

3.4. Microstructure Changes

Figure 10 compares the pore size distributions between cases with and without exposure to ESA for OCP samples at different exposure times. After one month of exposure to ESA, the effect of leaching is highlighted. The distribution of pore size after 2 months of ESA, however, is refined and tends towards the range from 10 nm to 100 nm (Figure 10a). Clearly, at an early age, hydration, leaching, and precipitation of sulfated species were competing processes, making it difficult to distinguish between them. Moreover, in the case of matured cement pastes (Figure 10b), the ESA appears to be more damaging for cementitious materials since the pore size distribution still tends to a smaller range (of the order of 10 nm), despite the quasi-complete hydration of the material, which suggests that the precipitation of the AFt and the gypsum occurs within a much smaller porosity than when the material is exposed at an early age.

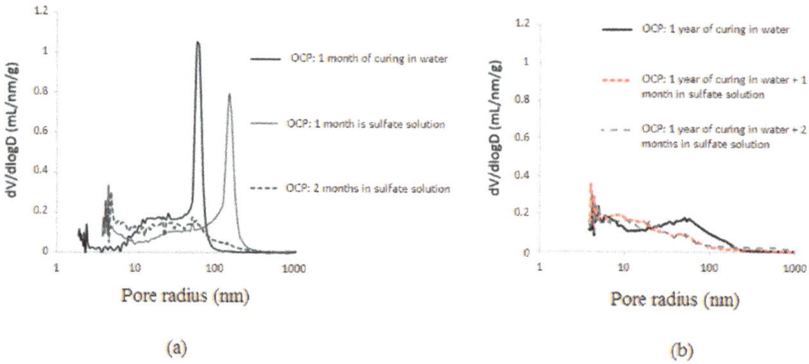

Figure 10. Evolution of the pore size distribution of OCP samples during ESA: (**a**) early-age exposure, (**b**) matured cement pastes.

A significant expansion and cracking of concrete structures is generally attributed to secondary ettringite and gypsum as a result of sulfate attack on hydrated alumina phases in the cement matrix. Visual examinations of samples exposed to sulfate solution after different exposure periods were conducted to confirm the role of sulfates in cementitious materials' damage. Figure 11 illustrates a comparison of samples at an early age (Figure 11a) and matured age (Figure 11b) after two months of exposure to sulfate solution. Early exposure did not show apparent degradation or cracking. After one year of contact with sodium sulfate solution, these observations remain valid. After just two months of exposure to sulfate, the mature samples have developed some cracks.

Figure 11. Observed damage of samples after two months of exposure to sulfate solution: (**a**) early-age exposure, (**b**) matured cement pastes.

Accordingly, it appears that AFt phases crystallize more readily in smaller pores and after a more advanced hydration reaction (saturated pores in matured samples) thereby resulting in greater cement paste risks (ettringite has approximately 2.28 times the volume of AFt phases; 707 cm^3 versus 309 cm^3). Additionally, mature samples are sufficiently rich in portlandite and C-S-H to precipitate gypsum upon dissolution. The situation is different if the exposure occurs at an early age. Moreover, gypsum has a 2.25 times greater molar volume than portlandite (74 cm^3 compared with 33 cm^3 per mole). As a result, the newly formed elements after an ESA (AFt and/or gypsum) could increase the expansion of cement paste samples when confined in porosity. Furthermore, SEM analysis of samples exposed to sulfate for early-age and mature cases has confirmed this observation. Based on Figure 12, the ettringite precipitates in confined spaces in mature samples (Figure 12a), confirming that the samples have expanded and cracked, while at an early age, the ettringite has longer spaces (Figure 12b).

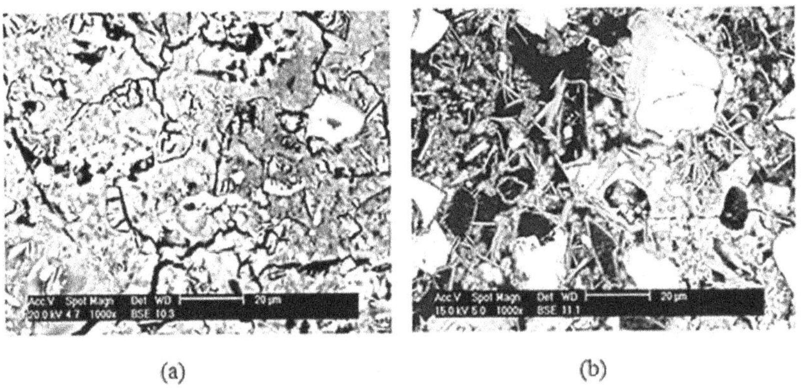

Figure 12. SEM images on OCP samples after two months of ESA: (**a**) compressed ettringite in matured samples, (**b**) ettringite in early-age samples.

4. Discussions

Experiments conducted within the framework of this study demonstrated firstly that the ESA is characterized by two major processes, a physical process of ionic transport and a chemical process of interaction with the microstructure. The first process is characterized by both a transport of sulfate ions towards the cementitious matrix and a leaching of calcium ions (resulting in the leaching of hydroxides from the portlandite) in the external solution. ICP was used to measure sulfate concentration profiles, which demonstrated that ESA is driven by diffusion and that chemical interactions (the second process) between hydrates and external sulfate supplies are more rapid than diffusion kinetics. In addition, sulfate ions have been demonstrated to self-slow during propagation. In the measured profiles, the sulfate ion migration front is rapidly stabilized followed by an accumulation of sulfate ions at the surface. Based on ATG/AD, DRX, and NMR analyses, this accumulation was explained by the fixation of sulfur ions by the precipitation of AFt as long as aluminum and calcium reserves were present, as well as the coexistence of gypsum and ettringite at times on the surface (especially in mature Portland cement pastes).

The effect of curing time on the degradation of the cementitious materials in contact with sodium sulfate solution was demonstrated in several steps. The first method is visual observations which show that Portland cement pastes resist ASE well when exposed early to sulfates (no degradation is observed after one year). The mature material, however, deteriorates rapidly and suddenly on the surface and then deeper towards the core after only two months of sulfate exposure. Hence, expansion is directly affected by porosity and pore size distribution. According to SEM visualization and MIP tests, samples for the early-age case have a greater amount of porosity, allowing them to accommodate expansive phases without sustaining damage. A long-term curing in water results in a rapid and sudden filling of the porosity, resulting in a sudden change in the sharpness of expansion curves for mature samples due to the fine microstructure and rigidity of the material.

In cement paste, the pore size distribution of capillary pores and hydrate pores has a significant effect on the transfer properties, particularly when the pores are interconnected. This observation is supported by the profiles of sulfate penetration that are much greater in the case of coarse porosity (early age). It should be noted that the chemical aspects of ESA are almost insensitive to these transport properties (especially in the case of an early-age exposure), but rather to the chemical reactions taking place within the material. These reactions are dependent on portlandite, the AFm and AFt phases, as well as the aluminum incorporated into the C-S-H prior to exposure to the sodium sulfate solution.

On the other hand, the quantification of precipitated ettringite in the two exposure cases demonstrated that degradation is not directly related to changes in molar volume. According to the hydrate distribution at an early age, ettringite was more abundant than in mature paste after two months of ASE. Nevertheless, these abundant precipitations were not the cause of the material's damage. In order to explain the cracks induced by tensile stresses, the crystallization pressure theory under supersaturation conditions was adopted in this study.

Based on the curing time of the material, the physical chemistry of ASE differs. As previously explained, this refers to chemical reactions rather than sulfate ions' transport, which is almost insensitive to curing duration. AFm/AFt ratios are very low when OPC is exposed at an early age. The investigation techniques used in this study indicated that ettringite continues to form and crystallize in a wide and evolving porosity with a hydration phenomenon occurring in the presence of sulfate, without causing damage to the matrix during the crystallization process. Conversely, when exposed to sulfates after 1 year of water hydration, the AFm/AFt ratio is significantly higher and portlandite is more available. Ettringite precipitates by dissolving CH and AFm and by decalcifying C-S-H. Gypsum is detectable by scanning electron microscopy and XRD as long as calcium and sulfate ion concentrations allow. These hydrates crystallize in a much more confined porous environment and a much more deleterious effect can be expected.

Based on this difference, it is evident that when concrete structures are exposed to sulfate-rich environments, such as marine environments or sulfate-rich soils, the curing time of these structures should be reduced. The structures would have a greater longevity if they were directly cured in a sulfate-rich environment. As previously explained, the material exposed directly to sulfate ions does not show any signs of degradation or cracking.

Therefore, this work has provided insight into the negative effects of long time curing in the case of Portland cement. However, these results are only valid for the CEM I used in this study. The cement, which is very rich in C_3A and C_4AF, was chosen despite its potential reactivity with sulfate ions. In Figure 13, the mechanisms of ESA development are shown step by step for both early and mature curing durations. Sulfate ions penetrate the cementitious matrix where they react directly with calcium ions resulting from the dissolution of the portlandite to produce gypsum and AFt. After these calcium ions have been consumed, the C-S-H decalcifies, thereby providing an additional quantity of calcium needed to form gypsum and AFt. This resulted in an increase in the C/S ratio. Another part of the sulfate ions reacts directly with the sources of aluminates (AFm, C3A, and C4AF) to form AFt. According to the previous results, curing duration affects the chemical mechanism. In mature exposures, these reagents are external sulfates, calcium derived from CH dissolution and C-S-H decalcification, and aluminate derived from AFm dissolution. However, the major reagents used at an early age are external sulfates and aluminates derived from anhydrous cement and part of the AFm.

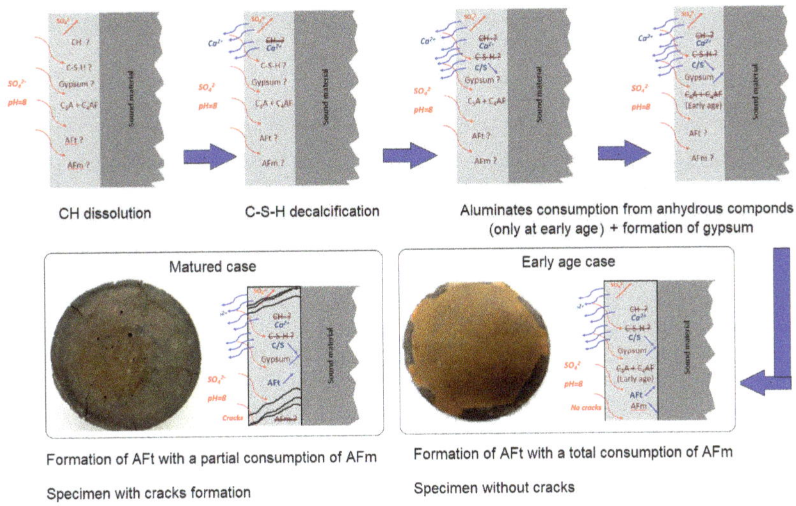

Figure 13. Mechanisms of ESA development for both early and mature curing durations.

5. Conclusions

The aim of these experimental investigations was to better understand the kinetics of sulfate diffusion and its effects on the microstructure of ordinary cement paste. In addition, the ESA mechanisms were studied using various and complementary investigation techniques for both early-age and mature curing conditions. Based on the obtained results, conclusions can be drawn as follows:

- The sulfate profiles in the two curing cases are similar, with a slight difference in the maximum sulfate content. There is a more significant effect in the first three millimeters when the exposure occurs at an early age. During this early curing, the material is highly porous and permeable. Chemical interactions with the cement matrix delay sulfate ingress.
- The physical and chemical interactions with cement paste hydrates appear to have faster kinetics than diffusion through concentration gradients and capillary adsorption. This finding confirms that ESA is characterized by both the diffusion and binding of sulfate ions in the cement matrix, with mainly a chemical fixation.
- Matured Portland cement paste showed rapid degradation. This is due to the presence of a significant quantity of compressed ettringite and gypsum, as highlighted by SEM analysis. On the other hand, Portland cement pastes that were exposed to sulfate solution early in the process did not develop cracks or spalls after one year of exposure.
- Although AFt formation is known to cause the degradation of cementitious materials when they are exposed to sulfate ions, the chemical mechanism varies with curing duration. At an early age, the main reagents are external sulfates and aluminates from the anhydrous cement and part of AFm. In the mature exposition case, these reagents are external sulfates, calcium derived from the dissolution of CH and decalcification of C-S-H, and aluminates derived from the total dissolution of AFm.

Author Contributions: Conceptualization, O.O.M., R.R., F.B. and J.-M.T.; methodology, O.O.M., R.R., L.D., J.-B.d.d.L. and N.R.; validation, O.O.M., R.R., F.B. and J.-M.T.; formal analysis, O.O.M., R.R., L.D., N.R. and J.-B.d.d.L.; investigation, O.O.M., R.R., J.-M.T. and J.-B.d.d.L.; resources, F.B.; writing—original draft preparation, O.O.M.; writing—review and editing, O.O.M. and J.-M.T.; visualization, O.O.M., R.R., F.B. and J.-M.T.; supervision, O.O.M., F.B. and J.-M.T.; project administration, F.B. and J.-M.T.; funding acquisition, O.O.M., R.R. and F.B. All authors have read and agreed to the published version of the manuscript.

Funding: This research was funded by Bouygues Construction (BYTP) grant number 2012/1460 And the APC was funded by the Bouygues Construction (BYTP).

Informed Consent Statement: Not applicable.

Data Availability Statement: All data are in this article. We don't have other data to propose.

Acknowledgments: The authors would like to acknowledge the Bouygues Society for financial and technical support. NMR equipment at ESPCI Paris is funded in part by the Paris Region.

Conflicts of Interest: The authors declare no conflict of interest.

References

1. Zhao, G.; Shi, M.; Guo, M.; Fan, H. Degradation Mechanism of Concrete Subjected to External Sulfate Attack: Comparison of Different Curing Conditions. *Materials* **2020**, *13*, 3179. [CrossRef]
2. Wagner, M.; Heisig, A.; Machner, A.; Beddoe, R.; Heinz, D. External Sulfate Attack on Cementitious Binders: Limitations and Effects of Sample Geometry on the Quantification of Expansion Stress. *Materials* **2022**, *15*, 3677. [CrossRef]
3. Liu, G.; Tang, Y.; Wang, J. Effects of carbonation degree of semi-dry carbonated converter steel slag on the performance of blended cement mortar—Reactivity, hydration, and strength. *J. Build. Eng.* **2023**, *63*, 105529. [CrossRef]
4. Flatt, R.J.; Roussel, N.; Cheeseman, C.R. Concrete: An eco-material that needs to be improved. *J. Eur. Ceram. Soc.* **2012**, *32*, 2787–2798. [CrossRef]
5. Planel, D.; Sercombe, J.; Le Bescop, P.; Adenot, F.; Torrenti, J.-M. Long-term performance of cement paste during combined calcium leaching–sulfate attack: Kinetics and size effect. *Cem. Concr. Res.* **2006**, *36*, 137–143. [CrossRef]
6. Kaddah, F.; Ranaivomanana, H.; Amiri, O.; Rozière, E. Accelerated carbonation of recycled concrete aggregates: Investigation on the microstructure and transport properties at cement paste and mortar scales. *J. CO2 Util.* **2022**, *57*, 101885. [CrossRef]
7. Xu, Z.; Ye, G. Understanding Chloride Diffusion Coefficient in Cementitious Materials. *Materials* **2023**, *16*, 3464. [CrossRef]
8. Martin, R.-P.; Omikrine-Metalssi, O.; Toutlemonde, F. Importance of considering the coupling between transfer properties, alkali leaching and expansion in the modelling of concrete beams affected by Internal Swelling Reactions. *Constr. Build. Mater.* **2013**, *49*, 23–30. [CrossRef]
9. Omikrine-Metalssi, O.; Kchakech, B.; Lavaud, S.; Godart, B. A new model for the analysis of the structural/mechanical performance of concrete structures affected by DEF—Case study of an existing viaduct. *Struct. Concr.* **2016**, *17*, 1104–1113. [CrossRef]
10. Al Shamaa, M.; Lavaud, S.; Divet, L.; Nahas, G.; Torrenti, J.-M. Influence of relative humidity on delayed ettringite formation. *Cem. Concr. Compos.* **2015**, *58*, 14–22. [CrossRef]
11. Santhanam, M.; Cohen, M.D.; Olek, J. Effects of gypsum formation on the performance of cement mortars during external sulfate attack. *Cem. Concr. Res.* **2003**, *33*, 325–332. [CrossRef]
12. Chen, X.; Gu, X.; Xia, X.; Li, X.; Zhang, Q. A Chemical-Transport-Mechanics Numerical Model for Concrete under Sulfate Attack. *Materials* **2021**, *14*, 7710. [CrossRef]
13. El-Hachem, R.; Rozière, E.; Grondin, F.; Loukili, A. Multi-criteria analysis of the mechanism of degradation of Portland cement based mortars exposed to external sulphate attack. *Cem. Concr. Res.* **2012**, *42*, 1327–1335. [CrossRef]
14. Neville, A. The confused world of sulfate attack on concrete. *Cem. Concr. Res.* **2004**, *34*, 1275–1296. [CrossRef]
15. Chen, W.; Huang, B.; Yuan, Y.; Deng, M. Deterioration Process of Concrete Exposed to Internal Sulfate Attack. *Materials* **2020**, *13*, 1336. [CrossRef]
16. Schmidt, T.; Lothenbach, B.; Romer, M.; Neuenschwander, J.; Scrivener, K. Physical and microstructural aspects of sulfate attack on ordinary and limestone blended Portland cements. *Cem. Concr. Res.* **2009**, *39*, 1111–1121. [CrossRef]
17. Jabbour, M.; Quiertant, M.; Baroghel-Bouny, V. A Critical Review of Existing Test-Methods for External Sulfate Attack. *Materials* **2022**, *15*, 7554. [CrossRef]
18. Santhanam, M.; Cohen, M.D.; Olek, J. Mechanism of sulfate attack: A fresh look Part 2. Proposed mechanisms. *Cem. Concr. Res.* **2003**, *33*, 341–346. [CrossRef]
19. Wagner, M.; Decker, M.; Kunther, W.; Machner, A.; Beddoe, R.E.; Heisig, A.; Heinz, D. Gypsum formation mechanisms and their contribution to crystallisation pressure in sulfate resistant hardened cement pastes during early external sulfate attack at low sulfate concentrations. *Cem. Concr. Res.* **2023**, *168*, 107138. [CrossRef]
20. Yu, C.; Scrivener, K. Mechanism of expansion of mortars immersed in sodium sulfate solution. *Cem. Concr. Res.* **2013**, *43*, 105–111. [CrossRef]
21. Zuo, X.-B.; Zheng, Z.-K.; Li, X.-N.; Zou, Y.-X.; Li, L. Mesoscale numerical simulation on the deterioration of cement-based materials under external sulfate attack. *Eng. Fail. Anal.* **2023**, *151*, 107419. [CrossRef]
22. Ran, B.; Omikrine-Metalssi, O.; Fen-Chong, T.; Dangla, P.; Li, K. Pore crystallization and expansion of cement pastes in sulfate solutions with and without chlorides. *Cem. Concr. Res.* **2023**, *166*, 107099. [CrossRef]
23. Ragoug, R.; Metalssi, O.O.; Barberon, F.; Torrenti, J.-M.; Roussel, N.; Divet, L.; de Lacaillerie, E. Durability of cement pastes exposed to external sulfate attack and leaching: Physical and chemical aspects. *Cem. Concr. Res.* **2019**, *116*, 134–145. [CrossRef]

24. Metalssi, O.O.; Touhami, R.R.; Barberon, F.; de Lacaillerie, E.; Roussel, N.; Divet, L.; Torrenti, J.-M. Understanding the degradation mechanisms of cement-based systems in combined chloride-sulfate attack. *Cem. Concr. Res.* **2023**, *164*, 107065. [CrossRef]
25. Shao, W.; Li, Q.; Zhang, W.; Shi, D.; Li, H. Numerical modeling of chloride diffusion in cement-based materials considering calcium leaching and external sulfate attack. *Constr. Build. Mater.* **2023**, *401*, 132913. [CrossRef]
26. Xiong, C.; Jiang, L.; Xu, Y.; Chu, H.; Jin, M.; Zhang, Y. Deterioration of pastes exposed to leaching, external sulfate attack and the dual actions. *Constr. Build. Mater.* **2016**, *116*, 52–62. [CrossRef]
27. Rozière, E.; Loukili, A.; El Hachem, R.; Grondin, F. Durability of concrete exposed to leaching and external sulphate attacks. *Cem. Concr. Res.* **2009**, *39*, 1188–1198. [CrossRef]
28. Bary, B.; Leterrier, N.; Deville, E.; Le Bescop, P. Coupled chemo-transport-mechanical modelling and numerical simulation of external sulfate attack in mortar. *Cem. Concr. Compos.* **2014**, *49*, 70–83. [CrossRef]
29. Wu, Q.; Ma, Q.; Huang, X. Mechanical Properties and Damage Evolution of Concrete Materials Considering Sulfate Attack. *Materials* **2021**, *14*, 2343. [CrossRef] [PubMed]
30. Wang, P.; Mo, R.; Zhou, X.; Xu, J.; Jin, Z.; Zhao, T. A chemo-thermo-damage-transport model for concrete subjected to combined chloride-sulfate attack considering the effect of calcium leaching. *Constr. Build. Mater.* **2021**, *306*, 124918. [CrossRef]
31. Feng, P.; Garboczi, E.J.; Bullard, C.M.J.W. Microstructural origins of cement paste degradation by external sulfate attack. *Constr. Build. Mater.* **2015**, *96*, 391–403. [CrossRef] [PubMed]
32. Biczok, I. *Concrete Corrosion Concrete Protection;* Chemical Publishing: New York, NY, USA, 1967.
33. Kanaan, D.; Soliman, A.M.; Suleiman, A.R. Zero-Cement Concrete Resistance to External Sulfate Attack: A Critical Review and Future Needs. *Sustainability* **2022**, *14*, 2078. [CrossRef]
34. Brown, P.W. An evaluation of sulfate resistance of cements in a controlled environment. *Cem. Concr. Res.* **1981**, *11*, 137–143. [CrossRef]
35. *ASTM International C1012-04;* Standard Test Method for Length Change of Hydraulic-Cement Mortars Exposed to a Sulfate Solution. American Society for Testing and Materials (ASTM) International: West Conshohocken, PA, USA, 2004.
36. *ASTM International C452-06;* Standard Test Method for potential Expansion of Portland-Cement Mortars Exposed to Sulphate. American Society for Testing and Materials (ASTM) International: West Conshohocken, PA, USA, 2006.
37. *ASTM International C150-09;* Standard Specification for Portland Cement. American Society for Testing and Materials (ASTM) International: West Conshohocken, PA, USA, 2009.
38. Massiot, D.; King, I.; Calvé, S.L.; Alonso, B.; Durand, J.O.; Bujoli, B.; Gan, Z.; Hoatson, G. Modelling one and two-dimensional solid-state NMR spectra, Magn. *Reson. Chem.* **2002**, *40*, 70–76. [CrossRef]
39. de Lacaillerie, J.-B.D.; Fretigny, C.; Massiot, D. MAS NMR Spectra of Quadrupolar Nuclei in Disordered Solids: The Czjzek Model. *J. Magn. Reson.* **2008**, *192*, 244–251. [CrossRef]
40. Richardson, I.G. Model structures for C-(A)-S-H(I), Acta Crystallographica Section B Structural Science. *Cryst. Eng. Mater.* **2014**, *70*, 903–923. [CrossRef]
41. Kunther, W.; Lothenbach, B.; Skibsted, J. Influence of the Ca/Si ratio of the C-S-H phase on the interaction with sulfate ions and its impact on the Ettringite crystallization pressure. *Cem. Concr. Res.* **2015**, *69*, 37–49. [CrossRef]

Disclaimer/Publisher's Note: The statements, opinions and data contained in all publications are solely those of the individual author(s) and contributor(s) and not of MDPI and/or the editor(s). MDPI and/or the editor(s) disclaim responsibility for any injury to people or property resulting from any ideas, methods, instructions or products referred to in the content.

materials

Article

The Evolution of Insulation Performance of Fiber-Reinforced Silica Aerogel after High-Temperature Treatment

Rui Gao [1], Zhangjian Zhou [1,*], Hongbo Zhang [1], Xiaoge Zhang [2] and Yuming Wu [2]

1 School of Materials Science and Engineering, University of Science and Technology Beijing, Beijing 100083, China; gao_rui95@163.com (R.G.)
2 Zhongfa Innovation (Beijing) Energy Conservation Technology Co., Ltd., Beijing 100086, China
* Correspondence: zhouzhj@mater.ustb.edu.cn

Abstract: Fiber-reinforced silica aerogel blankets (FRABs) are an important high-temperature thermal insulation material for industry applications that have emerged in recent years. In order to better understand the performance evolution of FRABs at high temperatures, the effect of heat treatment at different temperatures on the performance of FRABs as well as their base material, hydrophobic silica aerogel powder and glass wool, was investigated. The property evolution of the hydrophobic silica aerogel powder showed two stages with an increase in thermal treatment temperatures. The skeleton structure of the aerogel remained unchanged, but the residual organic chemicals, such as hydrophobic groups, were decomposed when the heat treatment temperature was lower than 400 °C. Above 400 °C, the skeleton began to shrink with the increase in temperature, which led to an increase in thermal conductivity. The structure and room-temperature thermal conductivity of the glass wool blanket were less affected by a heat treatment temperature under 600 °C. Therefore, the performance degradation of FRABs at high temperatures is mainly due to the change in the aerogel powder. The insulation performance of the glass wool and FRAB at high temperatures was studied using a heating table which was designed to simulate working conditions. The energy savings of using FRABs instead of glass fiber were calculated and are discussed here.

Keywords: fiber-reinforced silica aerogel; glass wool; heat treatment; thermal conductivity

1. Introduction

The energy crisis and carbon emissions have increasingly become the focus of international attention. One of the most important ways to achieve the goal of energy saving and emission reduction is to use efficient insulation materials in industry and architecture [1].

Hydrophobic silica aerogel has excellent properties such as extremely low density, very low thermal conductivity, and excellent waterproof performance [2,3]. Using hydrophobic silica aerogel as an insulation material to replace traditional insulation materials in the fields of building, high-temperature pipelines, and space exploration can greatly improve energy efficiency and therefore reduce carbon emissions [4–6].

The current commercially available hydrophobic silica aerogels are quite brittle; therefore, it is usually necessary to composite aerogel powder with other matrix materials (typically glass wool) for industrial applications as insulation materials. A number of aerogel-based composites have been developed and investigated, such as glass fiber–aerogel composites [7,8], expanded perlite–silica aerogel composites [9], polyurethane–silica aerogel composites [10], polypropylene–silica aerogel composites [11], etc. Among them, glass-fiber-reinforced aerogel blankets (FRABs) have been applied in industry [12,13]. For instance, FRABs have been applied to building and shown excellent characteristics, including very low thermal conductivity, a light weight, very good sound isolation, and excellent waterproof performance [14–16]. Furthermore, FRABs also have great application prospects in high-temperature environments, such as steam pipes in power plants [17,18] and petrochemical plants [19]. The typical operation temperature is usually between 200 °C and 600 °C. By using FRABs instead of traditional insulation materials (basalt fiber and glass fiber), it is expected that a significant amount of energy can be saved [20]. The investigation of the effect of high temperature on the microstructure stability and insulation performance of FRABs is crucial for engineering applications [21].

Many authors have studied the effect of heat treatment temperature on the microstructure changes of laboratory-made aerogels. Pradip B. Sarawade et al. [22] studied the changes in the microstructure of aerogels after heat treatment at temperatures up to 500 °C and found that the hydrophobicity of the silica aerogel spheres gradually decreased with the increase in temperature. After heat treatment at 500 °C, the aerogel microspheres became completely hydrophilic. As the temperature increased, the pore size of the aerogel microspheres increased. Gaosheng Wei et al. [23] measured the thermal conductivity of aerogels at different temperatures using the transient thermal strip method and found that the thermal conductivity increased significantly with an increase in temperature. The aerogel particles were partially sintered, and the pores were destroyed when the temperature rose to 950 °C, resulting in an increase in thermal conductivity and bulk density [24,25]. Huafei Cai et al. [26] investigated the thermal stability of silica aerogel using TEM with an in situ heating device. They found that the silica aerogel first shrank at the initial time (within 2 h) at a temperature between 600 °C and 1000 °C, and then the structure became stable during longer-term heating. Previous studies have shown that the pore structure of aerogel can be destroyed by high temperatures, especially when the temperature is higher than 600 °C, which will lead to a drastic decrease in the insulation performance. Moreover, when the heat treatment temperature is lower than 600 °C, the pore structure of the aerogel does not change significantly over a short period of time (less than 10 h). Nevertheless, the performance change due to structural changes in aerogel particles caused by the high temperatures is not representative of the overall FRAB performance change.

Although the effect of glass fiber content in fiber-reinforced aerogel composites (FRABs) on their properties has been widely studied [27,28], the current research has little involvement in the relationship between the service temperature and the insulation performance of FRABs. Most of the obtained results were based on calculations and simulations. For example, the total effective thermal conductivity of aerogels and their composites considering the combined effect of conduction and radiation was predicted [29,30]. Changes in the structure and properties of aerogel composites at different temperatures have also been investigated via calculations [31,32]. Miros et al. compared the thermal conductivity of FRABs and mineral wool at different temperatures using the guarded hot plate method [33]. Fang et al. used a model to predict the thermal conductivity of aerogel composites and then verified the accuracy of the model by measuring the high-temperature thermal conductivity using the hot disk method [34]. Furthermore, in recent studies, efforts were mainly focused on the development of new grades of aerogel materials which are still a long way from commercial application. For pure aerogel material, only a glass window combining silica aerogel has been reported as an engineering application [35,36]. FRABs have been widely used in high temperature industries, but their performance in use has rarely been evaluated, and the results of pure aerogel obtained in the laboratory cannot be used to accurately evaluate the service behavior of composite materials such as FRABs. In order to better understand the service behavior of FRABs, it is necessary to compare and evaluate aerogel, the matrix of glass wool, and FRABs under the same simulated actual operating conditions.

In this study, a novel heating table was designed and constructed to simulate actual operating conditions; therefore, the heat loss of the heat-treated FRAB and glass wool with different thicknesses at different temperatures was able to be measured and compared. These data, combined with the evolution of the microstructure and thermal conductivity of aerogel and FRABs after heat treatment, can help better guide the application of aerogel-based insulation materials.

2. Experimental Materials and Methods

2.1. Experimental Material

FRABs and their constituent material glass wool and silica aerogel were provided by JINNA TECH. The silica aerogel was prepared using water glass as the precursor. The wet gel was obtained through a sol-gel route, followed by solvent exchange with ethanol, surface modification with trimethylchlorosilane, and supercritical carbon dioxide drying. The specific surface area of the aerogel is 339.064 m^2/g, and the average pore size is 12.69 nm. The glass wool matrix used to make FRABs has a density of 160 kg/m^3. FRABs were prepared by infiltrating wet gel into glass wool and then supercritical drying. The density of the FRABs is 200 kg/m^3, of which the aerogel content is 20 wt.%.

2.2. Heat Treatment Method

In order to explore the change in the thermal conductivity of aerogels after thermal treatment at different temperatures, FRABs and their constituent material glass wool and silica aerogel were heat-treated in a box-type resistance furnace (SX2-12-12, Shanghai Zhetai Machinery Manufacturing Co., Ltd., Shanghai, China) with a heating rate of 3 °C/min. When the temperature reached the specified value (200 °C, 300 °C, 400 °C, 500 °C, 600 °C, and 700 °C), the furnace was maintained at that temperature for 2 h. Then, these samples were naturally cooled to room temperature.

2.3. Characterization Methods

2.3.1. Characterization of Apparent Properties

The bulk density of the aerogel samples was calculated by measuring their mass-to-volume ratio. The hydrophobic properties were tested using the following methods: Aerogel powders before and after heat treatment were spread on a polyethylene plate, respectively, as polyethylene does not absorb water. Then, water was dropped on it to observe the droplet spreading characteristics.

According to the "Determination of thermal conductivity of non-metallic solid materials—hot wire method", the TC-3000E thermal conductivity tester (Xiaxi Technology, Xi'an, China) was used to determine the room temperature thermal conductivity of the samples.

The thermal stability of the aerogel samples was analyzed by TG-DTA using an SDT Q600 (TA Instruments, New Castle, DE, USA). The samples were heat-treated under an air atmosphere from room temperature (~25 °C) to 1000 °C with a controlled heating rate of 10 °C/min.

2.3.2. Microstructural Characterization

In order to study the microstructure of the samples before and after heat treatment, a scanning electron microscope (SEM) (SIGMA 300, Carl Zeiss AG, Oberkochen, Germany) was used. Aerogels and FRABs were fixed on the sample holder using a carbon pad and subsequently coated with platinum for SEM analysis. The surface chemical modification of the aerogels was studied using Fourier transform infrared spectroscopy (FTIR) (TENSOR 37, Bruker, Karlsruhe, Germany). FTIR data for heat-treated (200 °C, 400 °C, 600 °C) aerogel powders were collected.

2.3.3. High Temperature Insulation Performance Test

In order to explore the high-temperature insulation performance of FRABs, we designed the following heating table, as shown in Figure 1. It contains two parts: the console and the heating device. The temperature of the heating element is adjusted by the control instrument; we also call it the working temperature. The heater power and, consequently, the heating pace are adjusted using the power knob. The temperature of the thermocouple is displayed by the display instrument. The testing process is divided into the following three steps: First, put the insulation material evenly into the heating device. Ensuring that there are no gaps between each contact surface is the key to a successful test. Second, turn on the power and set the temperature and then wait for a while until the heating power stabilizes. Third, record data obtained from the thermocouples and heating power.

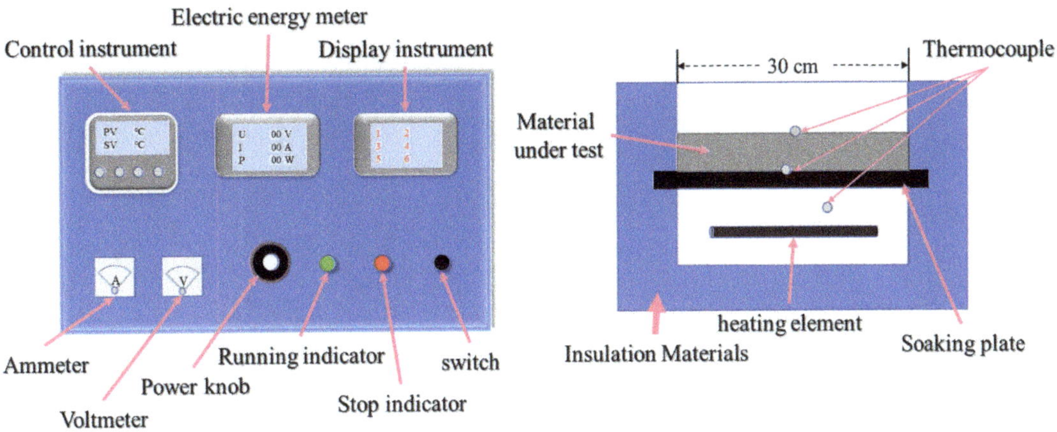

Figure 1. Schematic diagram of plate heater: console (**left**) and the heating device (**right**).

The temperature on the surface and the heating power of the insulation with different thicknesses and different working temperatures were tested at the same ambient temperature. The temperature on the surface is the temperature of the surface of the tested material. The heating power is the power consumption of the heating device during the test.

3. Results and Discussion

3.1. The Effect of Heat Treatment on Silica Aerogel

TG-DTA was used to explore the weight change and potential reaction of aerogel at high temperatures. The result is shown in Figure 2.

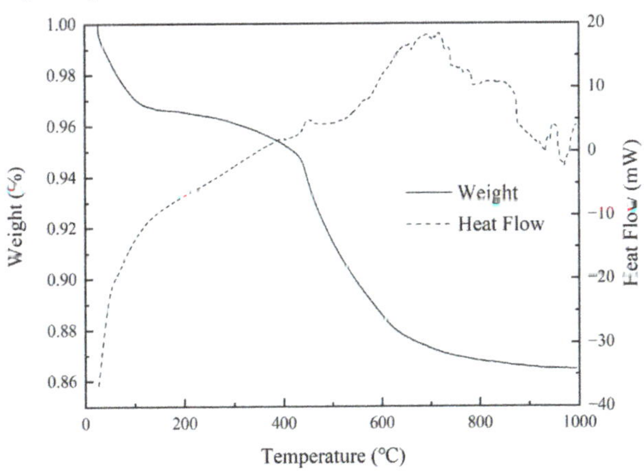

Figure 2. TG-DSC curves of the aerogel samples.

There is a strong endothermic peak before 100 °C with a weight loss of 4%. According to the preparation process of the aerogel, it should be caused by the evaporation of residual low-boiling organics inside the aerogel [37]. When the temperature was 400 °C~600 °C, a more obvious weight loss of 8% could be found, which can be judged due to the decomposition of hydrophobic groups in the hydrophobic aerogel at high temperatures [38]. This inference was further supported by Figure 3, which shows how heat treatment affected the hydrophobic properties of the aerogel. When the aerogel was heated at 200 °C, the hydrophobicity of the aerogel remained relatively good; however, when the temperature increased to 400 °C, the hydrophobicity completely disappeared. After 600 °C, the weight loss decreased obviously because the methyl group decomposition had been completed.

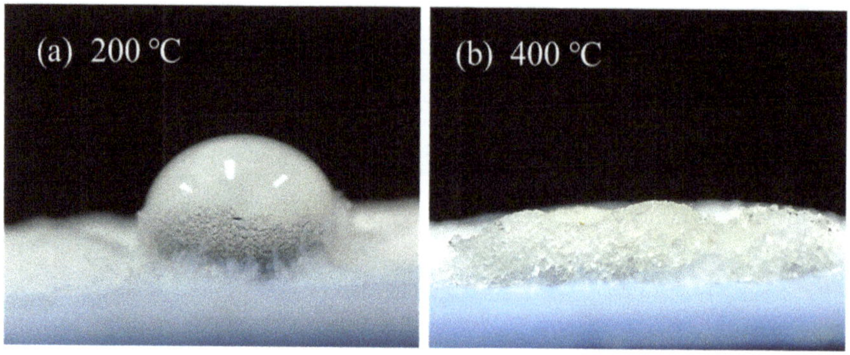

Figure 3. Droplet-spreading characteristics of aerogel after heat treatment at (**a**) 200 °C and (**b**) 400 °C.

Figure 4 shows the infrared spectroscopy results to further confirm the reason for hydrophobicity disappearance. The vibrations at 3666 cm^{-1} and 1650 cm^{-1} were due to residual –OH groups or adsorbed water. The background peak was caused by potassium bromide introduced into the test. The absorption peaks near 1085, 810, and 455 cm^{-1} represented the asymmetric stretching vibration, symmetric stretching vibration, and bending vibration of Si-O-Si [39]. The absorption peaks at 850 cm^{-1} and 2968 cm^{-1} could be attributed to S-C bending vibration and $-CH_3$ stretching vibration [40,41]. For the samples heat treated at 200 °C, pronounced absorption peaks were observed in the spectra. In contrast, almost no absorption peaks appeared at these two points in the spectra of the samples heat treated at 400 °C and 600 °C, respectively. There was additional absorption at 3200 cm^{-1} formed by −OH for the 200 °C-treated sample, but it was absent for samples treated at higher temperatures.

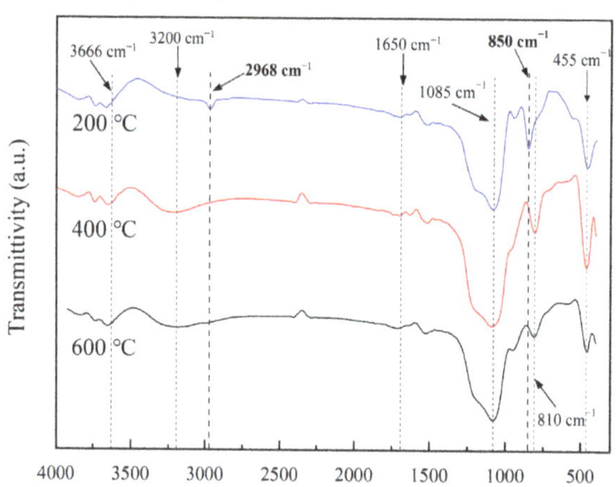

Figure 4. FT-IR spectra of the hydrophobic silica aerogel after heat treatment at 200 °C, 400 °C, and 600 °C.

The above analysis indicated that when heat treated above 400 °C, the methyl group is pyrolyzed, therefore destroying the hydrophobicity of aerogel.

The macroscopic morphology of aerogel after heat treatment at different temperatures is shown in Figure 5. It can be clearly seen that the aerogel shrank slightly after heat treatment at 400 °C, while significant volume shrinkage occurred when treated at temperatures higher than 600 °C.

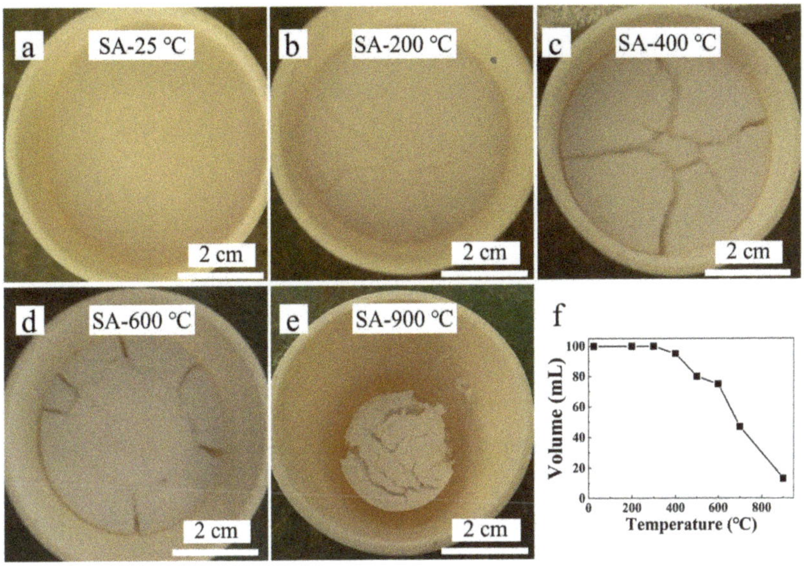

Figure 5. Photograph of the hydrophobic silica aerogel heat-treated at different temperatures (**a**) 25 °C, (**b**) 200 °C, (**c**) 400 °C, (**d**) 600 °C, (**e**) 900 °C, and (**f**) volume change.

Figure 6 shows the changes in the bulk density and RT thermal conductivity of silica aerogels after heat treatment at different temperatures. When the heat treatment temperature was lower than 300 °C, the bulk density of the aerogel decreased first and then remained stable. The decrease in the bulk density of the aerogel in this case is due to the decomposition of residual organic compounds. Therefore, it can be inferred that the skeleton structure of the aerogel will be stable at temperatures below 300 °C. After heating at 400 °C, the aerogel volume shrank slightly, the weight decreased due to the loss of hydrophobic groups, and the bulk density increased slightly. Then, the bulk density increased obviously along with the increase in heat treatment temperature, especially when the temperature was higher than 600 °C, as shown in Figure 5. Especially, when the temperature reached 900 °C, a severe shrinkage of the aerogel could be found. This indicates that the aerogel skeleton is severely damaged and the pores collapse when heat treated at high temperatures.

The change in thermal conductivity with heat treatment temperature was consistent with the bulk density change. When the heat treatment temperature was lower than 400 °C, the thermal conductivity even decreased slightly due to the volatilization of the organic chemicals. Then, the thermal conductivity increased along with the increase in heat treatment temperature due to the volume shrinkage, especially when the heat treatment temperature was higher than 600 °C.

Figure 7 shows the microstructure of aerogel powders after heat treatment at different temperatures. Untreated silica aerogel showed a highly porous and homogeneous nanostructure, which was similar to that of the samples heat treated at 400 °C. After heat treatment at 600 °C, a small part of the microscopic pore structure of the aerogel was destroyed. Particle diameter and pore sizes changed slightly, and the collapse of the framework was not obvious. For the samples heat-treated at 900 °C, the porous structure collapsed badly, and the particles aggregated and grew up.

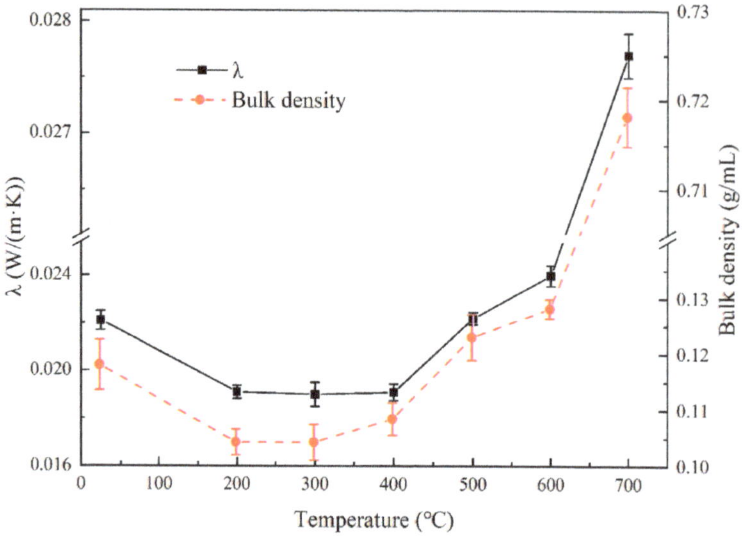

Figure 6. Bulk density and thermal conductivity of the silica aerogel samples after heat treatment from RT to 700 °C.

Figure 7. SEM images of silica aerogel after heat treatment at (**a**) 25 °C, (**b**) 400 °C, (**c**) 600 °C, (**d**) 900 °C.

According to the above test results, the change in silica aerogel after heat treatment can be divided into two stages. When the heat treatment temperature was lower than 400 °C in the first stage, the residual organic compound or hydrophobic groups introduced during the preparation process were decomposed, while the skeleton structure of the aerogel remained unchanged, resulting in a slight reduction in thermal conductivity. The second stage occurred when the heat treatment temperature was higher than 400 °C, which resulted in volume shrinkage and the destruction of the aerogel skeleton, therefore increasing the thermal conductivity. This phenomenon was more significant at temperatures above 600 °C.

3.2. The Effect of Heat Treatment on Glass Wool

The glass wool matrix was also heat-treated, and the weight loss and thermal conductivity were measured, as shown in Figure 8. When the heat treatment temperature exceeded 300 °C, the weight of the glass wool decreased by 2% due to the decomposition of residual organic additives added during the preparation of glass wool. According to Figure 8, room temperature thermal conductivity did not change significantly after heat treatment from 200 °C to 600 °C. Glass fibers had low solid-state thermal conductivity due to their low density and thin glass fibers. Meanwhile, fine glass fibers can capture and fix the air between the fibers, thereby preventing heat transmission by convection and limiting gaseous heat conduction by minimizing collisions between gas molecules. Therefore, the RT thermal conductivity of glass wool is very low and will not change with the heat treatment temperature.

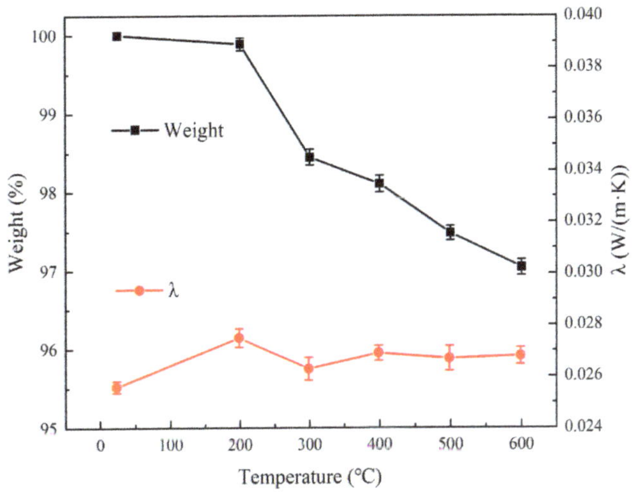

Figure 8. Weight loss and thermal conductivity of the glass wool after heat treatment from RT to 600 °C.

3.3. The Effect of Heat Treatment on FRAB

The relationship between the weight and thermal conductivity of FRABs after heat treatment from RT to 600 °C is shown in Figure 9. The mass loss under 400 °C was considered to be the decomposition of organic groups ($-CH_3$ groups) in the composites. This could be confirmed by the droplet-spreading characteristics of aerogel after heat treatment as shown in Figure 10. In Figure 10b, no droplets could be seen on the surface of the FRABs because droplets penetrate quickly into the materials. The FRABs experience a sharp weight loss from 400 °C to 600 °C, which is consistent with Liao's research [42]. The RT thermal conductivity of FRAB decreased first and then increased with the increase in heat treatment temperature, which had the same trend as aerogel shown in Figure 11. The structure and thermal conductivity of the glass wool were less affected by the heat treatment temperature under 600 °C in Section 3.2. Therefore, the performance change of FRABs at high temperatures was mainly due to the change in the aerogel.

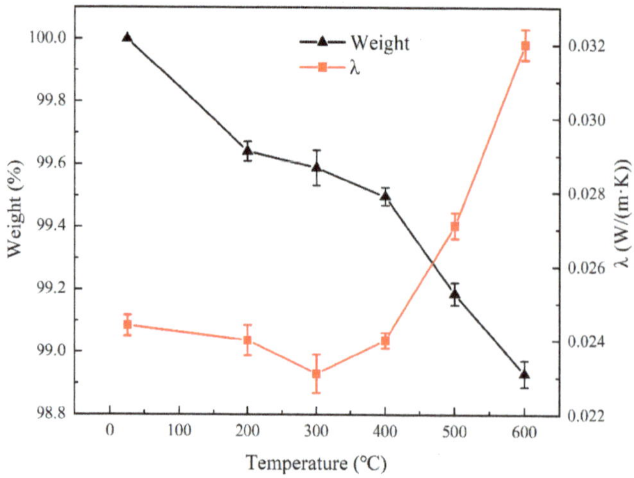

Figure 9. Weight loss and thermal conductivity of the FRAB after heat treatment from RT to 600 °C.

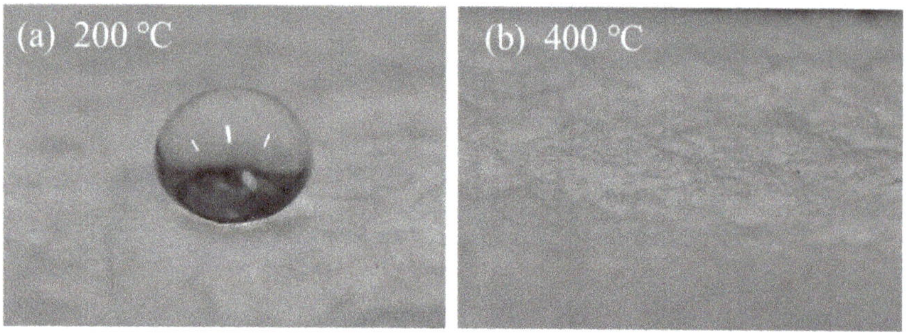

Figure 10. Droplet-spreading characteristics of FRAB after heat treatment at (**a**) 200 °C and (**b**) 400 °C.

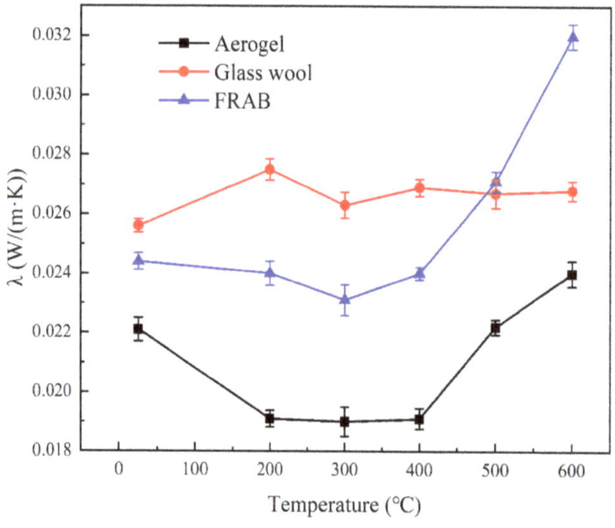

Figure 11. Thermal conductivity of the aerogel, glass wool, and FRAB after heat treatment from RT to 600 °C.

The microstructure of the FRAB samples was characterized through SEM images, as shown in Figure 12. At low magnification in Figure 12a,d,g, fibers were easily observed because of the high contrast. The fiber structure did not change with the increase in the heat treatment temperature. The change in the aerogel microstructure in the FRABs along with the heat treatment temperature is shown in Figure 12c,f,i. The pore structure of the aerogel in FRABs after heat treatment at 400 °C is similar to that of the untreated one. However, the pore size after heat treatment at 600 °C decreased slightly. After heat treatment at 600 °C, the aerogel particles coalesced to form larger particles. Meanwhile, the size of the pores surrounded by these particles was reduced. This was consistent with the aerogel changes after heat treatment. The state of the individual fibers was also observed in Figure 12b,e,h. The adhesion of the aerogel to the fiber and evidence of the sintering of aerogel particles can be found after heat treatment at a temperature of 600 °C, as shown in Figure 12h. Woignier et al. studied the sintering behavior of silica aerogel and found that sintering due to a diffusional process occurs in the temperature range of 500–700 °C [43]. The sintering behavior of aerogel was influenced by the contact between fiber and aerogel, which had an effect on the thermal insulation properties of FRABs at high temperatures. Therefore, it is important to test the thermal insulation properties of FRABs at high temperatures.

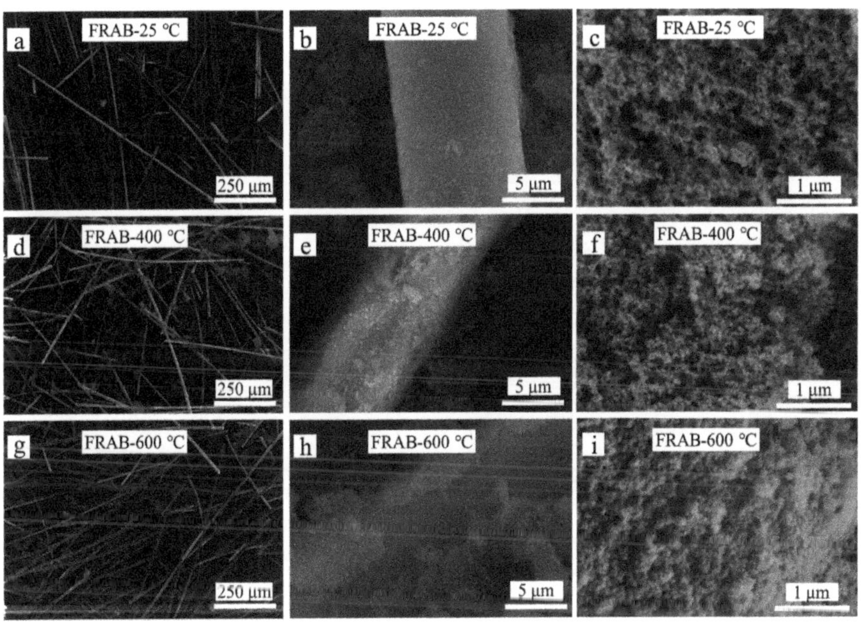

Figure 12. SEM images of FRABs before and after heat treatment. (**a–c**): untreated FRABs; (**d–f**): FRABs after heat treatment at 400 °C; (**g–i**): FRABs after heat treatment at 600 °C.

3.4. The Performance of FRABs at High Temperature

The RT thermal conductivity of the FRABs after heat treatment cannot truly reflect the thermal insulation performance of the FRABs at high temperatures. The data tested in the actual service temperature are more convincing. Therefore, a heating table was used to simulate the performance of FRABs and glass wool at high temperatures, as presented in Section 2.3.3. The value of the surface temperature was recorded at different operating temperatures for samples with a thickness of 5 cm. The variation in surface temperature with heating time is shown in Figure 13. The cold surface temperature reached a relatively stable state after 120 min of heating at 200 °C and 300 °C, respectively. The time required for the cold surface temperature to reach a steady state is 300 min, at which point, the system reached a steady state at 600 °C. This time is much longer than 2 h, which was reported by Huafei Cai [26]. In this experiment, 5 cm thick FRABs were used, and it took a long time for the heat flow to pass through the FRABs due to their excellent insulation capacity. As a result, the aerogel particles were affected and aggregated relatively slowly. Glass wool was also tested in the same way under the same conditions for comparison. The heating power of glass wool and FRABs were tested at different temperatures and sample thicknesses, and the results are shown in Figure 14.

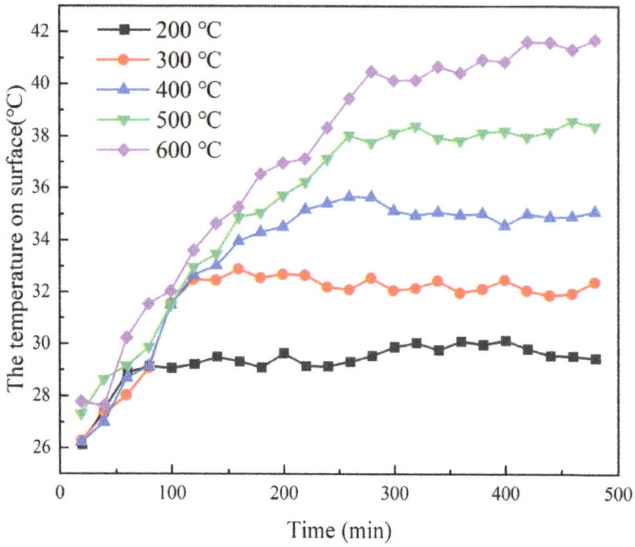

Figure 13. The temperature on surface changes with time at different temperatures.

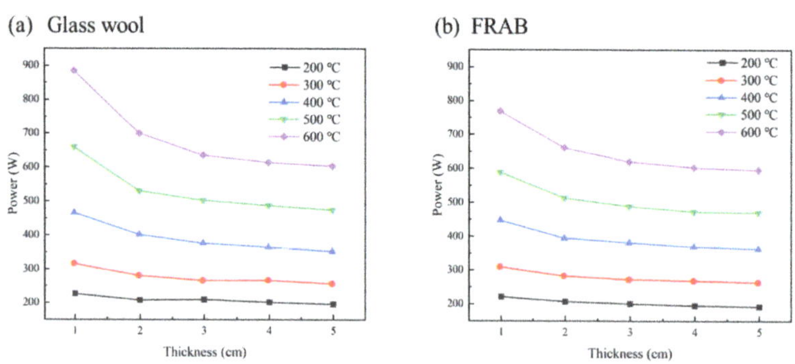

Figure 14. Heating power of (**a**) glass wool, and (**b**) FRABs changes with thickness at different temperatures.

It is obvious that the heating power increased with increasing temperature. The heating power gradually decreased to a constant value with the increasing thickness of the tested materials. When the thickness is large enough, it is considered that heat will not be lost through the tested materials. In this case, the heating power is equal to the heat loss power of the heating table.

The thermal conductivity of a homogeneous material does not change with direction. In the steady state without an internal heat source, it can be obtained according to Fourier's law:

$$\lambda = \frac{\varnothing}{A} \cdot \frac{d}{t_2 - t_1} \tag{1}$$

where d is the thickness of the tested material, A is the area of the tested material, t_2 is the operating temperature, t_1 is the temperature on the surface, and $\varnothing$ is the heat flow, i.e., the heat loss of the FRABs or glass wool.

The heat loss power φ of the plate heater is regarded as a fixed value at the same working temperature and has the following relationship with the heating power P:

$$P = \varnothing + \varphi \tag{2}$$

It is approximately considered that the temperature difference $t_2 - t_1$ is equal to the average value Δt at the same working temperature.

$$P = \lambda A \Delta t \cdot \frac{1}{d} + \varphi \tag{3}$$

There is a linear relationship between the heating power P and the reciprocal thickness $1/d$. The fittings are shown in Figure 15.

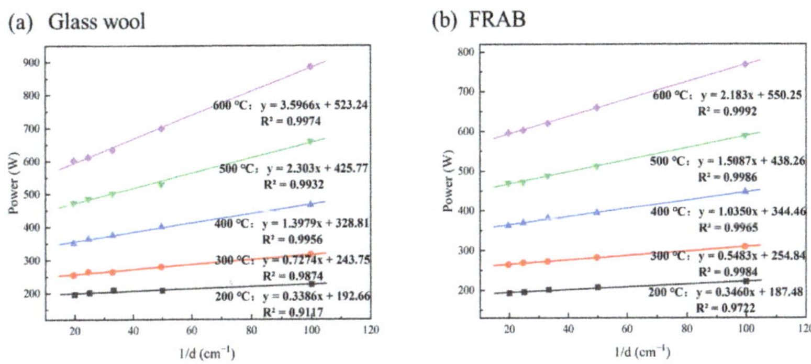

Figure 15. The relationship between heating power and the reciprocal thickness $1/d$.

The thermal conductivity of glass wool and FRABs at different temperatures can be calculated according to Figure 15. The thermal conductivity comparison of glass wool and FRABs at high temperatures is shown in Figure 16. The results are consistent with Miros's research which was measured on a two-plate guarded hot plate apparatus [33].

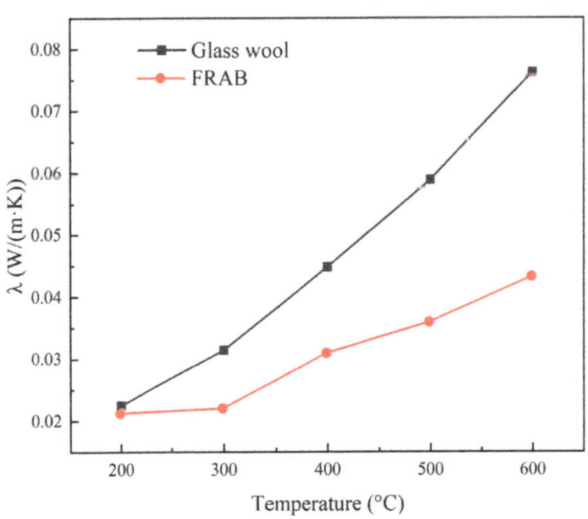

Figure 16. The thermal conductivity of glass wool and FRAB at high temperatures.

The thermal conductivity of FRABs is lower than that of glass fiber in the temperature range of 200 °C to 600 °C as shown in Figure 16. The pore size of FRABs is smaller than that of glass wool due to the filling of aerogel powders. According to the Knudsen effect [44], the smaller the pore size, the weaker the convective heat transfer of the gas and the lower the thermal conductivity. There is also an interfacial thermal resistance in the composite material [30,34,45]. The addition of aerogel creates more interfaces in FRABs, resulting in a higher thermal resistance.

In the practical application of insulation materials, heat loss is one of the most important parameters of interest to the user. Using the heating table designed in this experiment, not only the thermal conductivity of the material but also the corresponding heat loss can be calculated. According to the data in Figure 13 and Equation (2), the heat loss of FRABs and glass wool at different temperatures can be obtained using the following equation.

$$\varnothing = P - \varphi \tag{4}$$

where $\varnothing$ is the heat loss of FRABs or glass wool. P is the heating power. φ is the heat loss power of the heating table. The heat loss of FRABs and glass wool at an operating temperature of 600 °C is shown in Table 1. Heat loss refers to the heat passing through per square meter of insulation material per hour at the working temperature, so its unit is kJ/(h·m^2). Using FRABs instead of glass fiber saves 656 kJ per square meter per hour when the thickness of insulation is 5 cm, and 5056 kJ per square meter when the thickness of the insulation is 1 cm, as shown in Figure 17. The heat loss of FRABs is much lower than that of glass wool. Using FRABs instead of glass wool is a very effective way to save energy.

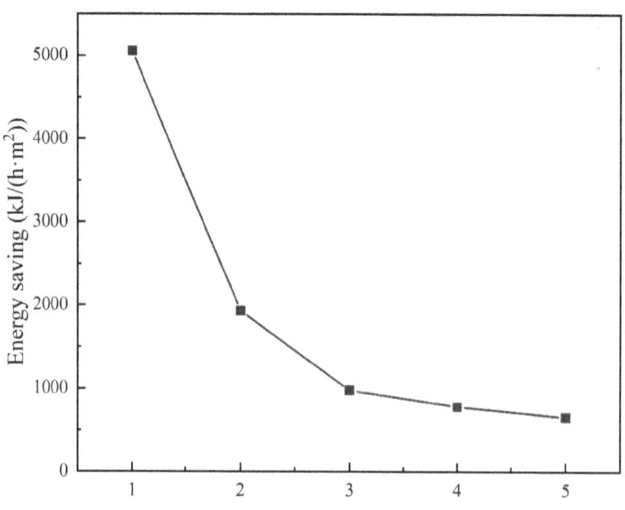

Figure 17. Energy savings change with thickness by using FRAB instead of glass fiber at 600 °C.

Table 1. Heat loss of FRAB and glass fiber at different thicknesses at 600 °C.

Thickness (cm)	Heat Loss (kJ/(h·m^2))	
	FRAB	Glass Fiber
1	9430.4	14,486.4
2	5126.4	7058.4
3	3506.4	4486.4
4	2834.4	3614.4
5	2542.4	3198.4

4. Conclusions

In this work, the performances of aerogels, glass wool, and FRABs after heat treatment in the temperature range from RT to 600 °C were investigated. A heating table was designed that could simulate the scene of the thermal insulation material in the actual working process. The thermal insulation properties of glass wool and FRABs were investigated considering the working environment. The main conclusions can be summarized as follows:

(1) The thermal conductivity of aerogel first decreased and then increased with the increase in heat treatment temperature, due to the change in the microstructure. The hydrophobicity of the aerogel disappeared after heat treatment at 400 °C.
(2) The RT thermal conductivity of the glass fiber was less affected by the heat treatment temperature under 600 °C. A scanning electron microscope observation showed that the microstructure of glass fibers in FRABs remained stable after heat treatment.
(3) The effect of heat treatment on the properties of FRABs was similar to that of aerogels. The contact of the aerogel with the fibers appeared to affect the sintering behavior of the aerogel particles at high temperatures, which may affect the performance of FRAB when in use.
(4) Heat losses of FRAB and glass fibers were calculated and compared in service by using the heating table. When the working temperature is 600 °C and the thickness of the insulation is 5 cm, using FRAB instead of glass fiber can save 656 kJ per square meter per hour. Additionally, when the thickness of the insulation is 1 cm, the energy saved per square meter per hour is 5056 kJ. Using FRAB instead of glass wool is a very effective way to save energy.

Author Contributions: Conceptualization, R.G. and Z.Z.; methodology, H.Z.; validation, R.G., Y.W. and Z.Z.; investigation, X.Z.; resources, X.Z.; writing—original draft preparation, R.G.; writing—review and editing, R.G. and Z.Z.; project administration, Z.Z.; funding acquisition, Z.Z. All authors have read and agreed to the published version of the manuscript.

Funding: This research was funded by [Innovation Center of Nuclear Materials, China Atomic Energy Authority] grant number [ICNM-2023-ZH-01].

Institutional Review Board Statement: Not applicable.

Informed Consent Statement: Not applicable.

Data Availability Statement: The data that support the findings of this study are available on request from the corresponding author, Z.Z., upon reasonable request.

Conflicts of Interest: The authors declare no conflict of interest.

References

1. Aditya, L.; Mahlia, T.M.I.; Rismanchi, B.; Ng, H.M.; Hasan, M.H.; Metselaar, H.S.C.; Muraza, O.; Aditiya, H.B. A Review on Insulation Materials for Energy Conservation in Buildings. *Renew. Sustain. Energy Rev.* **2017**, *73*, 1352–1365. [CrossRef]
2. Sarawade, P.B.; Quang, D.V.; Hilonga, A.; Jeon, S.J.; Kim, H.T. Synthesis and Characterization of Micrometer Sized Silica Aerogel Nanoporous Beads. *Mater. Lett.* **2012**, *81*, 37–40. [CrossRef]
3. Wei, G.; Liu, Y.; Zhang, X.; Yu, F.; Du, X. Thermal Conductivities Study on Silica Aerogel and Its Composite Insulation Materials. *Int. J. Heat Mass Transf.* **2011**, *54*, 2355–2366. [CrossRef]
4. Moretti, E.; Belloni, E.; Merli, F.; Zinzi, M.; Buratti, C. Laboratory and Pilot Scale Characterization of Granular Aerogel Glazing Systems. *Energy Build.* **2019**, *202*, 109349. [CrossRef]
5. Baetens, R.; Jelle, B.P.; Gustavsen, A. Aerogel Insulation for Building Applications: A State-of-the-Art Review. *Energy Build.* **2011**, *43*, 761–769. [CrossRef]
6. Hasan, M.A.; Rashmi, S.; Esther, A.C.M.; Bhavanisankar, P.Y.; Sherikar, B.N.; Sridhara, N.; Dey, A. Evaluations of Silica Aerogel-Based Flexible Blanket as Passive Thermal Control Element for Spacecraft Applications. *J. Mater. Eng. Perform.* **2018**, *27*, 1265–1273. [CrossRef]
7. Li, C.; Cheng, X.; Li, Z.; Pan, Y.; Huang, Y.; Gong, L. Mechanical, Thermal and Flammability Properties of Glass Fiber Film/Silica Aerogel Composites. *J. Non. Cryst. Solids* **2017**, *457*, 52–59. [CrossRef]

8. Siligardi, C.; Miselli, P.; Francia, E.; Lassinantti Gualtieri, M. Temperature-Induced Microstructural Changes of Fiber-Reinforced Silica Aerogel (FRAB) and Rock Wool Thermal Insulation Materials: A Comparative Study. *Energy Build.* **2017**, *138*, 80–87. [CrossRef]
9. Jia, G.; Li, Z.; Liu, P.; Jing, Q. Preparation and Characterization of Aerogel/Expanded Perlite Composite as Building Thermal Insulation Material. *J. Non. Cryst. Solids* **2018**, *482*, 192–202. [CrossRef]
10. Cho, J.; Jang, H.G.; Kim, S.Y.; Yang, B. Flexible and Coatable Insulating Silica Aerogel/Polyurethane Composites via Soft Segment Control. *Compos. Sci. Technol.* **2019**, *171*, 244–251. [CrossRef]
11. Motahari, S.; Motlagh, G.H.; Moharramzadeh, A. Thermal and Flammability Properties of Polypropylene/Silica Aerogel Composites. *J. Macromol. Sci. Part B Phys.* **2015**, *54*, 1081–1091. [CrossRef]
12. Leventis, N.; Sotiriou-Leventis, C.; Zhang, G.; Rawashdeh, A.M.M. Nanoengineering Strong Silica Aerogels. *Nano Lett.* **2002**, *2*, 957–960. [CrossRef]
13. Wang, W.; Pang, L.; Jiang, M.; Zhu, Y.; Wang, F.; Sun, J.; Qi, H. Fabrication of SiCN(O) Aerogel Composites with Low Thermal Conductivity by Wrapping Mesoporous Aerogel Structures over Mullite Fibers. *Materials* **2022**, *15*, 8811. [CrossRef]
14. Zhou, T.; Cheng, X.; Pan, Y.; Li, C.; Gong, L.; Zhang, H. Mechanical Performance and Thermal Stability of Glass Fiber Reinforced Silica Aerogel Composites Based on Co-Precursor Method by Freeze Drying. *Appl. Surf. Sci.* **2018**, *437*, 321–328. [CrossRef]
15. Shafi, S.; Navik, R.; Ding, X.; Zhao, Y. Improved Heat Insulation and Mechanical Properties of Silica Aerogel/Glass Fiber Composite by Impregnating Silica Gel. *J. Non. Cryst. Solids* **2019**, *503–504*, 78–83. [CrossRef]
16. Lakatos, Á. Stability Investigations of the Thermal Insulating Performance of Aerogel Blanket. *Energy Build.* **2019**, *185*, 103–111. [CrossRef]
17. Narasimhan, S.; Rajendran, V. Performance Improvement in Temperature Thermometry Using Data Analytics—A Nuclear Power Plant Perspective. *Prog. Nucl. Energy* **2021**, *134*, 103669. [CrossRef]
18. Petkov, M.P.; Chevalier, M.; Dean, D.; Cocks, A.C.F. Creep-Fatigue Interactions in Type 316H under Typical High-Temperature Power Plant Operating Conditions. *Int. J. Press. Vessel. Pip.* **2021**, *194*, 104500. [CrossRef]
19. Hara, T.; Semba, H.; Amaya, H. *Pipe and Tube Steels for Oil and Gas Industry and Thermal Power Plant*; Elsevier Ltd.: Amsterdam, The Netherlands, 2021; ISBN 9780128197264.
20. Fedyukhin, A.V.; Strogonov, K.V.; Soloveva, O.V.; Solovev, S.A.; Akhmetova, I.G.; Berardi, U.; Zaitsev, M.D.; Grigorev, D.V. Aerogel Product Applications for High-Temperature Thermal Insulation. *Energies* **2022**, *15*, 7792. [CrossRef]
21. He, S.; Huang, Y.; Chen, G.; Feng, M.; Dai, H.; Yuan, B.; Chen, X. Effect of Heat Treatment on Hydrophobic Silica Aerogel. *J. Hazard. Mater.* **2019**, *362*, 294–302. [CrossRef]
22. Sarawade, P.B.; Kim, J.K.; Hilonga, A.; Quang, D.V.; Jeon, S.J.; Kim, H.T. Synthesis of Sodium Silicate-Based Hydrophilic Silica Aerogel Beads with Superior Properties: Effect of Heat-Treatment. *J. Non. Cryst. Solids* **2011**, *357*, 2156–2162. [CrossRef]
23. Wei, G.; Wang, L.; Xu, C.; Du, X.; Yang, Y. Thermal Conductivity Investigations of Granular and Powdered Silica Aerogels at Different Temperatures and Pressures. *Energy Build.* **2016**, *118*, 226–231. [CrossRef]
24. Zhao, J.J.; Duan, Y.Y.; Wang, X.D.; Wang, B.X. Experimental and Analytical Analyses of the Thermal Conductivities and High-Temperature Characteristics of Silica Aerogels Based on Microstructures. *J. Phys. D Appl. Phys.* **2013**, *46*, 015304. [CrossRef]
25. Huang, D.; Guo, C.; Zhang, M.; Shi, L. Characteristics of Nanoporous Silica Aerogel under High Temperature from 950 °C to 1200 °C. *Mater. Des.* **2017**, *129*, 82–90. [CrossRef]
26. Cai, H.; Jiang, Y.; Feng, J.; Chen, Q.; Zhang, S.; Li, L.; Feng, J. Nanostructure Evolution of Silica Aerogels under Rapid Heating from 600 °C to 1300 °C via In-Situ TEM Observation. *Ceram. Int.* **2020**, *46*, 12489–12498. [CrossRef]
27. Patil, S.P.; Bachhav, B.S.; Markert, B. Thermal Conductivity of Glass Fiber-Reinforced Silica Aerogels Using Molecular Dynamics Simulations. *Ceram. Int.* **2022**, *48*, 2250–2256. [CrossRef]
28. Bi, C.; Tang, G.H.; He, C.B.; Yang, X.; Lu, Y. Elastic Modulus Prediction Based on Thermal Conductivity for Silica Aerogels and Fiber Reinforced Composites. *Ceram. Int.* **2022**, *48*, 6691–6697. [CrossRef]
29. Guo, J.F.; Tang, G.H.; Jiang, Y.G.; Cai, H.F.; Feng, J.; Feng, J.Z. Inhibited Radiation Transmittance and Enhanced Thermal Stability of Silica Aerogels under Very-High Temperature. *Ceram. Int.* **2021**, *47*, 19824–19834. [CrossRef]
30. He, Y.L.; Xie, T. Advances of Thermal Conductivity Models of Nanoscale Silica Aerogel Insulation Material. *Appl. Therm. Eng.* **2015**, *81*, 28–50. [CrossRef]
31. Liu, H.; Liu, J.; Tian, Y.; Wu, X.; Li, Z. Investigation of High Temperature Thermal Insulation Performance of Fiber-Reinforced Silica Aerogel Composites. *Int. J. Therm. Sci.* **2023**, *183*, 107827. [CrossRef]
32. Liu, H.; Hu, M.; Jiao, J.; Li, Z. Geometric Optimization of Aerogel Composites for High Temperature Thermal Insulation Applications. *J. Non. Cryst. Solids* **2020**, *547*, 120306. [CrossRef]
33. Miros, A.; Psiuk, B.; Szpikowska-Sroka, B. Aerogel Insulation Materials for Industrial Installation: Properties and Structure of New Factory-Made Products. *J. Sol-Gel Sci. Technol.* **2017**, *84*, 496–506. [CrossRef]
34. Fang, W.Z.; Zhang, H.; Chen, L.; Tao, W.Q. Numerical Predictions of Thermal Conductivities for the Silica Aerogel and Its Composites. *Appl. Therm. Eng.* **2017**, *115*, 1277–1286. [CrossRef]
35. Liu, S.; Wu, X.; Li, Y.; Cui, S.; Shen, X.; Tan, G. Hydrophobic In-Situ SiO2-TiO2 Composite Aerogel for Heavy Oil Thermal Recovery: Synthesis and High Temperature Performance. *Appl. Therm. Eng.* **2021**, *190*, 116745. [CrossRef]
36. Li, D.; Zhang, C.; Li, Q.; Liu, C.; Arıcı, M.; Wu, Y. Thermal Performance Evaluation of Glass Window Combining Silica Aerogels and Phase Change Materials for Cold Climate of China. *Appl. Therm. Eng.* **2020**, *165*, 114547. [CrossRef]

37. Soleimani Dorcheh, A.; Abbasi, M.H. Silica Aerogel; Synthesis, Properties and Characterization. *J. Mater. Process. Technol.* **2008**, *199*, 10–26. [CrossRef]
38. Li, Z.; Cheng, X.; Shi, L.; He, S.; Gong, L.; Li, C.; Zhang, H. Flammability and Oxidation Kinetics of Hydrophobic Silica Aerogels. *J. Hazard. Mater.* **2016**, *320*, 350–358. [CrossRef]
39. Cui, S.; Liu, Y.; Fan, M.H.; Cooper, A.T.; Lin, B.L.; Liu, X.Y.; Han, G.F.; Shen, X.D. Temperature Dependent Microstructure of MTES Modified Hydrophobic Silica Aerogels. *Mater. Lett.* **2011**, *65*, 606–609. [CrossRef]
40. Yu, Y.; Guo, D.; Fang, J. Synthesis of Silica Aerogel Microspheres by a Two-Step Acid–Base Sol–Gel Reaction with Emulsification Technique. *J. Porous Mater.* **2015**, *22*, 621–628. [CrossRef]
41. Pan, Y.; He, S.; Gong, L.; Cheng, X.; Li, C.; Li, Z.; Liu, Z.; Zhang, H. Low Thermal-Conductivity and High Thermal Stable Silica Aerogel Based on MTMS/Water-Glass Co-Precursor Prepared by Freeze Drying. *Mater. Des.* **2017**, *113*, 246–253. [CrossRef]
42. Liao, Y.; Wu, H.; Ding, Y.; Yin, S.; Wang, M.; Cao, A. Engineering Thermal and Mechanical Properties of Flexible Fiber-Reinforced Aerogel Composites. *J. Sol-Gel Sci. Technol.* **2012**, *63*, 445–456. [CrossRef]
43. Woignier, T.; Phalippou, J.; Prassas, M. Glasses from Aerogels—Part 2 The Aerogel-Glass Transformation. *J. Mater. Sci.* **1990**, *25*, 3118–3126. [CrossRef]
44. Nan, C.W.; Birringer, R.; Clarke, D.R.; Gleiter, H. Effective Thermal Conductivity of Particulate Composites with Interfacial Thermal Resistance. *J. Appl. Phys.* **1998**, *81*, 6692. [CrossRef]
45. Swartz, E.T.; Pohl, R.O. Thermal Boundary Resistance. *Rev. Mod. Phys.* **1989**, *61*, 605. [CrossRef]

Disclaimer/Publisher's Note: The statements, opinions and data contained in all publications are solely those of the individual author(s) and contributor(s) and not of MDPI and/or the editor(s). MDPI and/or the editor(s) disclaim responsibility for any injury to people or property resulting from any ideas, methods, instructions or products referred to in the content.

materials

MDPI

Article

Systematic Evaluation of Permeability of Concrete Incorporating Coconut Shell as Replacement of Fine Aggregate

Akram M. Mhaya [1], Hassan Amer Algaifi [2], Shahiron Shahidan [1,*], Sharifah Salwa Mohd Zuki [1], Mohamad Azim Mohammad Azmi [1], Mohd Haziman Wan Ibrahim [1] and Ghasan Fahim Huseien [3,*]

[1] Faculty of Civil Engineering and Built Environment, Universiti Tun Hussein Onn Malaysia, Parit Raja 86400, Batu Pahat, Johor, Malaysia
[2] School of Civil Engineering, Faculty of Engineering, Universiti Teknologi Malaysia, Skudai 81310, Johor Bahru, Johor, Malaysia
[3] Institute of Architecture and Construction, South Ural State University, Lenin Prospect 76, 454080 Chelyabinsk, Russia
* Correspondence: shahiron@uthm.edu.my (S.S.); eng.gassan@yahoo.com (G.F.H.)

Abstract: The concern about coconut shell disposal and natural fine aggregate depletion has prompted researchers to utilize coconut shell as aggregate in recent years. However, the majority of the present literature has focused on utilizing coconut shell as a coarse aggregate replacement in concrete via the traditional method. In this study, concrete incorporating coconut shell as a fine aggregate replacement (10–100%) was evaluated using permeability and water absorption tests in a systematic way. The response surface methodology (RSM) was first used to design the experimental works. In addition, an artificial neural network (ANN) and genetic expression programming (GEP) were also taken into account to mathematically predict the permeability and water absorption. Based on both experimental and theoretical modeling, three scenarios were observed. In the first scenario, high quality concrete was achieved when the replacement percentage of sand by coconut shell ranged from 0% to 10%. This is because both the permeability and water absorption were less than 1.5×10^{-11} m and 5%, respectively. In the second scenario, an acceptable and reasonable low permeability (less than 2.7×10^{-11} m/s) and water absorption (less than 6.7%) were also obtained when the replacement percentage increased up to 60%. In contrast, the high content coconut shell, such as 90% and 100%, developed concrete with a high permeability and water absorption and was defined in the third scenario. It was also inferred that both the experimental and mathematical models (ANN, GEP, and RSM) have consistent and accurate results. The correlation statistics indicators (R^2) were greater than 0.94 and the error was less than 0.3, indicating a strong correlation and minimum error. In conclusion, coconut shell could act as a good alternative material to produce cleaner concrete with an optimum value of 50% as a fine aggregate replacement.

Keywords: waste materials; cleaner concrete; durability; coconut shell; systematic experimental work; informational modeling

1. Introduction

Coconut shell is one of the agricultural wastes that becomes a contributor to environmental pollution [1–5]. It is grown in more than 90 countries in areas measuring 14.231 million hectares with a total copra production equivalent to 11.04 million tonnes. Indonesia (25.63%), the Philippines (23.91%), and India (19.20%) are the major coconut-producing countries in the world [6,7]. In 2018, it was reported that a total of 504,700 metric tonnes of coconuts in Malaysia were produced in a year with a coverage plant area of 84,600 hectares [8,9]. As a result, researchers shifted their attention to solve this problem. Similarly, many researchers were also prompted to exploit other waste materials such as tire rubber [10–12], coal bottom ash [13–15], and glass powder [16–18] as a replacement of concrete aggregate. However, the target waste materials must satisfy the requirement of both the mechanical and durable qualities of concrete during the lifespan of concrete structures [19–21]. Permeability and water absorption tests are important indicators of concrete durability that must be taken into account [22–24]. This is because high permeability and water absorption could accelerate the deterioration of the concrete matrix. In particular, it promotes the easy path of aggressive ions to penetrate concrete and ultimately destroy the physical and structural integrity of concrete. For example, Shaaban and Rizzuto [25] evaluated the air permeability of concrete incorporating waste tire extracts. The study outcome revealed that the inoculation of 1% steel fibers and 10% crumb rubber led to the improved air permeability of concrete exposed to oven-drying compared to that of the air-dried specimens [26]. The improvement of permeability was attributed to pores filling due to rubber expanding at high temperatures. Moreover, the permeability of self-compacting concrete minimized with the addition of natural waste perlite powders (260 kg/m^3) in a study reported by El Mir and Nehme [27]. Similarly, exploiting polished granite waste as a partial replacement of coarse aggregate exhibited a better result for both water permeability and water absorption [28,29]. In contrast, the water permeability of concrete incorporating waste glass as a partial replacement of sand increased owing to the development of extra voids between cement paste and waste glass particles at the interface [30]. In the same context, the combination of both coconut shell and fly ash as partial replacements of coarse aggregate and cement, respectively, improved the resistance against water absorption and permeability in a study reported by Prakash and Thenmozhi [31]. Similarly, the replacement of coarse aggregate by coconut shell in self-compacting concrete exhibited positive results up to 75% in the presence of silica fume and rice husk ash [32–34].

From another perspective, the majority of the existing literature relies on the traditional lab work, while the mathematical and systematic experimental work of concrete containing coconut shell is still in its infancy. In this spirit, extensive theoretical research was carried out to predict and optimize the concrete durability using response surface methodology (RSM), artificial neural networks (ANN), and genetic expression programing (GEP) in the existing literature [35–38]. Indeed, mathematical modeling is regarded as one of the important methods not only to minimize the number of experiments but also to carefully assess the relationship and interaction between variables. For instance, Abbas [39] developed a non-linear model to predict and optimize the permeability of high performance concrete incorporating silica fume and natural pozzolan. Similarly, Güneyisi and Gesoğlu [40] improved the optimum content of metakaolin and fly ash that achieved the best performance of chloride permeability and water absorption of concrete using the response surface method. Moreover, the statistical algorithms method was also adopted to predict the chloride permeability of self-compacting concrete in a study reported by Kumar and Rai [41]. In addition, the evolved support vector regression method proved its ability to effectively predict the permeability coefficient of pervious concrete in a study reported by Sun and Zhang [42]. Based on the study's findings, a high correlation coefficient (R) and a low root-mean-square error (RMSE) were achieved indicating that the model was reliable and efficient [43].

It can be inferred that very limited experimental research was conducted to examine the permeability and water absorption of concrete containing coconut shell as fine aggregates. Accounting for that in this present study, and to fill the gap of the existing literature, both experimental and mathematical modeling were considered to predict and obtain the optimum content of coconut shell as the replacement of fine aggregate using genetic expression programming (GEP), artificial neural networks (ANN), and response surface methodology (RSM). In addition, ANOVA, error and correlation statistical validation methods were considered to evaluate the performance of the proposed models.

2. Experimental and Informational Modeling

As shown in Figure 1, phase I dealt with determining the experimental results of the proposed concrete in terms of water absorption and permeability. Phase II dealt with the assessment of the theoretical analysis of the proposed concrete.

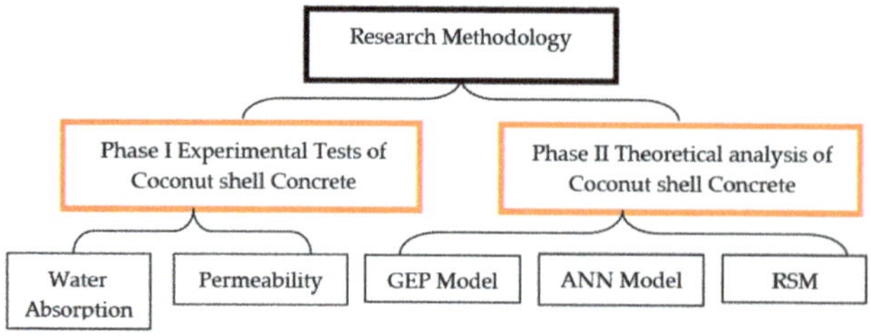

Figure 1. Diagram of representation of empirical program.

2.1. Experiment Design

Experiment design is an optimization algorithm that involves both mathematical and statistical procedures to analyze the interaction and relationship between the output or, namely, the responses (dependent variables) and input or influential factors (independent variables) [44]. In addition, the two main ideas behind the exploit of experiment design are (1) determine a minimum number of experimental tests and (2) obtain the optimum values of the influential factors that achieve the best performance of the target output. The central composite design (CCD) of response surface methodology is a type of experiment design approach that is used for optimization purpose in the present study. Indeed, CCD is composed of three types of points including center points, $2n$ axial points and 2^n factorial points where n is the number of independent variables. Figure 2 shows the required experimental tests on the basis that two independent variables were used. It is interesting to note that the coded values +1 and -1 refer to the high and low limit of each parametrize, while α is the distance from center of the cube which equal to 1.414 in the present study. It should be also noted that five center points were adopted to examine and assess the prediction error. Similarly, the number of experimental tests (Q) was found to be thirteen using Equation (1) where m represented the number of center points [45]. Moreover, Equation (2) was used to convert the coded values to real values where Z and Zc are the real value of the independent value and real value of independent variables at the center point, respectively [46]. Furthermore, L denoted the coded value of the independent variable.

$$Q = 2^n + 2n + m \tag{1}$$

$$L = \frac{Z - Z_c}{\alpha} \tag{2}$$

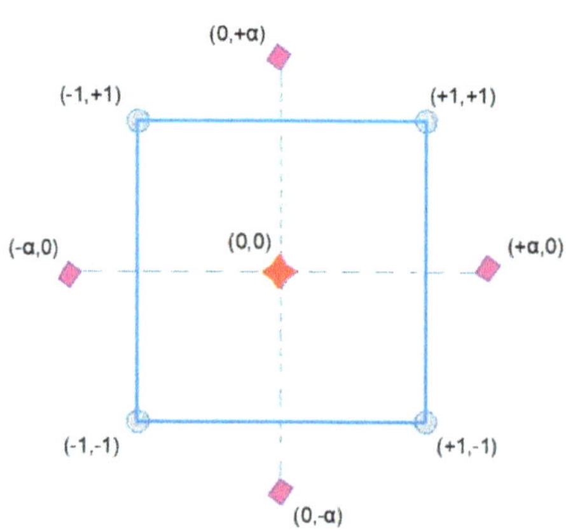

Figure 2. Basic concept of experimental design use.

A second order polynomial equation was considered to represent the permeability of CA-based concrete. The general form of the second order polynomial equation is presented in Equation (3) [47] where β_{ii} is the quadratic coefficients, β_o corresponds to intercept of the model, and β_i denotes to the linear coefficients. Furthermore, X_1 and X_2 represent the input data involving CA content and time, while Y is the response (permeability and water absorption). For the purpose of verification of the proposed equation, analysis of variance (ANOVA) was taken into account. In particular, R^2 was calculated to evaluate the closeness between the response and real results as shown in Equation (4) where SS_T represents the total sum of square error, SS_E is the sum of square error based on the predicted results, and $\overline{Y}_A$ denote the mean value of actual value. In addition, Y_P and Y_A are the predicted and real values. In the same context, $R^2_{adj.}$ was also calculated to evaluate the effect of the number of independent variables on the correlation between the real and predicted results as shown in Equation (5) [48] where DF is the degree of freedom and SS_R represents the sum square of differences error between the actual and predicted values. The predicted $R^2_{pred.}$ was also determined according to Equation (6) [49]. The differences between $R^2_{pred.}$ and $R^2_{adj.}$ should be less than 0.2 to ensure that the equation has the ability to predict for more data [50] where W refers to the estimated residual sum of square without the ith. Meanwhile, the signal-to-noise ratio was evaluated using the adequate precision (SN) as shown in Equation (7) in which its value should be greater than 4 [51] where σ^2 denote to the residual mean square. In addition, the p-value and F-value were also taken into account to validate the significance of the proposed equation. The achievement of the high F-value and p-value less than 0.005 led to the equation being considered as significant [52].

$$Y = \beta_o + \sum \beta_i X_i + \sum \beta_{ii} X_i^2 + \sum \beta_{ij} X_i X_j \tag{3}$$

$$R^2 = \frac{SS_E}{SS_T} = \frac{\sum_{i=1}^{n} \left(Y_P - \overline{Y}_A\right)^2}{\sum_{i=1}^{n} \left(\overline{Y}_A - Y_A\right)^2} \tag{4}$$

$$R^2_{adj} = 1 - \frac{SS_R/DF_R}{SS_T/DF_T} \tag{5}$$

$$R^2_{pred} = 1 - \frac{W}{SS_T} \tag{6}$$

$$SN = \frac{\max(Y_p) - \min(Y_p)}{\sqrt{\frac{p\sigma^2}{n}}} \tag{7}$$

2.2. Preparation of Concrete Mix Design

Prior to obtaining the ingredients of the concrete mixture, the materials are first collected and prepared to satisfy the specification. For example, the crushed granite was exploited as coarse aggregate, while the fine aggregate was collected for the natural sand. Both the coarse and fine aggregate met the requirement of international standard EN 933-1:2012. Meanwhile, the ordinary Portland cement is produced (OPC). Another crucial element in the mixing of concrete is water. Water helps to bind the cement and aggregates to produce concrete. BS EN 1008: 2002 defines sources of water and provides requirements as well as testing frequencies for qualifying individual or combined water sources. Apart from that, the pH value of water is also taken into consideration. In this research, the source of water was tap water used for water absorption and permeability tests. Regarding coconut shell, it was collected from a local supplier in Malaysia. After that, the CA is exposed to grinding via an impact pulverizer to achieve the suitable target size (Figure 3a). The sieve analysis was also considered to ensure that all CA particles passed a sieve size of 4.75 mm. In addition, a sufficient amount of CA particles passing the sieve size of 600 µm was also checked (Figure 3b). Consequently, the Department of Environment (DoE) method (British Standard) was adopted to calculate the proportions of the control concrete mixture that has strength of 30 MPa at 28 days. The obtained results from DOE revealed that the amount of cement, water, coarse aggregate, and fine aggregate were 435 kg, 195 L, 1251 kg/m^3 and 536 kg/m^3, respectively. It is interesting to note that the CA was added as a partial replacement of fine aggregate according to the suggested array experiment design of CCD as shown in Table 1. Rheobuild 1100 superplasticising admixture (up to 1.2% by cement weight) was also used to acquire a slump in the range of 100–140 mm. Table 2 exhibits the mix proportion of the proposed concrete.

Figure 3. Preparation of coconut shell: (**a**) chips (20 mm) and (**b**) fine powder.

Table 1. The required experimental run according to CCD.

Run NO.	Coded Value		Real Value		CCD Division
			Replaced Coconut Shell (%)	Time (Days)	
1	−1	−1	10	7	Factorial points (2^n)
2	1	−1	100	7	
3	−1	1	10	28	
4	1	1	100	28	
5	1	0	100	17	Axial points ($2n$)
6	−1	0	10	17	
7	0	−1	55	7	
8	0	1	55	28	
9	0	0	55	17	Centre points
10	0	0	55	17	
11	0	0	55	17	
12	0	0	55	17	
13	0	0	55	17	

Table 2. The actual experimental data used for ANN and GEP models.

Name of Specimens	Percentage of Fine Coconut Shell	Permeability m/s		Water Absorption %	
		7 Days	28 Days	7 Days	28 Days
FCSC10	10%	8.92×10^{-12}	1.2×10^{-11}	4.65	5.21
FCSC20	20%	1.37×10^{-11}	1.53×10^{-11}	5.12	5.48
FCSC30	30%	1.48×10^{-11}	1.73×10^{-11}	5.34	5.53
FCSC40	40%	1.81×10^{-11}	1.96×10^{-11}	5.96	6.28
FCSC50	50%	2.02×10^{-11}	2.3×10^{-11}	5.99	6.63
FCSC60	60%	2.17×10^{-11}	2.7×10^{-11}	6.37	7.28
FCSC70	70%	2.58×10^{-11}	3.38×10^{-11}	7.78	8.17
FCSC80	80%	3.59×10^{-11}	3.56×10^{-11}	8.45	8.77
FCSC90	90%	4.52×10^{-11}	4.69×10^{-11}	9.54	10.23
FCSC100	100%	5.47×10^{-11}	6.43×10^{-11}	11.63	13.85

FCS a coconut shell as partial replacement of sand, FCSC10 (10% fine aggregates replacement), FCSC100 (100% fine aggregates replacement).

2.3. Water Absorption Test

Aiming to explore the impact and extent to which coconut shell could affect the capillary and concrete voids, a water absorption (*WA*) test was carried out in compliance with BS 1881-122:2011. Three concrete cubes with the dimension of (100 × 100 × 100) mm were considered for each experiment run as shown in Figure 4. All cubes were tested after 28 days of curing age. Moreover, the value of water absorption was calculated using a traditional mathematical formula as described in Equation (8) where W_2 and W_1 represent the wet weight of the concrete cube and the dry weight of the concrete cube, respectively. It is also interesting to note that the correction factor (*CF*) was taken into account to tackle the sample length variation, which was in line with the study reported by Kwan and Ramli [53].

$$WA = \frac{W_2 - W_1}{W_1} \times 100 \times CF \quad (8)$$

Figure 4. Water penetration tests set up: (**a**) water absorption and (**b**) water permeability.

2.4. Permeability Test

To further analyze the evolution of water transport in concrete with and without coconut shell, a water permeability test was conducted in accordance with BS EN-12390-8:2009. In particular, water permeability coefficients (K_w) were obtained to evaluate the performance of concrete incorporating CA as shown in Equation (9) where d is the depth of water penetration, T refers to time under pressure, and h represent the hydraulic head. In addition, the porosity (v) is the function of the area of the cubes (A), water density (ρ), depth of penetration (d), and the differences of mass sample (m) as shown in Equation (10). To implement the test, the concrete cubes in the dimension of (100 × 100 × 100) mm were tested after 28 days for each experiment run. After the test arrangement was pressurized with 5 bars for 3 days as shown in Figure 4b, the concrete cubes were split into half and the water penetration depth measured.

$$K_w = \frac{d^2 v}{2Th} \tag{9}$$

$$v = \frac{m}{Ad\rho} \tag{10}$$

2.5. Prediction Model Using ANN

In recent years, extensive theoretical research was adopted to quickly analyze the behavior of concrete properties under different influential parameters. Herein, a data-driven model using an artificial neural network was utilized to provide a predictive mathematical equation that is able to examine and evaluate the effect of coconut shell on permeability as well as the absorption of concrete. Table 2 shows the collected real results that were inserted as input data in MATLAB to develop the ANN model. In particular, 25% of the input data were used for validation purposes, while 75% were used for training. Meanwhile, the TanH sigmoid function is taken into account as an activation function shown in Equation (11). Moreover, three layers were used to develop the data-driven model. The input data were inserted, weighted, and summed in the first layer, which simulates the collecting and receiving of the information by a human brain. After that, the input data were exposed to mathematical processes through the TanH sigmoid activated function in the second layer or namely the hidden layer. This step is line with the human process in which the received information is processed and converted to electrical signals. In the third layer, namely output layer, the action is taken either to accept or reject the output results (signal electric). In the human brain, the acceptance or rejection is dependent on the strength of the electrical signal, while it depends on the satisfaction of statistical indicators of the ANN model. In this study, the root average squared error (RASE) and R^2 were adopted as statistic validation indictors to verify the accuracy and strength of the proposed equation as shown in Equations (12) and (13), respectively.

$$f_{(x)} = \frac{e^{2x} - 1}{e^{2x} + 1} \tag{11}$$

$$\begin{aligned} RASE &= \sqrt{\frac{SSR}{n}} \\ &= \sqrt{\frac{\sum_{i=1}^{n} (Y_A - Y_P)^2}{n}} \end{aligned} \tag{12}$$

$$R^2 = 1 - \frac{\sum_{i=1}^{n} (Y_A - Y_P)^2}{\left(\sum_{i=1}^{n} (Y_A)^2 - \frac{\left(\sum_{i=1}^{n} Y_A \right)^2}{n} \right)} \tag{13}$$

2.6. Prediction Model Using GEP

The theoretical framework of concrete properties has become a hot trend in the scientific community, specifically for civil structural engineers. Moreover, many researchers have recognized the GEP model as a reliable and robust model. In this study, two mathematical equations were developed to simulate the permeability and water absorption of CA-based concrete. As shown in Table 2, the collected input data were used to develop the GEM model. Two independent variables were considered involving coconut shell content and time, while two dependent variables represent the permeability and water absorption of concrete in the proposed GEP models separately. In addition, the GeneXpro Tools 5.0 software was used to achieve this goal. In a similar manner to the ANN model, the collected data were divided into groups. The first group was defined as a training step that use 75% of the data, while 25% was related to the validation step. For validation, the statistic indicators are divided into correlation and error statistics indicators. Regarding the correlation statistics indicators, R and R^2 were calculated to assess the relationship between the real and predicted results as descried in Equations (14) and (15). The closer the R to one, the strength and closeness results could be achieved. For the error statistic indicators, four methods were taken into account involving the mean absolute error (MAE), mean root relative squared error ($RRSE$), relative absolute error (RAE), and root mean square error ($RMSE$) as show in Equations (14)–(19).

$$R^2 = 1 - \frac{\sum_{i=1}^{n} (Y_A - Y_P)^2}{\sum_{i=1}^{n} (Y_A - \overline{Y_A})^2} \tag{14}$$

$$R = \sqrt{1 - \frac{\sum_{i=1}^{n} (Y_A - Y_P)^2}{\sum_{i=1}^{n} (Y_A - \overline{Y_A})^2}} \tag{15}$$

$$MAE = \frac{1}{n}\sum_{i=1}^{n} |Y_A - Y_P| \tag{16}$$

$$RRSE = \sqrt{\frac{\sum_{i=1}^{n} (Y_P - Y_A)^2}{\sum_{i=1}^{n} (Y_A - \overline{Y_A})^2}} \tag{17}$$

$$RAE = \frac{\sum_{i=1}^{n} |Y_A - Y_P|}{\sum_{i=1}^{n} \left| Y_A - \frac{1}{n}\sum_{i=1}^{n} Y_A \right|} \tag{18}$$

$$RMSE = \sqrt{\frac{1}{n}\sum_{i=1}^{n} (Y_A - Y_P)^2} \tag{19}$$

3. Result and Discussion

3.1. Parametric Analysis

The evolution of water absorption and permeability of concrete under different replacements of fine aggregate by coconut shell were also investigated and evaluated using both experiments and modeling involving RSM, ANN, and GEP models.

3.1.1. Water Absorption

As is well known, the water absorption (WA) of concrete is one way to assess its durability and it is related to the movement of liquid such as water into the concrete matrix owing to surface tension acting in the capillaries. BS 1881-122:1983 defined the WA as the weight's increment by absorbed water that might be occurred compared to that of the dry concrete samples. This increment is attributed to the high pressure inside the pores when the dry concrete samples are exposed to and submerged in water. Therefore, a high percentage of pore volume and interconnectivity pores inside the concrete could be indicated by the value of the water absorption of concrete samples. Less water absorption refers to dense and high-quality concrete, while high absorption could be considered as a concern of concrete deterioration. An excellent water absorption of concrete could be achieved when its value is lower than 5% according to ASTM C 642-06 [54].

In the present study, an excellent water absorption (less than 5%) was obtained when the replacement of fine aggregate by coconut shell was 10% compared to the normal concrete as shown in Figure 5. In the same context, it can be seen that the *WA* increased with the increase in the sand replacement level through two scenarios. In the first scenario, the increment of water absorption continued and slightly increased up to 55% of the sand replacement percentage by coconut shell. After that, the *WA* has a sharp slop and significantly increased in comparison with the normal concrete, which is defined as the second scenario. For example, the increment of water absorption was in the range of 5.48–6.63% when the sand was replaced by the coconut shell from 20% to 50%. Such results could indicate a good water absorption. This is in line with Mo and Thomas [55] who highlighted that the concrete could be classified as good quality concrete if the *WA* is less than 10%. Similarly, Shafigh and Nomeli [56] defined the concrete incorporating the partial replacement of fine aggregate (37.5%) by oil palm shell as a good quality concrete due to the reasonable value of water absorption obtained (less than 10%). Back to the second scenario, the 100% replacement of fine aggregate by coconut shell lead to an increase in the *WA* up to 13.85%, which is considered as high and not desirable water absorption. Such an increment might be attributed to the capacity of coconut shell to absorb water itself and create a high osmosis pressure inside the concrete matrix. Moreover, the coconut shell is not dense compared to that of the natural aggregate, such as granite, in which it somewhat has pores that correspond to high concrete absorption. In particular, it could increase the chance of interconnectivity pores inside the concrete microstructure when the higher replacement of sand by coconut shell was used.

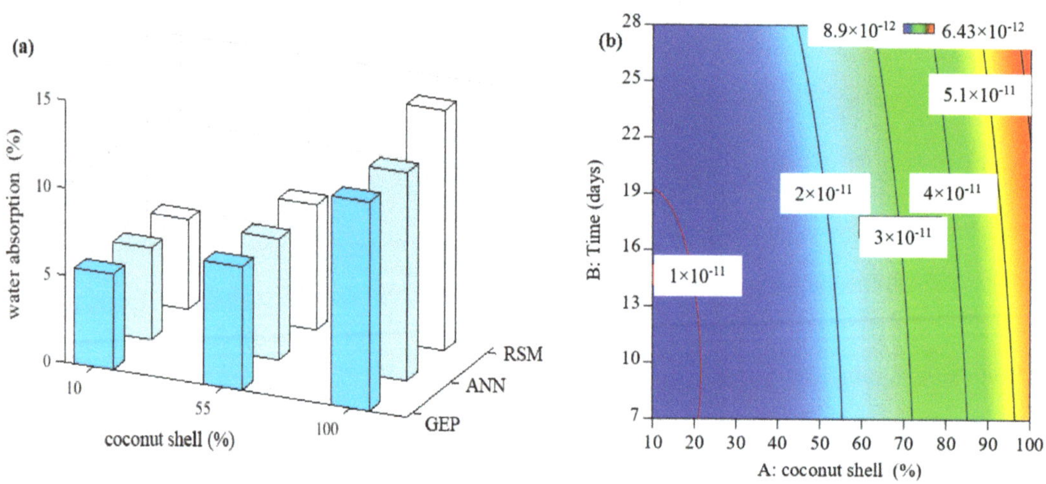

Figure 5. *Cont.*

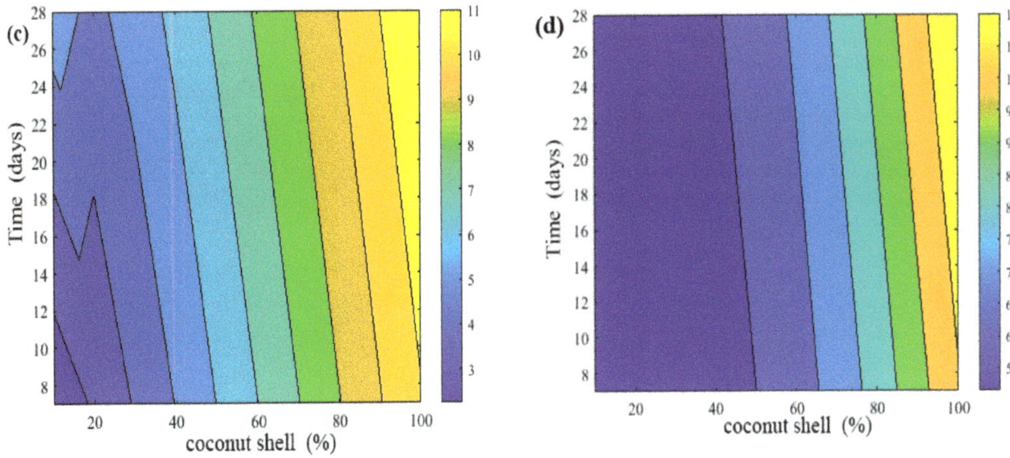

Figure 5. Evolution of water absorption of CA-based concrete: (**a**) all models (**b**), RSM (**c**) GEP, and (**d**) ANN model.

3.1.2. Water Permeability

Despite both the permeability and absorption of concrete being experimental indicators to assess the quality of concrete in terms of water penetration, absorption tests differ from concrete permeability tests and do not necessarily fully relate the absorption to the permeability [57]. As is well known, the microstructure of concrete is composed of a complex pore system, such as pores and interconnectivity pores, as well as invertible microcracks. Both the pores and interconnectivity are attributed to the water cement ratio, degree of compaction, and degree of hydration, while the occurrence of microcracks is invertible due to external and internal stress such as shrinkage, bleeding, and other factors that cause volumetric changes [58]. The pores are classified into three types according to its size (i.e., gel pores, capillary pore, and air voids). The size of the gel pores and capillary pores are 0.5–10 nm [59] and 50 nm–10 μm, respectively [60], whereas the size of air voids ranges from tens of micrometers to a millimeter [61]. These pores will later connect with each other to form an interconnectivity pore system that is considered as the main idea behind the permeability concept. In other words, the continuity of pores, their size, and distribution are related to the permeability. As such, it can be inferred that the permeability of concrete is important as it negatively causes the lifespan of cement based structures to be exposed to water or aggressive chemical ions.

In the present study, the water permeability of control concrete without the coconut shell was 1.1×10^{-11} m/s at 28 days, which is in a good agreement with the present literature. For example, Cuadrado-Rica and Sebaibi [62] concluded that the range of normal concrete permeability was 1.0×10^{-11} m/s to 5.0×10^{-11} m/s. In addition, according to ACI standard 301-89, a high-quality concrete is obtained when the water permeability is lower than 1.5×10^{-11} m/s. In our study, the water permeability of concrete containing 10% of the coconut shell as a replacement of the fine aggregate was found to be 1.2×10^{-11} m/s confirming that the concrete quality is still high when it met the specifications. In the same context, with the increase in the coconut shell up to 60%, the water permeability of the concrete slightly increased up to 2.7×10^{-11} m/s, which also can be considered a low and reasonable permeability. This fact is in line with Amriou and Bencheikh [63], who defined a low water permeability of concrete as when its value was located in the range between 8×10^{-12} m/s and 3.2×10^{-11} m/s. This result was consistent with all the data sets obtained from the RSM, ANN, and GEP models as shown in Figure 6a. It can be seen that the permeability of concrete containing coconut shell up to 55% was lesser than 2×10^{-11} m/s in indicating that the concrete quality is good. After that, the slop of water permeability increased with the increases in coconut shell content, specifically, 90% and 100%. This result is also presented using a counter plot as shown in Figure 6b–d in which the blue color reflects low permeability, whereas the yellow and red colors relate to high permeability. This mean that the zone of low permeability was located between 0 and 60% of coconut shell replacement, while the incorporation of high content coconut shell (more than 60%) would weaken the resistance of the concrete permeability. In addition, it might be attributed to the increment of pores and interconnectivity pores of concrete incorporating the high content of coconut shell.

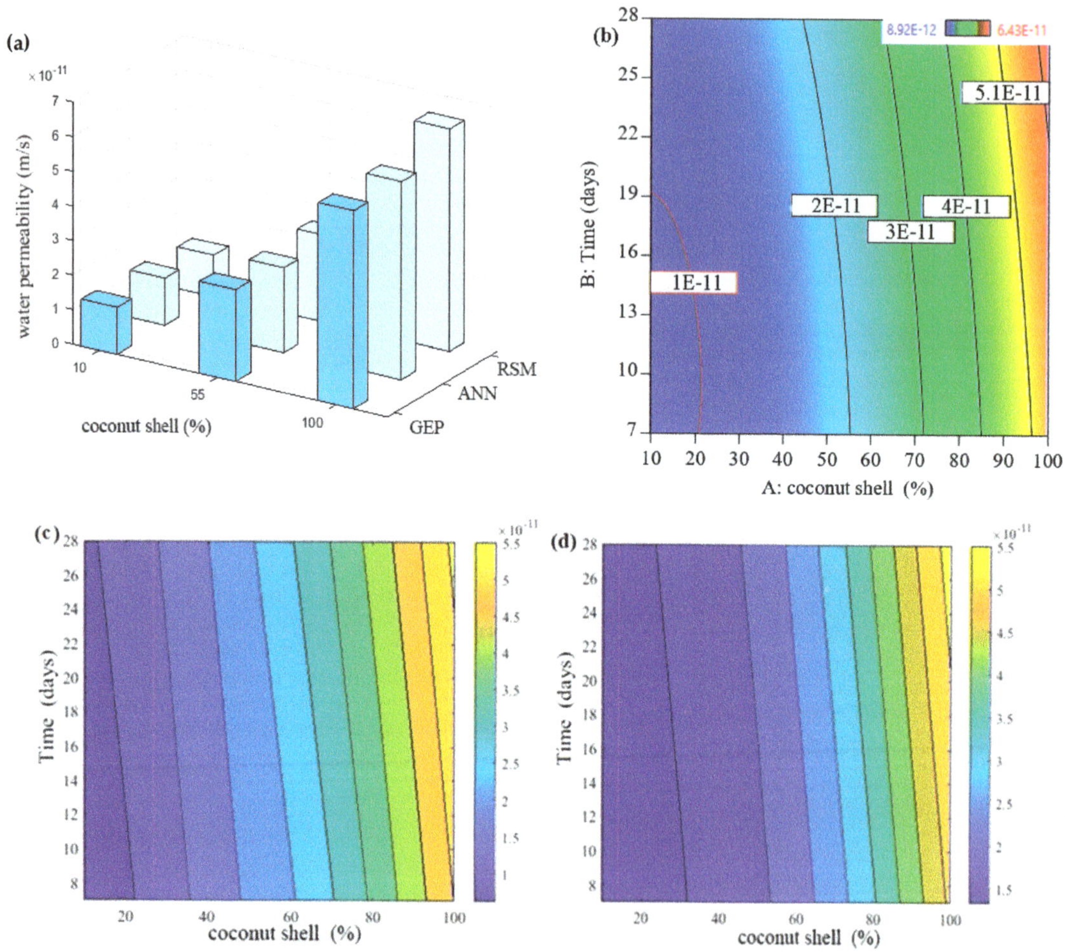

Figure 6. Evolution of water permeability of CA-based concrete: (**a**) all models, (**b**) RSM, (**c**) GEP, and (**d**) ANN model.

3.2. Informational Modeling Using RSM

The evolution of the water permeability and absorption of CA-based concrete were estimated and evaluated using quadratic equations as shown in Table 3. It can be seen that both equations are functions of time and coconut shell content. Based on the ANOVA results, both the permeability and water absorption equations were found to be significant in which the F-values were 384.41 and 300.17, while the p-values were 0.0026 and 0.0033, respectively. This is consistent with the previous studies that determined that the model could be considered as significant if the *p*-value is lesser than 0.005 and F-value is high [64]. The accuracies of both equations were also tested using *RMSE*. It was found that the value of the RMSE is minimum, specifically the RMSE value of *WA* and *WP* were 0.1492 and 8.2×10^{-13}, indicating that the model was able to estimate accurate results. This fact is in line with Algaifi and Alqarni [65] who used RMSE to prove the adequacy of the predicted equation of bacterial concrete strength. Based on their outcomes, the RSME was 2.04, confirming that the predicted and actual results were close and accurate. In the same contest, R^2 proved the closeness and correlation between the predicted and actual results in which the value of R^2 was high. According to Huseien and Sam [66], a good correlation could be obtained when R^2 is greater than 0.7. Herein, the R^2 value of *WA* and *WP* were 0.9993 and 0.9995, thus highlighting that the predicted results were acceptable. In addition, the capability of these quadratic equations to accurately predict further data was also proved using $R^2_{predicted}$ and R^2_{adj}. In particular, it was found that the differences between the $R^2_{predicted}$ and R^2_{adj} were less than 0.2. This fact is in good agreement with the existing literature. For example, Jitendra and Khed [67] developed an RSM model to predict and optimize the water absorption, chloride ions penetration, and compressive strength of concrete blocks containing foundry sand and fly ash as partial replacements of natural sand and cement, respectively. The outcome of their study revealed that the model was reliable and could be used for further prediction. This is because a reasonable difference (less than 0.2) between $R^2_{predicted}$ and R^2_{adj} was achieved for all data sets.

Table 3. Verification of the permeability and water absorption equations (RSM model).

Item	Second Polynomial Equations and the Involved Statistics Parameters				
Water Permeability (*WP*)	$R^2 = 0.999$	$R^2_{adj} = 0.9964$	$R^2_{predicted}$ 0.9813	Adeq. Precision 44.47	RMSE 8.2×10^{-13}
	$WP = 10E - 12(0.219 + 0.24d_0 + 2.78d_1 + 1.63d_0d_1 + 0.12d_0^2 + 0.1d_1^2)$				
Water Absorption (*WA*)	$R^2 = 0.9987$	$R^2_{adj} = 0.9953$	$R^2_{predicted}$ 0.976	Adeq. Precision 40.757	RMSE 0.1492
	$WA = 6.21 + 3.88d_0 + 0.59d_1 + 0.42d_0d_1 + 2.27d_0^2 + 0.355d_1^2$				

For the same regards, the significance of each parameter involving time and coconut shell on the permeability and water absorption of concrete was investigated and evaluated. It can be inferred that the coconut shell content has three scenarios in terms of slop gradient as shown in Figure 7a,b. In the first scenario, the line is almost horizontal indicating that the there is no effect. This is because the replacement percentage of CA is still low (less than 25%). After that, the line slop started to increase up with the increase in the CA content up to 50% which was classified and defined in the second scenario. In the third scenario, a sharp slop was observed indicating that the CA content (higher than 50) has a great significance on both the water absorption and permeability. In particular, with the increase in coconut shell content, both the permeability and water absorption were also increased. This fact is also confirmed by ANOVA in which the *p*-value of the coconut shell content was 0.0007 and 0.0006 based on the water absorption and permeability, respectively, while the F-value of *CA* was 1352.83 and 1787.27.

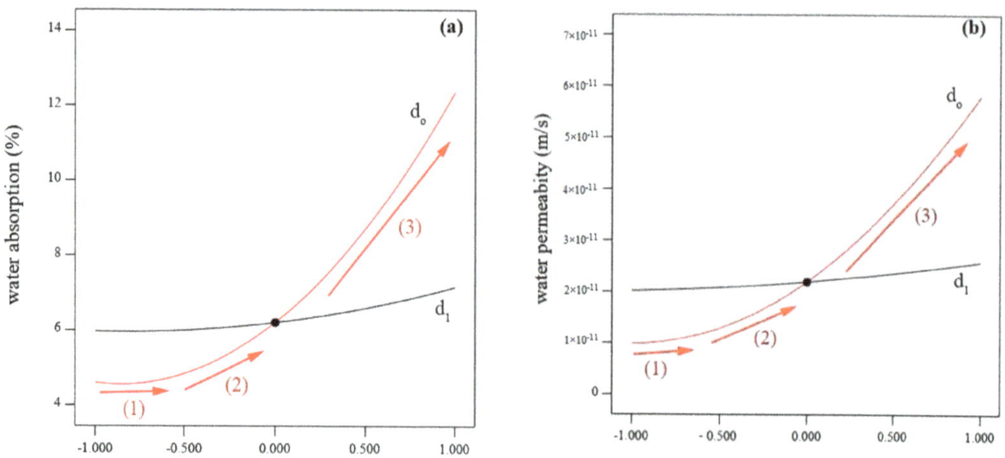

Figure 7. Significance of the involved influential parameters on: (**a**) water absorption and (**b**) permeability.

Moreover, the desirability functions were also used to optimize the optimum value of coconut shell content in CA-based concrete. Equation (19) represents the general mathematical form for the optimization purpose using RSM. Herein, the optimization equation is a function of two independent variables involving time and CA content. It was found that 78 solutions were obtained using the optimization equation. In addition, the optimal content of the CA content was considered as 53% in the present study as shown in Figure 8a. This is because, beyond this value, the increment of water absorption and permeability greatly increased. This fact is also in line with the illustrated results in Figure 8b,c. For example, two different slops were recorded. In the first slop, the increment of WA and WP of concrete incorporating CA up to 55% was reasonable and almost acceptable, while the high replacement percentage of CA (greater than 55%) significantly increased the WA and WP. This is also consistent with BS 1881-122 (2011) in which a good quality concrete could be considered when the concrete absorption is lower than 10%. The second reason to consider 53% as an optimum value is that a significant reduction on the mechanical properties of concrete was recorded when the CA content is greater than 53%, which was explained and discussed in our previous published paper. From another point of view, our attention focused to produce normal concrete incorporating CA. Beyond 53% of the CA, the concrete could be classified as lightweight concrete in which its density was lower than 2240 kg/m^3. This fact is line with Khoshkenari and Shafigh [68] and Nowak and Rakoczy [69], who demonstrated that the density of lightweight concrete has a density in the range of 1440–1840 kg/m^3, while the density of ordinary or normal concrete is in the range between 2240 and 2400 kg/m^3. Herein, the concrete density is divided into two zones as shown in Figure 8d. The first zone represents the normal concrete that has a density greater than 2240 kg/m, while the second zone represent the lightweight concrete that has a lower a density.

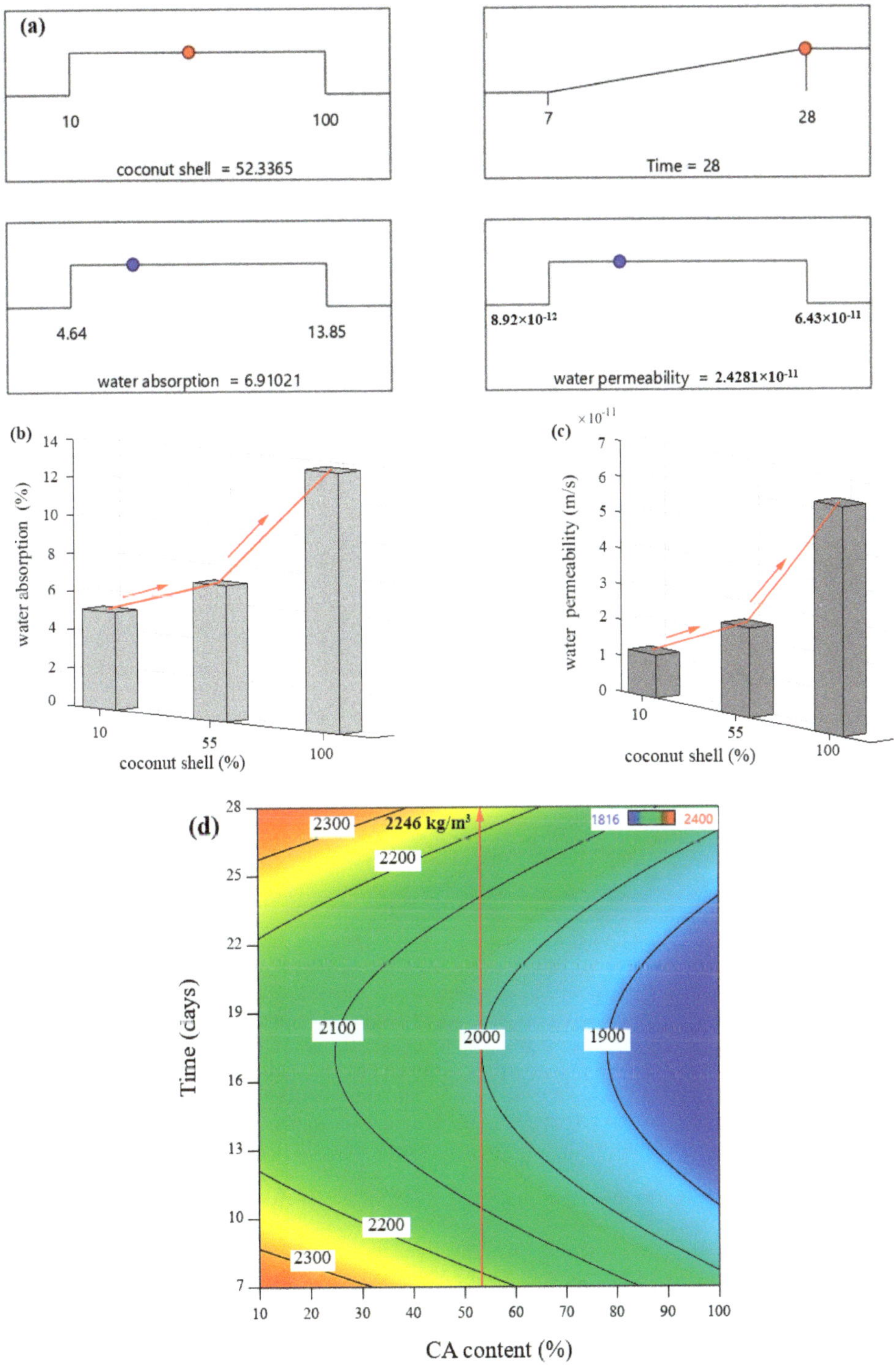

Figure 8. Optimization of coconut shell: (**a**) desirability functions, (**b**) predicted WA evolution, (**c**) predicted WP evolution, and (**d**) CA-based density evolution.

3.3. Informational Modeling Using GEP and AMM

The evolution of water absorption and permeability were also theoretically predicted using GEP. As shown in Figure 9a,b, the WA and WP equations were first developed and expressed using structural trees, respectively. It can be seen that each equation has one chromosome that, in turn, composed of one gene involving (/, +, Sqrt, Avg. and constants (c_o)). The structural tree was later converted into a mathematical expression using the Karva language.

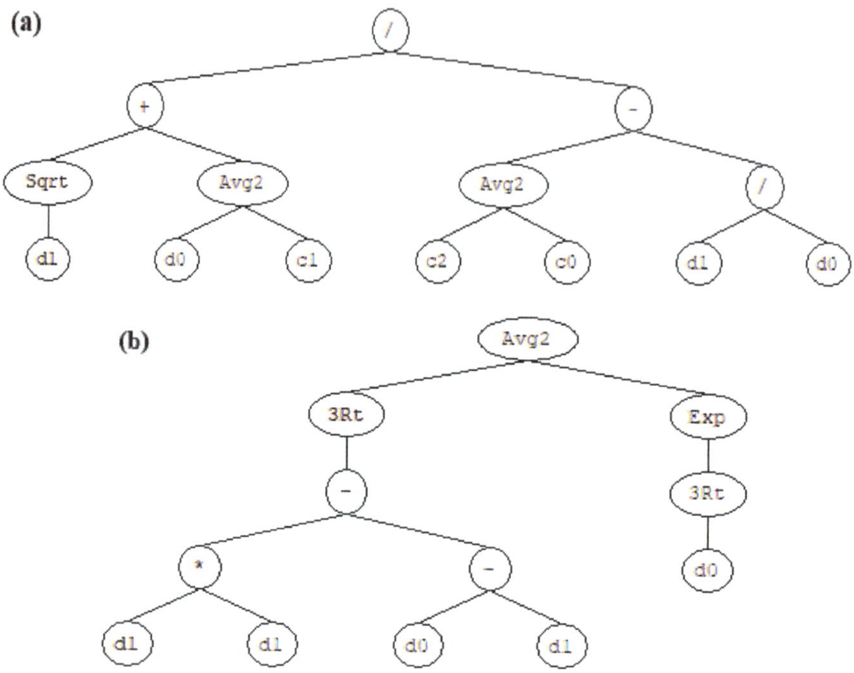

Figure 9. Structural tree of the predicated behavior of CA-based concrete: (**a**) water absorption and (**b**) water permeability.

The accuracy and reliability of the predicted GEP equations were also examined and assessed using both correlation and error statistics validations methods. It was found that the differences between actual and estimated results were found to be minimum and acceptable. According to Ali Khan and Zafar [70], the accuracy of the predicted equation of geopolymer concrete was proved using RMSE and MAE values lower than 5.971 and 5.832 for both training and validation. Herein, the values of RMSE and MAE were less than 1.037 for both WA and WP equations, indicating that the predicted GEP results could be considered as accurate and reliable. Similar trends were also observed using an ANN model. It was found that the RSME value of WA and WP of CA-based concrete was lower than 0.4447 for both training and validation data as shown in Table 4. Such results confirmed that both ANN and GEP equations could be used for further prediction with less errors. In the same context, high correlation and closeness between the actual and predicted results were obtained for both ANN and GE models. Based on Alabduljabbar and Huseien [71], the good correlation between the predicted and actual results of the alkali activated concrete's compressive strength were obtained using R^2 values of 0.991 and 0 and 9878 for training and validation. This outcome was similar to the present findings. In particular, the value of R^2 of the proposed water permeability equation using GEP and ANN were 0.9519 and 0.9719 for training, while its value was higher than 0.967 for validation. These positive results are almost in line with the findings from the predicted equation of water absorption using ANN and GEP. In particular, the R^2 were higher than 0.856 for both training and validations. As such, it can be inferred that both GEP and ANN models showed its ability to predict with high correlation and minimum error.

Table 4. The developed ANN and GEP equations and their performance.

Model	Item	Mathematical Equation and Related Statistics Validation Parameters				
GEP	WA	Training	MAE = 0.817	RMSE = 1.037	$R = 0.925$	$R^2 = 0.856$
		Validation	MAE = 0.848	RMSE = 0.959	$R = 0.982$	$R^2 = 0.964$
		$WA = \frac{2\sqrt{d_1}+d_0+c_1}{-2d_1d_0^{-1}+c_2+c_0}$				
	WP	Training	MAE = 2.73×10^{-12}	RMSE = 3.2×10^{-12}	$R = 0.9756$	$R^2 = 0.9519$
		Validation	MAE = 2.6×10^{-12}	RMSE = 2.8×10^{-12}	$R = 0.995$	$R^2 = 0.991$
		$WP = 10^{-12} \times \left(0.5\sqrt[3]{d_1^2 + d_1 - d_0} + 0.5EXP\left(\sqrt[3]{d_0}\right)\right)$				
ANN	WA	Training	MAE = 0.32	RMSE = 0.4447	$R = 0.979$	$R^2 = 0.9598$
		Validation	MAE = 0.123	RMSE = 0.221	$R = 0.983$	$R^2 = 0.967$
		$WA = 10.16 - 5.15\mathrm{Tan}H(0.5(4.91 - 0.05d_0 - 0.019d_1))$				
	WP	Training	MAE = 2.0×10^{-12}	RMSE = 2.4×10^{-12}	$R = 0.985$	$R^2 = 0.9719$
		Validation	MAE = 6×10^{-13}	RMSE = 7×10^{-13}	$R = 0.99$	$R^2 = 0.9981$
		$WP = 10^{-12} \times (45.9 - 33.7\mathrm{Tan}H(0.5(4.91 - 0.05d_0 - 0.019d_1)))$				

4. Conclusions

In recent years, the utilization of waste agricultural material as an alternative of aggregate in concrete was rapidly increased in order to address the environmental problems. In the present study, coconut shell was taken into account as a replacement of fine aggregate. Both the permeability and water absorption of CA-based concrete tests were assessed in a systematic way using RSM, ANN and GEP models. It can be concluded that utilizing coconut shell as a fine aggregate is helpful to address environmental problems, however, our work was focused on conventional concrete. Pozzolanic-based material such as fly ash could be very useful to enhance and produce CA-concrete. In addition, further study is encouraged to assess the thermal conductivity of concrete using coconut shell.

In the same context, based on the experimental and predicted results, it can be concluded that:

1. All mathematical models of RSM, ANN, and GEP proved their ability to evaluate the behavior of CA-based concrete, in which the predicted data and the actual data were consistent.
2. Based on ANN, GEP, and RSM models, the replacement percentage of fine aggregate by coconut shell up to 50% produce a good quality concrete in which the permeability and water absorption were less than 2.7×10^{-11} m/s and 5%, receptively.
3. The ANN, RSM, and GEP also revealed that the high replacement of fine aggregate by coconut shell produced a concrete with high permeability (greater than 4.5×10^{-11} m/s) and high water absorption (greater than 10%).

Author Contributions: A.M.M. verify the manuscript structure and write the final draft; H.A.A.: conceptualization, writing—original draft, methodology, investigation, and formal analysis; S.S.: supervision, validation, visualization, and reviewing; S.S.M.Z.: reviewing and editing; M.A.M.A.: reviewing and editing. M.H.W.I.; G.F.H. verification of the manuscript structure, reviewing, and editing: supervision, reviewing, and editing. All authors have read and agreed to the published version of the manuscript.

Funding: This research was supported and funded by Universiti Tun Hussein Onn Malaysia (UTHM) and PLUS MALAYSIA BERHAD through industrial grant (no. of grant M106). The authors also thank the Ministry of Higher Education.

Institutional Review Board Statement: The study did not require ethical approval.

Informed Consent Statement: Not applicable.

Data Availability Statement: Data sharing not applicable.

Conflicts of Interest: The authors declare no conflict of interest.

References

1. Aziz, W.; Aslam, M.; Ejaz, M.F.; Ali, M.J.; Ahmad, R.; Raza, M.W.U.H.; Khan, A. Mechanical properties, drying shrinkage and structural performance of coconut shell lightweight concrete. In *Structures*; Elsevier: Amsterdam, The Netherlands, 2022.
2. Tangadagi, R.B.; Manjunatha, M.; Preethi, S.; Bharath, A.; Reshma, T.V. Strength characteristics of concrete using coconut shell as a coarse aggregate—A sustainable approach. *Mater. Today: Proc.* **2021**, *47*, 3845–3851. [CrossRef]
3. Tiwari, A.; Singh, S.; Nagar, R. Feasibility assessment for partial replacement of fine aggregate to attain cleaner production perspective in concrete: A review. *J. Clean. Prod.* **2016**, *135*, 490–507. [CrossRef]
4. Janani, S.; Kulanthaivel, P.; Sowndarya, G.; Srivishnu, H.; Shanjayvel, P.G. Study of coconut shell as coarse aggregate in light weight concrete—A review. *Mater. Today Proc.* **2022**, *65*, 2003–2006. [CrossRef]
5. Alengaram, U.J.; Jumaat, M.Z.; Mahmud, H. Ductility behaviour of reinforced palm kernel shell concrete beams. *Eur. J. Sci. Res.* **2008**, *23*, 406–420.
6. Verma, D.; Gope, P. The use of coir/coconut fibers as reinforcements in composites. In *Biofiber Reinforcements in Composite Materials*; Elsevier: Amsterdam, The Netherlands, 2015; pp. 285–319.
7. Gunasekaran, K.; Kumar, P.; Lakshmipathy, M. Mechanical and bond properties of coconut shell concrete. *Constr. Build. Mater.* **2011**, *25*, 92–98. [CrossRef]
8. Nor, M.; A'liah, N.A.; Engku Ariff, E.E.; Nik Omar, N.R.; Zainol Abidin, A.Z.; Muhammad, R.M.; Rahim, H.; Nazmi, M.S.; Sulaiman, N.H. Total productivity and technical efficiency of coconuts in Malaysia. *Econ. Technol. Manag. Rev.* **2020**, *15*, 11–22.
9. Gunasekaran, K.; Annadurai, R.; Kumar, P. Long term study on compressive and bond strength of coconut shell aggregate concrete. *Constr. Build. Mater.* **2012**, *28*, 208–215. [CrossRef]
10. Bušić, R.; Miličević, I.; Šipoš, T.K.; Strukar, K. Recycled rubber as an aggregate replacement in self-compacting concrete—Literature overview. *Materials* **2018**, *11*, 1729. [CrossRef] [PubMed]
11. Li, Y.; Zhang, S.; Wang, R.; Dang, F. Potential use of waste tire rubber as aggregate in cement concrete—A comprehensive review. *Constr. Build. Mater.* **2019**, *225*, 1183–1201. [CrossRef]
12. Mhaya, A.; Abidin, A.R.Z.; Sarbini, N.N.; Ismail, M. Role of crumb tyre aggregates in rubberised concrete contained granulated blast-furnace slag. In *IOP Conference Series: Earth and Environmental Science*; IOP Publishing: Bristol, UK, 2019; pp. 1–12.
13. Singh, N.; Mithulraj, M.; Arya, S. Influence of coal bottom ash as fine aggregates replacement on various properties of concretes: A review. *Resour. Conserv. Recycl.* **2018**, *138*, 257–271. [CrossRef]
14. Ibrahim, M.H.B.W.; Shahidan, S.; Algaifi, H.A.; Hamzah, A.F.B.; Jaya, R.P. CBA Self-compacting Concrete Exposed to Chloride and Sulphate. In *Properties of Self-Compacting Concrete with Coal Bottom Ash under Aggressive Environments*; Springer: Berlin/Heidelberg, Germany, 2021; pp. 33–57.
15. Ibrahim, M.H.B.W.; Shahidan, S.; Amer Algaifi, H.; Bin Hamzah, A.F.; Putra Jaya, R. CBA Self-compacting Concrete Exposed to Seawater by Wetting and Drying Cycles. In *Properties of Self-Compacting Concrete with Coal Bottom Ash under Aggressive Environments*; Springer: Berlin/Heidelberg, Germany, 2021; pp. 59–75.
16. Małek, M.; Łasica, W.; Jackowski, M.; Kadela, M. Effect of waste glass addition as a replacement for fine aggregate on properties of mortar. *Materials* **2020**, *13*, 3189. [CrossRef] [PubMed]
17. Harrison, E.; Berenjian, A.; Seifan, M. Recycling of waste glass as aggregate in cement-based materials. *Environ. Sci. Ecotechnol.* **2020**, *4*, 100064. [CrossRef] [PubMed]
18. Mhaya, A.M.; Baharom, S.; Huseien, G.F. Improved strength performance of rubberized Concrete: Role of ground blast furnace slag and waste glass bottle nanoparticles amalgamation. *Constr. Build. Mater.* **2022**, *342*, 128073. [CrossRef]
19. Medina, C.; Frías, M.; De Rojas, M.S.; Thomas, C.; Polanco, J.A. Gas permeability in concrete containing recycled ceramic sanitary ware aggregate. *Constr. Build. Mater.* **2012**, *37*, 597–605. [CrossRef]
20. Jiao, H.; Chen, W.; Wu, A.; Yu, Y.; Ruan, Z.; Honaker, R.; Chen, X.; Yu, J. Flocculated unclassified tailings settling efficiency improvement by particle collision optimization in the feedwell. *Int. J. Miner. Metall. Mater.* **2022**, *29*, 2126–2135. [CrossRef]
21. Chen, F.; Xu, B.; Jiao, H.; Chen, X.; Shi, Y.; Wang, J.; Li, Z. Triaxial mechanical properties and microstructure visualization of BFRC. *Constr. Build. Mater.* **2021**, *278*, 122275. [CrossRef]

22. Pan, X.; Shi, C.; Jia, L.; Zhang, J.; Wu, L. Effect of inorganic surface treatment on air permeability of cement-based materials. *J. Mater. Civ. Eng.* **2016**, *28*, 04015145. [CrossRef]
23. Jia, L.; Shi, C.; Pan, X.; Zhang, J.; Wu, L. Effects of inorganic surface treatment on water permeability of cement-based materials. *Cem. Concrete Composites* **2016**, *67*, 85–92. [CrossRef]
24. Baghban, M.H.; Mhaya, A.M.; Faridmehr, I.; Huseien, G.F. Carbonation Depth and Chloride Ion Penetration Properties of Rubberised Concrete Incorporated Ground Blast Furnace Slag. In *Solid State Phenomena*; Trans Tech Publications Ltd.: Bach, Switzerland, 2022; Volume 329, pp. 101–108.
25. Shaaban, I.G.; Rizzuto, J.P.; El-Nemr, A.; Bohan, L.; Ahmed, H.; Tindyebwa, H. Mechanical properties and air permeability of concrete containing waste tires extracts. *J. Mater. Civ. Eng.* **2021**, *33*, 04020472. [CrossRef]
26. Mhaya, A.M.; Baghban, M.H.; Faridmehr, I.; Huseien, G.F. Performance Evaluation of Modified Rubberized Concrete Exposed to Aggressive Environments. *Materials* **2021**, *14*, 1900. [CrossRef]
27. El Mir, A.S.; Nehme, G.; Assaad, J.J. Durability of self-consolidating concrete containing natural waste perlite powders. *Heliyon* **2020**, *6*, e03165. [CrossRef] [PubMed]
28. Sharma, N.K.; Kumar, P.; Kumar, S.; Thomas, B.S.; Gupta, R.C. Properties of concrete containing polished granite waste as partial substitution of coarse aggregate. *Constr. Build. Mater.* **2017**, *151*, 158–163. [CrossRef]
29. Mhaya, A.M.; Huseien, G.F.; Abidin, A.R.Z.; Ismail, M. Long-term mechanical and durable properties of waste tires rubber crumbs replaced GBFS modified concretes. *Constr. Build. Mater.* **2020**, *256*, 119505. [CrossRef]
30. Bisht, K.; Ramana, P. Sustainable production of concrete containing discarded beverage glass as fine aggregate. *Constr. Build. Mater.* **2018**, *177*, 116–124. [CrossRef]
31. Prakash, R.; Thenmozhi, R.; Raman, S.N.; Subramanian, C.; Divyah, N. An investigation of key mechanical and durability properties of coconut shell concrete with partial replacement of fly ash. *Struct. Concr.* **2021**, *22*, E985–E996. [CrossRef]
32. Palanisamy, M.; Kolandasamy, P.; Awoyera, P.; Gobinath, R.; Muthusamy, S.; Krishnasamy, T.R.; Viloria, A. Permeability properties of lightweight self-consolidating concrete made with coconut shell aggregate. *J. Mater. Res. Technol.* **2020**, *9*, 3547–3557. [CrossRef]
33. Mathew, S.P.; Nadir, Y.; Arif, M.M. Experimental study of thermal properties of concrete with partial replacement of coarse aggregate by coconut shell. *Mater. Today Proc.* **2020**, *27*, 415–420. [CrossRef]
34. Nadir, Y.; Sujatha, A. Durability properties of coconut shell aggregate concrete. *KSCE J. Civ. Eng.* **2018**, *22*, 1920–1926. [CrossRef]
35. Liu, Q.-F.; Iqbal, M.F.; Yang, J.; Lu, X.Y.; Zhang, P.; Rauf, M. Prediction of chloride diffusivity in concrete using artificial neural network: Modelling and performance evaluation. *Constr. Build. Mater.* **2021**, *268*, 121082. [CrossRef]
36. Liu, Q.-F.; Hu, Z.; Lu, X.Y.; Yang, J.; Azim, I.; Sun, W. Prediction of chloride distribution for offshore concrete based on statistical analysis. *Materials* **2020**, *13*, 174. [CrossRef]
37. Boubekeur, T.; Boulekbache, B.; Aoudjane, K.; Ezziane, K.; Kadri, E.H. Prediction of the durability performance of ternary cement containing limestone powder and ground granulated blast furnace slag. *Constr. Build. Mater.* **2019**, *209*, 215–221. [CrossRef]
38. Mhaya, A.M.; Huseien, G.F.; Faridmehr, I.; Abidin, A.R.Z.; Alyousef, R.; Ismail, M. Evaluating mechanical properties and impact resistance of modified concrete containing ground Blast Furnace slag and discarded rubber tire crumbs. *Constr. Build. Mater.* **2021**, *295*, 123603. [CrossRef]
39. Abbas, Y. Simplex-lattice strength and permeability optimization of concrete incorporating silica fume and natural pozzolan. *Constr. Build. Mater.* **2018**, *168*, 199–208. [CrossRef]
40. Güneyisi, E.; Gesoğlu, M.; Algın, Z.; Mermerdaş, K. Optimization of concrete mixture with hybrid blends of metakaolin and fly ash using response surface method. *Compos. Part B Eng.* **2014**, *60*, 707–715. [CrossRef]
41. Kumar, S.; Rai, B.; Biswas, R.; Samui, P.; Kim, D. Prediction of rapid chloride permeability of self-compacting concrete using Multivariate Adaptive Regression Spline and Minimax Probability Machine Regression. *J. Build. Eng.* **2020**, *32*, 101490. [CrossRef]
42. Sun, J.; Zhang, J.; Gu, Y.; Huang, Y.; Sun, Y.; Ma, G. Prediction of permeability and unconfined compressive strength of pervious concrete using evolved support vector regression. *Constr. Build. Mater.* **2019**, *207*, 440–449. [CrossRef]
43. Mhaya, A.M.; Baharom, S.; Baghban, M.H.; Nehdi, M.L.; Faridmehr, I.; Huseien, G.F.; Algaifi, H.A.; Ismail, M. Systematic Experimental Assessment of POFA Concrete Incorporating Waste Tire Rubber Aggregate. *Polymers* **2022**, *14*, 2294. [CrossRef]
44. Oyebisi, S.O.; Ede, A.N.; Olutoge, F.A. Optimization of design parameters of slag-corncob ash-based geopolymer concrete by the central composite design of the response surface methodology. *Iran. J. Sci. Technol. Trans. Civ. Eng.* **2021**, *45*, 27–42. [CrossRef]
45. Habibi, A.; Ramezanianpour, A.M.; Mahdikhani, M.; Bamshad, O. RSM-based evaluation of mechanical and durability properties of recycled aggregate concrete containing GGBFS and silica fume. *Constr. Build. Mater.* **2021**, *270*, 121431. [CrossRef]
46. Shahmansouri, A.A.; Nematzadeh, M.; Behnood, A. Mechanical properties of GGBFS-based geopolymer concrete incorporating natural zeolite and silica fume with an optimum design using response surface method. *J. Build. Eng.* **2021**, *36*, 102138. [CrossRef]
47. Ferdosian, I.; Camões, A. Eco-efficient ultra-high performance concrete development by means of response surface methodology. *Cem. Concr. Compos.* **2017**, *84*, 146–156. [CrossRef]
48. Foroughi, M.; Rahmani, A.R.; Asgari, G.; Nematollahi, D.; Yetilmezsoy, K.; Samarghandi, M.R. Optimization of a three-dimensional electrochemical system for tetracycline degradation using box-behnken design. *Fresenius Environ. Bull.* **2018**, *27*, 1914–1922.
49. Mukhopadhyay, T.; Dey, T.K.; Chowdhury, R.; Chakrabarti, A. Structural damage identification using response surface-based multi-objective optimization: A comparative study. *Arab. J. Sci. Eng.* **2015**, *40*, 1027–1044. [CrossRef]

50. Mohammed, B.S.; Yen, L.Y.; Haruna, S.; Seng Huat, M.L.; Abdulkadir, I.; Al-Fakih, A.; Liew, M.S.; Abdullah Zawawi, N.A.W. Effect of Elevated Temperature on the Compressive Strength and Durability Properties of Crumb Rubber Engineered Cementitious Composite. *Materials* **2020**, *13*, 3516. [CrossRef] [PubMed]
51. Dan, S.; Banivaheb, S.; Hashemipour, H. Synthesis, characterization and absorption study of chitosan-g-poly (acrylamide-co-itaconic acid) hydrogel. *Polym. Bull.* **2021**, *78*, 1887–1907. [CrossRef]
52. Oyebisi, S.; Ede, A.; Owamah, H.; Igba, T.; Mark, O.; Odetoyan, A. Optimising the Workability and Strength of Concrete Modified with Anacardium Occidentale Nutshell Ash. *Fibers* **2021**, *9*, 41. [CrossRef]
53. Kwan, W.H.; Ramli, M.; Kam, K.J.; Sulieman, M.Z. Influence of the amount of recycled coarse aggregate in concrete design and durability properties. *Constr. Build. Mater.* **2012**, *26*, 565–573. [CrossRef]
54. Gurumoorthy, N.; Arunachalam, K. Durability studies on concrete containing treated used foundry sand. *Constr. Build. Mater.* **2019**, *201*, 651–661. [CrossRef]
55. Mo, K.H.; Thomas, B.S.; Yap, S.P.; Abutaha, F.; Tan, C.G. Viability of agricultural wastes as substitute of natural aggregate in concrete: A review on the durability-related properties. *J. Clean. Prod.* **2020**, *275*, 123062. [CrossRef]
56. Shafigh, P.; Nomeli, M.A.; Alengaram, U.J.; Mahmud, H.B.; Jumaat, M.Z. Engineering properties of lightweight aggregate concrete containing limestone powder and high volume fly ash. *J. Clean. Prod.* **2016**, *135*, 148–157. [CrossRef]
57. Bhardwaj, B.; Kumar, P. Waste foundry sand in concrete: A review. *Constr. Build. Mater.* **2017**, *156*, 661–674. [CrossRef]
58. Banthia, N.; Biparva, A.; Mindess, S. Permeability of concrete under stress. *Cem. Concr. Res.* **2005**, *35*, 1651–1655. [CrossRef]
59. Kumar, R.; Bhattacharjee, B. Porosity, pore size distribution and in situ strength of concrete. *Cem. Concr. Res.* **2003**, *33*, 155–164. [CrossRef]
60. Sidiq, A.; Gravina, R.J.; Setunge, S.; Giustozzi, F. High-efficiency techniques and micro-structural parameters to evaluate concrete self-healing using X-ray tomography and Mercury Intrusion Porosimetry: A review. *Constr. Build. Mater.* **2020**, *252*, 119030. [CrossRef]
61. Gallucci, E.; Scrivener, K.; Groso, A.; Stampanoni, M.; Margaritondo, G. 3D experimental investigation of the microstructure of cement pastes using synchrotron X-ray microtomography (μCT). *Cem. Concr. Res.* **2007**, *37*, 360–368. [CrossRef]
62. Cuadrado-Rica, H.; Sebaibi, N.; Boutouil, M.; Boudart, B. Properties of ordinary concretes incorporating crushed queen scallop shells. *Mater. Struct.* **2016**, *49*, 1805–1816. [CrossRef]
63. Amriou, A.; Bencheikh, M. New experimental method for evaluating the water permeability of concrete by a lateral flow procedure on a hollow cylindrical test piece. *Constr. Build. Mater.* **2017**, *151*, 642–649. [CrossRef]
64. Hou, D.; Chen, D.; Wang, X.; Wu, D.; Ma, H.; Hu, X.; Zhang, Y.; Wang, P.; Yu, R. RSM-based modelling and optimization of magnesium phosphate cement-based rapid-repair materials. *Constr. Build. Mater.* **2020**, *263*, 120190. [CrossRef]
65. Algaifi, H.A.; Alqarni, A.S.; Alyousef, R.; Bakar, S.A.; Ibrahim, M.W.; Shahidan, S.; Ibrahim, M.; Salami, B.A. Mathematical prediction of the compressive strength of bacterial concrete using gene expression programming. *Ain Shams Eng. J.* **2021**, *12*, 3629–3639. [CrossRef]
66. Huseien, G.F.; Sam, A.R.M.; Algaifi, H.A.; Alyousef, R. Development of a sustainable concrete incorporated with effective microorganism and fly Ash: Characteristics and modeling studies. *Constr. Build. Mater.* **2021**, *285*, 122899. [CrossRef]
67. Jitendra, K.; Khed, V.C. Optimization of concrete blocks with high volume fly ash and foundry sand. *Mater. Today Proc.* **2020**, *27*, 1172–1179. [CrossRef]
68. Khoshkenari, A.G.; Shafigh, P.; Moghimi, M.; Mahmud, H.B. The role of 0–2 mm fine recycled concrete aggregate on the compressive and splitting tensile strengths of recycled concrete aggregate concrete. *Mater. Des.* **2014**, *64*, 345–354. [CrossRef]
69. Nowak, A.; Rakoczy, A. Statistical model for compressive strength of lightweight concrete. *Archit. Civ. Eng. Env.* **2011**, *4*, 73–80.
70. Ali Khan, M.; Zafar, A.; Akbar, A.; Javed, M.F.; Mosavi, A. Application of Gene Expression Programming (GEP) for the prediction of compressive strength of geopolymer concrete. *Materials* **2021**, *14*, 1106. [CrossRef]
71. Alabduljabbar, H.; Huseien, G.F.; Sam, A.R.M.; Alyouef, R.; Algaifi, H.A.; Alaskar, A. Engineering Properties of Waste Sawdust-Based Lightweight Alkali-Activated Concrete: Experimental Assessment and Numerical Prediction. *Materials* **2020**, *13*, 5490. [CrossRef]

materials

MDPI

Article

A New Model Supporting Stability Quality of Materials and Industrial Products

Dominika Siwiec and Andrzej Pacana *

Faculty of Mechanical Engineering and Aeronautics, Rzeszow University of Technology, al. Powstancow Warszawy 12, 35-959 Rzeszow, Poland; d.siwiec@prz.edu.pl
* Correspondence: app@prz.edu.pl

Abstract: Stabilizing the quality of industrial product materials remains a challenge. This applies mainly to new or significantly modified materials. It also refers to special processes. The tests of product quality can stabilize the quality of industrial product materials. The popular method for this is using the non-destructive testing (NDT). The NDT identifies incompatibility but does not determine the cause of its occurrence. Hence, it was necessary to support the process of identifying causes of incompatibilities in products. The purpose of the article was to develop a model based on a new approach to determine the ranking of actions that are possible as part of the process of stabilizing the quality of industrial products. The model was developed to improve quality through sequential and systematic methods of identification (and reduce) and incompatibility. The quality management techniques and decision method were applied and combined in this model, i.e., SMART(-ER) the method, method of selecting a team of experts, brainstorming (BM), Ishikawa diagram with the 5M rule, Likert scale validation technique, arithmetic average, and Grey Relational Analysis (GRA). The test of this model was carried out to find cracks in the outer hull of 418 alloy four-point bearing (CPW-S 5616), which was identified by NDT (magnetic-powder method). As a result, a ranking of activities was obtained to stabilize the quality of the product and the main cause of incompatibility was indicated, i.e., the cause which can influence to the most degree influence on occurrence the incompatibility. The originality of the proposed model is an application in the right order of specially selected and combined qualitative methods and supporting decision methods. The finding of causes of incompatibility of products is the basis of product improvement in the area of stabilizing the quality of materials, mainly by the occurrence of special processes. The universality of the model refers to the possibility of its application for any material, processes of its formation, and processes of products, and any incompatibilities where the model can be integrated with quality control.

Keywords: grey relational analysis; Ishikawa diagram; multi-criteria decision methods; quality management tools; production engineering; mechanical engineering

1. Introduction

The special processes of materials are difficult to stabilize. Special processes are realized based on the validated process (method or algorithm). The identification and removal of these incompatibilities improve manufacturing quality or a quality of the processes as the method. The research of the quality of materials and products is mainly non-destructive testing [1]. However, this research does not determine the causes of incompatibility. Therefore, it is necessary to take future actions which rely on determining the causes of these incompatibilities [2–4]. In the process of achieving high quality of products [5], it is necessary to identify and eliminate possible incompatibilities of products [6,7]. It refers to determining the main causes of the problem in a precise way to carry out adequate improvement actions [8,9]. The right analysis of the quality of products and the implementation of supporting actions to reduce incompatibilities also contribute to stabilizing the production process [10,11]. Furthermore, it is possible to repeat results under industrial conditions [5,12,13]. Despite that, achieving stable production is still problematic [14,15]. The initial production series is carried out mainly in the case of new products or significantly modified [16].

The literature review shows that to verify incompatibilities of industrial products, the most frequently used methods were brainstorming (BM), Ishikawa diagram [17], Pareto-Lorenz [18], and the 5Why method [19]. The Ishikawa diagram and the Pareto-Lorenz diagram were used to verify the causes of incompatibility of industrial products [18–21]. For example, in study [18], the structure of the pulley defects was assessed, where the Pareto diagram was used to identify the most important incompatibilities, and the Ishikawa diagram was used to determine the causes of these incompatibilities. In turn, the study [22] presented a combination of methods, i.e., the Ishikawa diagram and Pareto analysis, to reduce defects of capacitors. Firstly, all defects were verified by causes and effects diagrams, and then the most important defects were identified by Pareto analysis. In a similar way, the authors of the study [19] analyzed the laser cutting process, where the 5Why method was used to verify the main causes of incompatibility in this process. However, in study [20], the general purpose technological analysis (GPT) was verified using the Ishikawa diagram. The universalmodel was carried out by authors of study [12], in which quality management tools were combined, that is, the SMART (-ER) method, the method of selecting a team of experts, brainstorming (BM), Ishikawa diagram, and the 5Why method. The model was carried out by mechanical seal from 410 alloy. Another example is in this study [21], in which the Ishikawa diagram was applied to analyze the causes of errors in assessment accuracy in part of the construction of the machine. The authors of this study presented another approach to the analysis of the incompatibility of the product [23], where the incompatibility of metal inclusions in the product from AMS6514 alloy was verified. The techniques were combined, i.e., brainstorming, Ishikawa diagram, and DEMATEL method. The idea was to verify cause-and-effects relations of incompatibility of product. In the study [24], the method was tested that was a combination technique, that is, brainstorming the Ishikawa diagram and the DEMATEL method. The purpose was to develop combination methods to verify complex cause-and-effect problems. These problems refer to the quality of products. The number of causes of these incompatibilities was large. These methods were shown in the study [25–27], where the techniques were combined: brainstorming (BM), cause and effect diagram, FAHP method (Fuzzy Analytic Hierarchy Process), and 5Why method (Why-Why method). The idea of this combination was to reduce inconsistencies and uncertainties in the expert team, where these evaluations refer to a large number of causes of product incompatibility. In turn, in this study [28], practical examples of using the basic quality tools (7QC tools) were analyzed. Other causes were shown in the study [29], in which there were analysis groups (categories) of the causes of problem according to the Ishikawa rule (5M+E). This tool was combined with the FTOPSIS method (Fuzzy Technique for Order Preference by Similarity to Ideal Solution) and the FAHP method. The test method was carried out to determine cracks in the pin connecting discs in the engine gear when grinding. The purpose of this combination was to verify the causes of various types of incompatibility of products.

It was concluded that quality management tools were used to verify the incompatibilities of industrial products, e.g., [12,18,21]. Their applications include verifying the potential causes of incompatibilities and determining the main causes [22,23]. Despite that, these analyses were not a destination of sequential reducing causes of incompatibility by its importance, i.e., determining the impact (importance) of these causes on occurring incompatibility. Therefore, these studies include a gap in the lack of methods that are applied to create a ranking of causes of incompatibility, where this ranking would allow for ranking actions that support the stabilization of product quality. Therefore, the objective of the article was to develop a new model that supports the stabilization quality of industrial products. During the development of this model, the hypothesis was assumed:

Hypothesis 1 (H1). *It is possible to support the stability quality of materials, processes of their formation and industrial products by determining the ranking of causes that have an impact on the incompatibility of the product; it is realized by determining all potential causes and then by their sequential and constructive reduction to reduce the main causes.*

The purpose of the study is to develop a model based on a new approach to determine the ranking of actions that can be taken as part of the process of stabilizing the quality of industrial products. It refers to the sequential and coherent analysis of the causes of problems with the quality of products, where the analysis can be realized even for four causes of incompatibility. It means that it is possible to analyze a small number of causes of incompatibility, where according to the GRA method, the minimum number of causes is equal to 4. However, the maximum number of causes of incompatibility for analysis is unlimited. Moreover, it is possible to reduce the causes of the problem from potential causes to the most important causes, which possibly have the largest degree of influence on the occurrence of the problem.

The model has a universal character. However, in view of the specificity of special processes occurring in the mechanical industry, it seems adequate to realize the initial test, for example, for the casting process. That resulted from a review of the literature on the subject [18–21,24,26,29], after which it was proved that the most frequent errors occur due to foundry processes. Therefore, the modeling was tested for cracks in the outer hull of the four-point bearing of alloy 418 (CPW-S 5616). This incompatibility was identified in the Polish company by non-destructive testing (magnetic powder method).

2. Model

2.1. Concept of Model

The concept of the model refers to the verification of incompatibility of industrial products, where efforts were made on systematic verification of incompatibilities as part of continuous improvement of products. The idea was to support the production process of products, i.e., mainly new products or modified products for which often the trial (initial) production series are realized. The main purpose of the model was to determine the ranking of actions that stabilize the quality of products. The general concept of the model is shown in Figure 1.

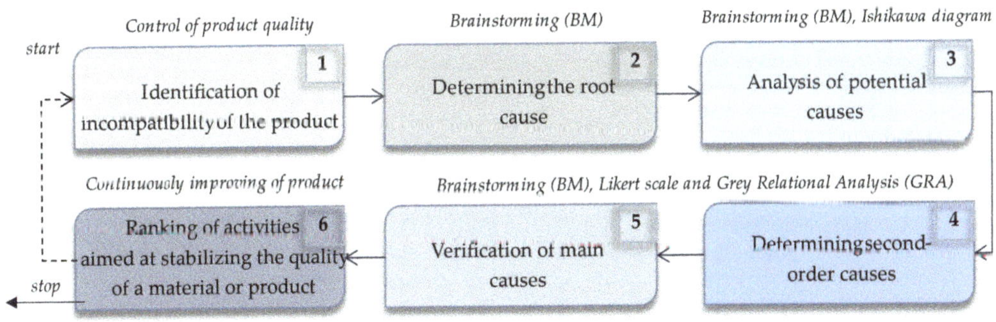

Figure 1. General concept of model.

The model was developed as a modified method and suited to the specification of searching for the importance of incompatibilities in industrial products. The mentioned support was a destination in the process of identification of incompatibility causes and their importance in determining the main causes of the problem, after which it is possible to make the right improvement actions. It refers to verification of all potential causes of incompatibility, their sequential analysis, and reducing to identify the main causes (i.e., having the most impact on the emergence of noncompliance). After determining the main causes of incompatibility, adequate improving actions are determined, i.e., actions which can reduce or mostly reduce incompatibility.

The originality of the model is the possibility to analyze, in sequential and coherent ways, the causes of incompatibilities of the product, where the number of these incompatibilities can be equal even to 4. Moreover, it is possible to reduce causes from potential causes to the most important causes, where this process is supported by a calculation process and a simple but simultaneously effective GRA method. Additionally, the model combines techniques: teamwork, visualization, and calculation methods, which support and realize each step of the model to identify the main cause (root) of problems.

2.2. Conditions and Justification for Choice of Methods for Model

The concept of the model is based on the integration of selected instruments of quality management and decision support methods in a fuzzy decision-making environment. These techniques were: SMART(-ER) method [30], method of choice of the team of experts [12,31], brainstorming (BM) [32,33], Ishikawa diagram (causes and effects) with the 5M rule [6–9], the technique of importance in Liker scale [5,7], average arithmetic, and Grey Relational Analysis (GRA) [34–39].

Firstly, the purpose of the analysis is determined. The purpose is determined by the SMART(-ER) method (S—specific, M—measurable, A—achievable, R—relevant or realistic or reward, T—based on timeline or timebound, E—exciting or evaluated, R—recorded or reward) [30].

Next, the team of experts is selected. The team is selected according to the method shown in the study [12,31]. The idea was to achieve an effective analysis of a problem by the team of experts. Therefore, the expert team should have knowledge and experience in the analysis of incompatibility and the ability to solve the problem. The appropriate choice of the team of experts has an impact on achieving the objective, as shown in the study [9].

Next, brainstorming (BM) is realized among a team of experts. The BM method allows for coherent and effective verification of any kind of problem that requires in-depth analysis [32,33]. This method is used in all stages of the proposed model, for example, to identify the potential causes of incompatibility (i.e., the causes which probably cause incompatibility of product). To determine these causes, it is necessary to answer the question "What has happened that this incompatibility occurred?".

Next, all causes are grouped according to the 5M rule (man, method, machine, material, and management). This rule is preferred to analyze the quality of industrial products [19,20]. The 5M is used in the Ishikawa diagram to group the causes of the problem. The aim of 5M is the simple visualization of the causes of incompatibility of products [21]. Therefore, the team of experts needs to understand each of the causes generated during brainstorming (BM).

Then, the team of experts determines the weights of incompatibility causes, i.e., the impact of these causes on the occurrence of incompatibility. This is achieved as part of the next part of brainstorming (BM) and by using the technique with the Likert scale to determine the importance of causes [5,7]. According to these assessments, the second-order causes of incompatibility are determined, that is, the ones that have the most impact on the occurring incompatibility from all potential causes. It refers to the estimate of the weights of causes according to arithmetic average from assessments of potential causes. The presentation of the weights of causes as average values of the assessments of the team of experts resulted from the need to combine all the evaluations as a single value. It is difficult and not precise to compare the causes when they are marked by a large number of different assessments.

Later, in a combined way techniques are used such as brainstorming (BM), importance technique with Likert scale, and GRA method. The purpose is to determine the main causes, i.e., having the maximum weight, the most impact on incompatibility. The GRA method has application for a small number of data (that is, even 4 data), where it is a common phenomenon during the analysis of causes of incompatibility [38,39].

2.3. Model Assumptions and Conditions Ensuring Its Novelty

The assumptions of the model were made after making the concept of the model and determining the conditions of the selected techniques. Moreover, these assumptions have resulted from literature review, e.g.:

- a lack of limitations for the number of potential causes [2,19];
- the minimum number of potential causes determined for a single category (5M) in the Ishikawa diagram is equal to 4 causes [36,37,39];
- the minimum number of all second-order causes in the Ishikawa diagram should be equal to 4 causes [34,35,39];
- Second-order causes are causes having the greatest impact on incompatibility of all potential causes [12,38];
- the main causes are the causes that have the most impact on incompatibility from all second-order causes [20,23];
- the causes of potential incompatibility may or may not be of equal importance to each other, i.e., they may have the same or different influence (severity) on the occurrence of incompatibility [12,39];
- verification of the impact of causes on occurred the incompatibility is supported by the process of importance of causes on the Likert scale by selected teams of experts [12,38].

These assumptions were detailed in stages of model, which was characterized in the next part of the study. The novelty of the model is possibilities of its application for any product, e.g., the new production process (new products) or a significantly modified product, where this production is not stable, for example, starting a trial (initial) series. Additionally, the model has applications for any kind of incompatibilities identified as part of special processes. Therefore, the model can be used for any entity (e.g., production enterprise). Furthermore, the model can be integrated with any quality control after which incompatibility was identified [12]. The novelty of the model resulted from the character of the implemented quality management tools and decision methods, e.g., the possibilities to analyze even four incompatibilities [34,35]. Therefore, the model can be used as part of continually improving products, and for the sustainable development of industrial products.

2.4. Characteristics of Model

The purpose of the proposed model is to support the stabilization of the quality of industrial products. The model was developed in eight main stages, as shown in Figure 2.

Detailed characteristics of the model stages are presented in the next part of the study.

Stage 1. Determine the main incompatibility and purpose of the analysis

The incompatibility to analyze should be the main incompatibility, i.e., incompatibility which is the most often occurring in the enterprise and has the biggest cost or affected waste of resources. This incompatibility is determined according to the control sheet or using Pareto analysis [18]. Then, for the chosen incompatibility, the purpose of the analysis is determined. The SMART(-ER) method is used for that [30]. The purpose is determined by the entity, e.g., an expert (for example, a quality control manager or a company owner). The incompatibility should be characterized by considering, e.g., the type of incompatibility, the product in which this incompatibility was identified, and the number of incompatibilities. This information about incompatibility is often available in the catalogue (specification) of incompatibility.

Stage 2. Choice of the team of experts

The purpose of choosing a team of experts is to determine the people responsible for executing the model and achieving the purpose of the model. This incompatibility of the product (selected at stage 1) should be analyzed by the expert team to precisely determine the main cause of this incompatibility. Therefore, the expert team should have knowledge and experience in the analysis of incompatibility and the ability to solve the problem. The appropriate choice of the team of experts has an impact on achieving the objective, as shown in the study [9]. The team of experts should be chosen according to the method shown in the study, i.e., [12,31].

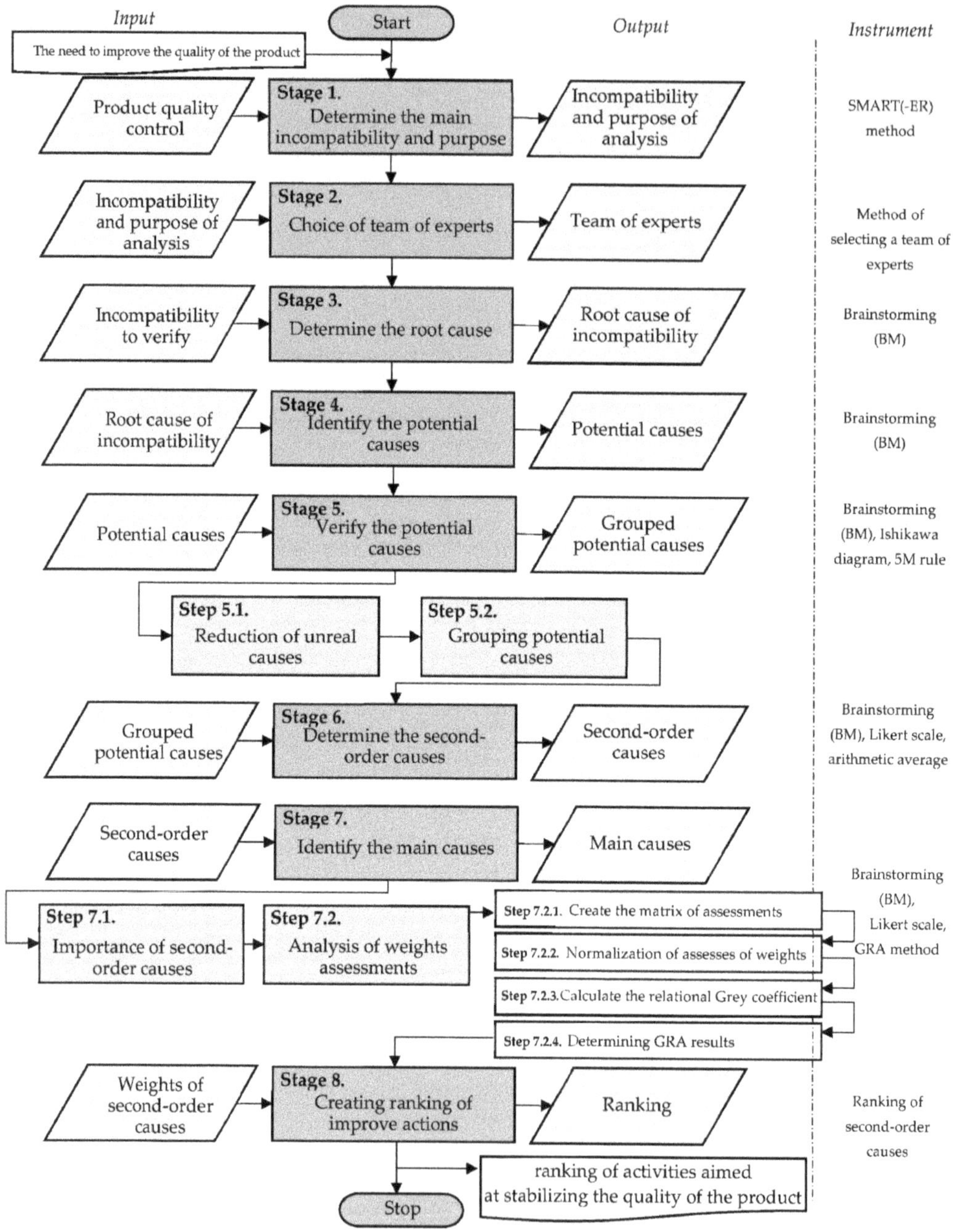

Figure 2. Model supporting stabilization quality of industrial products.

Stage 3. Determine the root cause

Determining the root cause refers to determining the incompatibility at the place of occurrence. To determine the root cause brainstorming (BM), it is used, which is shown among the team of experts. The Pareto rule (20/80) can be used for a large number of root causes, as shown in the study [18,20].

Stage 4. Identify the potential causes

Identification of potential causes (initial causes) includes determining causes that could have an impact on occurring incompatibility. At this stage, all potential causes are determined. Their impact (weight) on the incompatibility will be determined in the next stage of the model. To determine these causes, it is necessary to answer the question: "What has happened that this incompatibility occurred?". To identify potential causes, brainstorming is conducted among the team of experts. The brainstorming is carried out according to the method, which is shown in the study, e.g., [32,33]. It is necessary to generate (indication) all potential causes based on the root of the incompatibility (from stage 3). All potential causes are noted in a place visible to the team, e.g., a blackboard. The BM should end after about 30 min. The result of this stage is the highest number of potential causes of incompatibility.

Stage 5. Verify the potential causes

At this stage, verification of all potential causes is done. The purpose of this stage is to determine the potential causes, which could have an impact on the occurred incompatibility, and then show these causes in a standardized way. The result of this stage is the Ishikawa diagram for potential (initial) causes, which are grouped according to the 5M rule [18,19,21]. At this stage, all potential causes have the same weights, i.e., impact into incompatibility.

Step 5.1. Reduction of unreal causes

Firstly, it is necessary to delete potential causes which are unreal, i.e., that probably do not have an impact on incompatibility. For this purpose, the leader analyzes all potential causes (generated on stage 4) and, from among them, removes unreal causes.

Step 5.2. Grouping potential causes

Potential causes are grouped according to their categories. It was assumed to use the 5M rule, i.e., man, method, machine, materials, and management. However, it is possible to use any category which will be adequate for determining potential causes, for example, personnel measurement, environment. Brainstorming is used to group these causes, which is done among the selected team of experts. In turn, the Ishikawa diagram is used to group potential causes [18,19]. Next, for each category (5M), it is necessary to note appropriate potential causes. According to the concept of the model, in the Ishikawa diagram, it is necessary to note a minimum of four potential causes in each category of 5M [35,37].

Stage 6. Determine the second-order causes

At this stage, the weights (importance) of the potential causes were determined. It refers to determining the impact (importance) of these causes on incompatibility. Therefore, it is determined which potential causes could have the most likely to cause the incompatibility. Determining second-order causes relies on analyzing potential causes in each group of these causes (i.e., 5M) and determining their importance (impact) on the occurrence of the incompatibilities. For this purpose, it is necessary to use, in a combined way, brainstorming (BM), Likert scale, and arithmetic average from weight assessment. As a result, the ranking of potential causes is obtained, where the maximum weight in each group of 5M is the second-order cause. The team of experts assesses potential causes, i.e., determines importance (weight) of the impact of potential causes of incompatibility. In this aim, the brainstorming is carried out during which a team of experts assesses causes on the Likert scale [5,7], where 1—the cause has little influence on incompatibility (low importance), 5—the cause significantly influences the occurrence of incompatibility (high importance).

All potential causes shown on the Ishikawa diagram (from the 5 stage) should be assessed. The assessments are noted directly on the Ishikawa diagram by potential causes. The assessments are noted directly on the Ishikawa diagram by potential causes. After assessing all potential causes, it is necessary to estimate the weights of these causes. The arithmetic averages of all assessments of the weights are calculated (1):

$$w_i = \frac{\sum_{i=1}^{n} w_i}{n} \tag{1}$$

where: w—weight i-th potential cause, n—number of evaluates for i-th potential cause.

Second-order causes are chosen on the average weights of potential causes. It is necessary to choose a single cause in each group of causes (i.e., from each of 5M groups). It is a cause that has a maximum average weight value. It is useful to note these causes in the Ishikawa diagram. The second-order causes are verified in the next stage of the model.

Stage 7. Identify the main causes

At this stage, second-order causes are verified. The purpose is to identify the main causes, i.e., causes that have the most impact on the occurrence of incompatibility. It refers to reverifying second-order causes. Hence, these causes are verified simultaneously (without including their groups). In this purpose, it was assumed to be used in a combined way: brainstorming (BM) [40,41], Likert scale [5,7], and GRA method [35,38]. As a result, the second-order causes are achieved, where the maximum weight is the main cause. If some second-order causes will have the same weight, these causes are considered equally important. It is shown in three steps.

Step 7.1. Importance of second-order causes

The team of experts assess second-order causes, i.e., determines importance (weight) impact of causes on the occurrence of incompatibility. In this aim, the brainstorming is carried out during which a team of experts assess causes on the Likert scale [5,7], where 1—the cause has little influence on incompatibility (low importance), 5—the cause significantly influences the occurrence of incompatibility (high importance).

Step 7.2. Analysis of weight assessments of second-order causes

Based on weight assessments of second-order causes, it is necessary to verify their importance. The GRA method is used for this [37,38,42,43]. The choice of the GRA method resulted from its application to supporting decisions in the fuzzy (uncertain) area [34,35], where it is adequate to verify the importance of causes of incompatibility causes (determined subjectively by team of experts [36,37]. Furthermore, the GRA method has application for a small number of data (i.e., even 4 data), where it is a common phenomenon during analysis of causes of incompatibility [38,39], except that it conditioned versatility of the proposed model, where it is possible to verify both a large and a small number of causes. Ultimately, it is possible to determine adequate improvement actions. The method is shown in four steps. The result of this stage is weights of second-order causes.

Step 7.2.1. Create the matrix of assessments

First, the matrix is created, i.e., $M = m \times n$, where m—alternative (that is, group of causes), n—criterion (that is, cause). This matrix should be filled with weights of second-order causes [34,35].

Step 7.2.2. Normalization of assessment of weight of incompatibility causes

Then, it is necessary to process (normalize) the weights of causes to achieve assessments in the range from 0 to 1. According to the concept of a model, it was assumed that "the higher the rating, the greater the impact of the cause on the incompatibility". Therefore, according to the GRA method, Formula (2) is used [34,37]:

$$x_i^*(k) = \frac{x_i^{(O)}(k) - min\, x_i^{(O)}(k)}{max\, x_i^{(O)}(k) - min\, x_i^{(O)}(k)} \tag{2}$$

It was assumed that $x_0^{(O)}(k)$ and $x_i^{(O)}(k)$ are appropriately original and comparable sequence; $i = 1, 2, \ldots, m$; $k = 1, 2, \ldots, n$; and m—alternative (i.e., group of causes), n—criterion (i.e., cause) [3,43].

Step 7.2.3. Calculate the relational Grey coefficient

On the basis of normalized sequences, the relational Grey coefficient is calculated. Formula (3) is used for that [36,37]:

$$\gamma[x_0^*(k),\ x_i^*(k)] = \frac{\Delta_{min} + \xi\Delta_{max}}{\Delta_{0i}(k) + \xi\Delta_{max}}, \text{and } 0 < \gamma[x_0^*(k),\ x_i^*(k)] \leq 1 \tag{3}$$

where: $\Delta_{0i}(k)$ represents a sequence of variations between the original sequence $x_0^*(k)$ and comparison sequence $x_i^*(k)$, which is calculated from Formula (4) [39]:

$$\Delta_{0i}(k) = |x_0^*(k) - x_i^*(k)| \tag{4}$$

Similarly, the largest (5) and the smallest (6) deviations are calculated [34–36]:

$$\Delta_{max} = \max_{\forall j \in i} \max_{\forall k} \left| x_0^*(k) - x_j^*(k) \right| \tag{5}$$

$$\Delta_{min} = \min_{\forall j \in i} \min_{\forall k} \left| x_0^*(k) - x_j^*(k) \right| \tag{6}$$

In turn, the factor ξ from Formula (3) has values [0,1]. Most often, it is assumed that $\xi = 0.5$ [35,38].

Step 7.2.4. Determining Grey Relational Assessments

The relationship assessment of Grey is the weighted sum of Grey's coefficients, as shown in Formula (7) [36]:

$$\gamma(x_0^*,\ x_i^*) = \sum_{k=i}^{n} \beta_k \gamma[x_0^*(k),\ x_i^*(k)] \tag{7}$$

where: $\gamma(x_0^*,\ x_i^*)$, *tj*. a grey relational score that shows the level of correlation between the original sequence and the comparable sequence, as if they were identical.

A correctly defined grey relational score should be 1 (8) [34,40]:

$$\sum_{k=1}^{n} \beta k = 1 \tag{8}$$

As a result, on the basis of the GRA results, it is possible to determine the degree of influence of the second-order causes on the occurrence of incompatibility. This is shown in the next step of the model.

Step 7.3. Choice of main cause

Based on calculated values, it is possible to choose the main cause which has the largest impact on incompatibility. It is the cause that has the maximum value according to the GRA method [36]. It is useful to mark this cause in the Ishikawa diagram.

Stage 8. Creating a ranking of actions as part of stabilizing the quality of the product

This stage relies on ordering the weight values of second-order causes (estimated in Step 7.2). The cause with the maximum GRA value is the main cause (the first position in the ranking). The lower the value of GRA, the lower the cause is. According to the rules for continuous improvement of products [2–4], in the first position, it is necessary to propose improvement actions for the main cause. Following the authors of [18], eliminating the root cause in the first place can ensure that the incompatibility is reduced or eliminated with a certain probability. After improvement actions, it is necessary to verify their efficiency [40,41]. Then, it is possible to take actions for other incompatibilities of ranking. For that, it is necessary to use the sequence developed in the proposed model. Actions implemented in this way (in a repeatable and sequential manner) will help stabilize the quality of the product as part of the continuous improvement process.

3. Test of Model

The model was tested for cracks on the outer hull of the 418 alloy four-point bearing (CPW-S 5616), which were identified relatively often in the Polish company. The Polish industry was chosen because the foundry industry is the most developed in Poland. Additionally, in the foundry industry, special processes are the most needed, as shown in the studies [18–21,24,26,29]. Cracks occur as a result of the loss of ductility of the material (i.e., exceed the tensile strength). In enterprises, the methods supporting accurate determination are the main causes of incompatibility. So far, the bearing verification of the incompatibility has been based mainly on the experience and decisions of the quality manager. The choice of the outer hull of the 418 alloy four-point bearing was based on the individual needs of the enterprise. The incompatibilities of this product were verified relatively often, and the main causes of its occurrence were not precisely identified. In addition, in the general approach, the four-point bearing is popular and often used, for example, in the engineering industry. Therefore, the results obtained for this product can be useful in different applications. For this reason, it was considered justified to propose a model that supports the stabilization of the quality of this product.

Stage 1. Determine the main incompatibility and purpose of the analysis

The main incompatibility was a crack in the outer hull of 418 alloy four-point bearing (CPW-S 5616). This incompatibility was identified by non-destructive tests (magnetic powder method), as shown in this study [2]. The crack was a flat incompatibility, where stresses occur at the ends of it. As a result, a notch will form, causing the crack to develop further [42]. The example of the crack in the outer hull is shown in Figure 3.

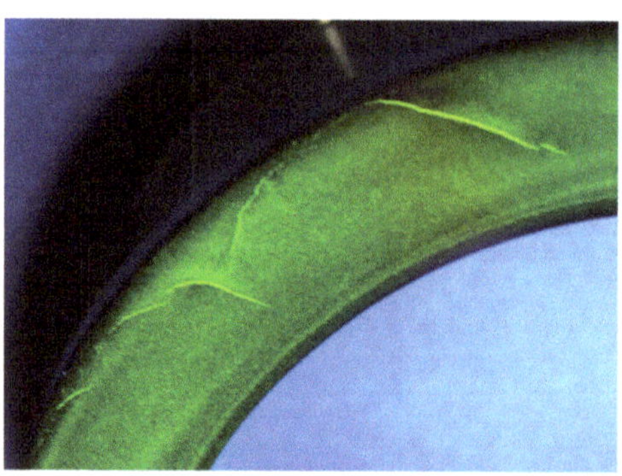

Figure 3. Crack on the outer hull of four-point bearing.

The outer hull of the four-point bearing was specially designed for oil-free screw compressors. They have a high-strength outer cage ring and can be operated accurately. These products can reach high speeds under high operating temperatures and high loads.

In addition, they provide reduced heat, vibration, and noise generation. The outer four-point bearing hull is used, among others, mounted on the trunnion of the first rotor disc of the compressor [43,44].

The outer hull of bearing was made as forging from 418 (CPW-S 5616) alloy. It is an alloy of modification with higher strength from the family of 12% chrome martensitic stainless steels. It is a precipitation hardening stainless steel with shear strength. Its chemical composition is as follows: Fe (81%), Cr (13%), W (3%), Ni (2%), and C (1%). The mechanical and physical strengths are presented in Table 1.

Table 1. Mechanical and physical strengths of 418 alloy. Own study based on [45].

Mechanical and Physical Strengths	Value (21 °C)
Ultimate Tensile Strength (MPa)	965
0.2% yield point (MPa)	760
Elongation (%)	15
Brinell hardness	302–352

Based on the selected incompatibility, it was possible to determine the purpose of the analysis. The purpose was to determine the rank of actions to stabilize quality of the outer hull of bearing from 418 alloy. This concerned the sequential reduction and the importance of the causes of fracture in the bearing housing until the main causes of this problem were identified and the sequence of appropriate improvement actions could be determined.

Stage 2. Choice of the team of experts

According to the expert second stage, the team was selected. The team included a quality control manager, NDT manager, and authors of the article. The team of experts had knowledge and experience in this type of incompatibility and methods used in the proposed model.

Stage 3. Determine the root cause

At this stage, the root cause of the crack on the outer hull of the four-point bearing was determined. For this purpose, brainstorming was conducted among the experts' team. It was assumed that root cause of the crack is a state of stress (deformation). As a result, the tensile strength of the material is locally exceeded. This is called loss of ductility and is followed by the formation of a notch that generates this incompatibility. This defect may cause the product to crack during operation.

Stage 4. Identify the potential causes

Then, the potential causes (initial) of the crack were identified. For this purpose, brainstorming was conducted among the experts' team. Potential causes such as:

1. Too fast welding speed;
2. Flow of liquid weld pool too fast;
3. Lack of clean welded layer;
4. Small width in relation to depth (joint proportion);
5. Stresses due to high thermal expansion;
6. Inappropriate selection of material;
7. Unprepared metal surface;
8. High carbon content in the weld;
9. Dirt inside the weld;
10. Employee rush;
11. No periodic training;
12. No TMP (Total Productive Maintenance);
13. Lack of up-to-date procedures;
14. Distraction;
15. Dirty tools;

16. Lack of unit controls;
17. Broken tools;
18. Failure to use the manual;
19. Moisture of the electrode;
20. Short work experience of the employee;
21. Inadequate lighting;
22. Noise;
23. Contamination of the site;
24. Uncalibrated tools;
25. Psychophysical condition of the worker (e.g., severe nervous tension, exhaustion, physical or mental malaise).

Twenty-five potential causes were identified and subjected to further verification in subsequent stages of the model.

Stage 5. Verify the potential causes

At this stage, all potential (initial) causes were verified. The purpose was to determine potential causes that could have an impact on incompatibility.

Step 5.1. Reduction of unreal causes

First, the impossible cause was deleted from all 25 potential causes, i.e., dirty tools. The result was the general overview of the tools used, which were kept in an orderly manner. Other causes were determined as possible causes of the outer hull of the bearing, which was made as forgings from alloy 418 (CPW-S 5616).

Step 5.2. Grouping potential causes

Then, the potential causes were grouped according to Rule 5M, i.e., man, method, machine, material, and management. Causes were grouped during brainstorming (BM) among a team of experts. The Ishikawa diagram was created, as shown in Figure 4.

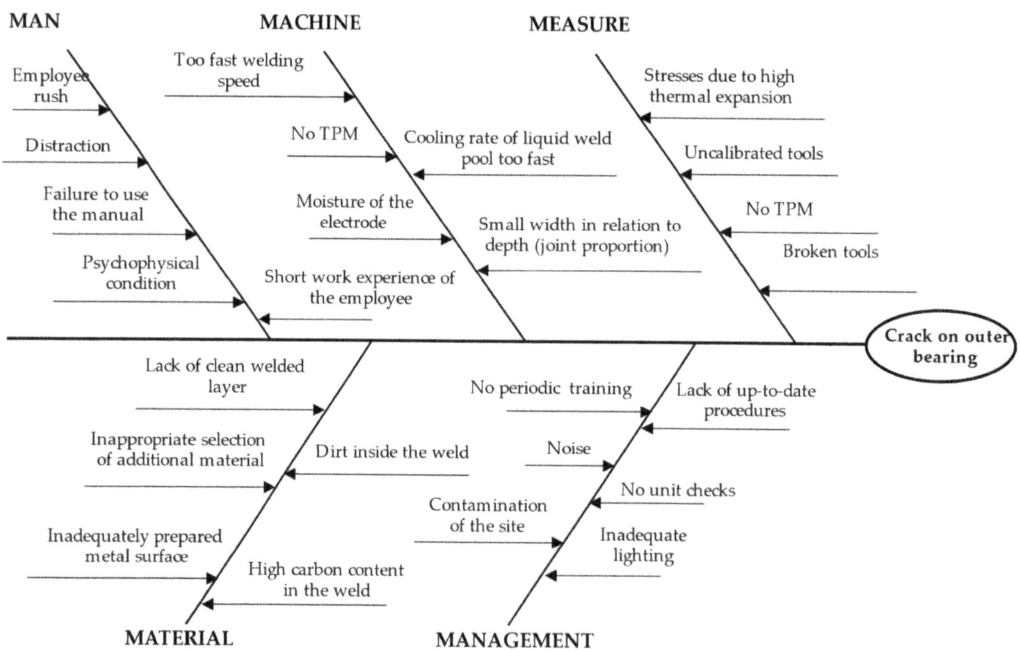

Figure 4. Ishikawa diagram for a crack in the bearing housing.

The developed Ishikawa diagram was modified after the next stages of the model were implemented.

Stage 6. Determine the second-order causes

At this stage, the team of experts determined the impact (importance) of potential causes. For this purpose, all causes from the Ishikawa diagram were analyzed. The impact (weight) of potential causes was determined in each 5M group. The brainstorming scale, Likert scale, and arithmetic average were used in a combined way.

The team of experts assessed the potential causes, i.e., the importance of potential causes of cracks on the outer hull of bearing. During brainstorming (BM), causes were assessed on the Likert scale. These assessments are shown in Figure 5.

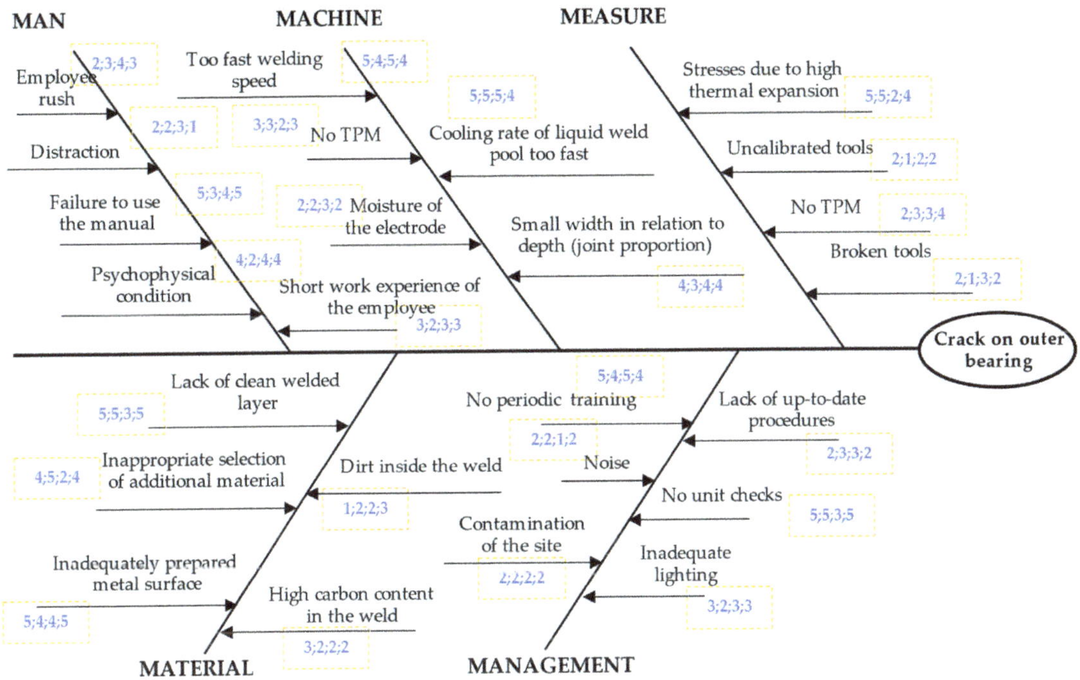

Figure 5. Ishikawa diagram for crack on the outer hull of four-point bearing to assess the importance of potential causes.

Then, it was possible to verify the causes and their assessments. The analysis included assessments of the weights of causes in the 5M group. The average of the assessments was calculated as the average weight of potential causes. The result is shown in Table 2.

Table 2. Average weights of potential causes.

Category 5M	No.	Potential Causes	Assessment of Importance	Average Weight
man	1	employee rush	2;3;4;3	3.00
	2	dissociation	2;2;3;1	2.00
	3	not following the manual	5;3;4;5	4.25
	4	psychophysical state	4;2;4;4	3.50
	5	short work experience of the employee	3;2;3;3	2.75
machine	6	too high cutting speed	5;4;5;4	4.50
	7	no TPM	3;3;2;3	2.75
	8	electrode humidity	2;2;3;2	2.25
	9	too high cooling rate of the weld pool liquid	5;5;5;4	4.75
	10	small width in relation to the depth	4;3;4;4	3.75
measure	11	stresses	5;5;2;4	4.00
	12	uncalibrated tool	2;1;2;2	1.75
	13	no TPM	2;3;3;4	3.00
	14	damaged tools	2;1;3;2	2.00
material	15	no cleaned top layer	5;5;3;5	4.50
	16	inappropriate selection of additional material	4;5;2;4	3.75
	17	inadequately prepared metal surface	5;4;4;5	4.50
	18	debris inside the weld	1;2;2;3	2.00
	19	high carbon content in the weld	3;2;2;2	2.25
management	20	no periodic training	5;4;5;4	4.50
	21	noise	2;2;1;2	1.75
	22	environment pollution	2;2;2;2	2.00
	23	lack of up-to-date procedures	2;3;3;2	2.50
	24	no unit checks	5;5;3;5	4.50
	25	inadequate lighting	3;2;3;3	2.75

It is not certain that we have identified all causes, therefore, the sum of average values should not exceed 100%. In this case, the value was equal to about 80% and is acceptable.

Second-order causes were selected based on the basis of average weights of potential causes. Second-order causes were selected in each group of causes (i.e., from each group of 5M). These were causes with the highest value of the arithmetic average. Second-order causes were not following the manual (C3), too high cooling rate of the liquid weld pool (C9), stresses (C11), no cleaned surface layer (C15), inadequately prepared metal surface (17), no periodic training (20), and no unit check (C24). The average values resulting from the assessments awarded by a team of experts (during the brainstorming method before step) were marked in the Ishikawa diagram (Figure 6).

The main causes from second-order causes were selected as the main causes. It is shown in the next stage of the model.

Stage 7. Identify the main causes

Initially, the team of experts assessed second-order causes, i.e., determined the importance (weights) of the impact the causes on cracks on the outer hull of four-point bearings. Brainstorming was used for that. The importance of second-order causes was assessed using the Likert scale. The result is shown in Table 3.

For the assessment of weights of second-order causes, it was achieved according to the GRA method. This included phases 1 to 4 of the model. First, the average assessments of Table 4 were normalized. Formula (2) was used for this. Then, with normalized sequences, the relational Grey coefficient was calculated. Formula (3) is used for those and adequate Formulas (4)–(6). The Grey Relational Assessment (GRA) is the weighted sum of the Grey coefficients, as shown in Formula (7). Based on GRA values, the ranking was created. The maximum value is determined as the main cause. The results are shown in Table 4.

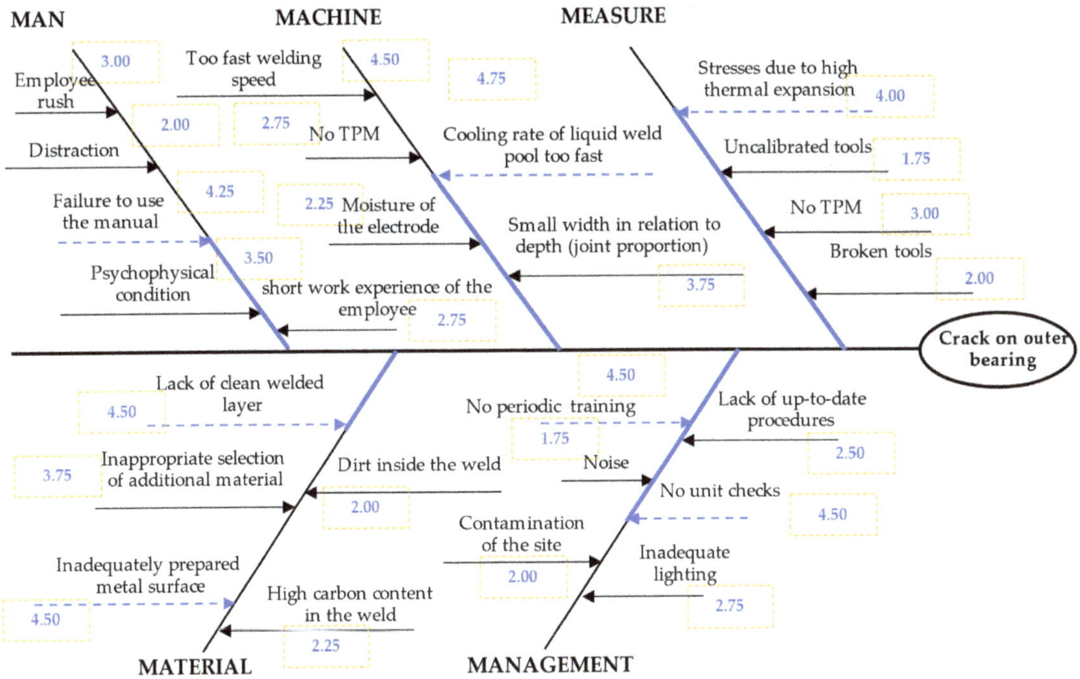

Figure 6. Ishikawa diagram for crack on the outer hull of four-point bearing to choose the second-order causes.

Table 3. Assessment of weights of second-order causes.

Category 5M	No.	Potential Causes	Average Weight
Man	C3	not following the manual	3;5;3;2
Machine	C9	too high cooling rate of the liquid weld pool	5;4;5;4
Measure	C11	stresses	5;4;3;5
Material	C15	no cleaned top layer	4;5;3;3
Material	C17	inadequately prepared metal surface	2;4;4;3
Management	C20	no periodic training	1;3;3;2
Management	C24	no unit checks	2;3;5;2

Table 4. Results from GRA to choose the main causes.

5M	No.	Normalization				Grey Relational Coefficient				GRA $\gamma(x_0^*, x_i^*)$	Ranking	Results
Man	C3	0.50	1.00	0.50	0.25	0.50	1.00	0.50	0.40	0.60	4	
Machine	C9	1.00	0.75	1.00	0.75	1.00	0.67	1.00	0.67	0.83	1	main cause
Measure	C11	1.00	0.75	0.50	1.00	1.00	0.67	0.50	1.00	0.79	2	
Material	C15	0.75	1.00	0.50	0.50	0.67	1.00	0.50	0.50	0.67	3	
Material	C17	0.25	0.75	0.75	0.50	0.40	0.67	0.67	0.50	0.56	6	
Management	C20	0.00	0.50	0.50	0.25	0.33	0.50	0.50	0.40	0.43	7	
Management	C24	0.25	0.50	1.00	0.25	0.40	0.50	1.00	0.40	0.58	5	

where: C3—not following the manual, C9—too high cooling rate of the liquid weld pool, C11—stresses, C15—no cleaned top layer, C17—inadequately prepared metal surface, C20—no periodic training, C24—no unit checks.

The main cause of the crack in the outer hull of four-point bearing was the too high cooling rate of the liquid weld pool. This cause had the highest value of the GRA method. This cause was marked in the Ishikawa diagram (Figure 7).

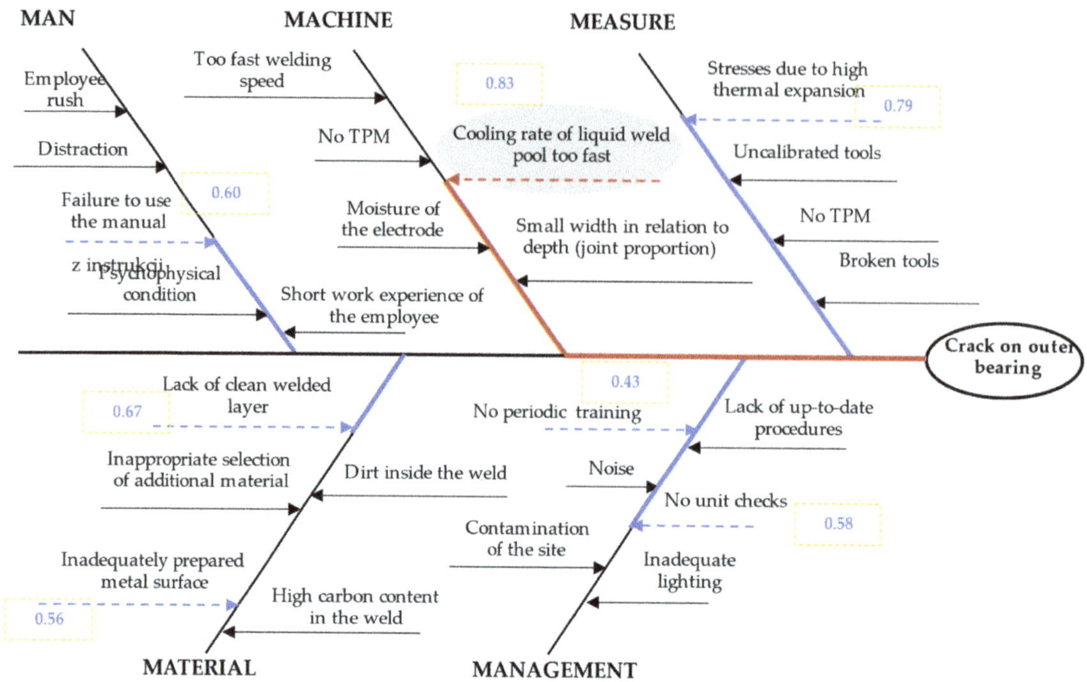

Figure 7. Ishikawa diagram for crack on the outer hull of four-point bearing to choose the main cause.

According to the main cause, the ranking of improving actions was created to stabilize the quality of the outer hull of four-point bearing.

Stage 8. Creating a ranking of actions as part of stabilizing the quality of the product

At this stage, the second-order causes were sorted. These causes were sorted according to the GRA values. The ranking of second-order causes is presented in Table 5.

Table 5. Ranking of activities to stabilize the quality of the bearing housing.

5M	No.	Second-Order Causes	GRA $\gamma(x_0^*, x_i^*)$	Ranking
Man	C3	not following the manual	0.60	4
Machine	C9	too high cooling rate of the liquid weld pool	0.83	1
Measure	C11	stresses	0.79	2
Material	C15	no cleaned top layer	0.67	3
Material	C17	inadequately prepared metal surface	0.56	6
Management	C20	no periodic training	0.43	7
Management	C24	no unit checks	0.58	5

As shown in the previous stage, the main cause was the cause with the maximum GRA value, that is, the cooling rate of the liquid weld pool was too high (0.83). According to the proposed concept, improvement actions should be taken for this reason in the first place. According to [42], the post-weld inspection includes the provision of appropriate geometrical dimensions and the development of surface and volume tests. After implementing improvement actions for these causes, one should propose actions to reduce stresses, then clean the surface layer, then the need to follow the instructions and introduce more frequent checks.

4. Discussion

Stabilization of the quality of products remains a challenge [6,12,23]. It results from the need to identify the incompatibility of the product, and then its effective verification to determine the main cause of its occurrence [5,7,14,21]. Later, it is possible to identify adequate improvement actions [46–49]. In addition, for the lack of stable quality of product, the problem consists of a large number of potential causes [50]. On the other hand, determining the most important (main) cause is difficult.

The GRA results were not verified with other MCDM (Multi-Criteria Decision-Making Method) methods. However, we tested other approaches for the possibility of including MCDM in the process of identifying causes of incompatibility [24–29]. The test of the combination method MCDM (or FMCDM—Fuzzy Multi-Criteria Decision-Making Method) with the quality management tools was effective in identifying the precise causes of the incompatibilities of product. However, the research so far has focused mainly on analyzing a large number of causes of incompatibility. Therefore, there have been no studies that would combine decision support methods and quality management tools to analyze even a small number of causes of noncompliance. Therefore, this is the main originality of the study, which uses the GRA method. This proposed model has been applied for both a large and small number of causes of incompatibility (even 4 causes). It is important to mention that, applying different MCDM techniques, the ranks are different for the same problem. Therefore, future research will focus on the comparison of existing methods developed [24–29] with the new proposed approach.

The problem of simultaneously determining the main cause of the problem was identified in the Polish company. The model was carried out by searching and ranking incompatibilities to identify the incompatibility of cracks on the outer hull of the four-point bearing of alloy 418 (CPW-S 5616). This made it difficult to start effective improvement actions for the incompatibility that generates the largest source of waste. During brainstorming, a team of experts determined the root cause, i.e., state of stress (deformation). Next, the team of experts determined 25 potential causes, which were grouped and visualized in the Ishikawa diagram with the 5M rule. Causes were evaluated on a Likert scale, where the team of experts assessed the impact of causes based on incompatibility. The arithmetic average was calculated from experts' assessments to determine the weights of potential causes. On the basis of the results, second-order cause was determined. These causes were analyzed using the GRA method. As a result, the ranking of the importance of second-order causes was prepared. The main cause was the cooling rate being too high of the liquid weld pool (0.83). According to the proposed concept, improvement actions should be taken for this reason in the first place. Then, the effectiveness of the actions taken should be verified, and further improvement actions should be taken for the next reason from the ranking.

Therefore, the objective of the study was to develop a new model that supports the stabilization quality of industrial products. After the model, it was shown that it is possible to support the stability quality of industrial products by determining the ranking of causes that have an impact on the incompatibility of the product, where it is realized by determining all potential causes and then by their sequential and constructive reduction to reduce the main causes. The initial test of the model shows that the model can be effective in improving products in the casting industry. On the basis of the proposed model for the problem analyzed, it was possible to show the main cause, which was the high cooling rate of the liquid weld pool. According to the model, for this reason, improvement actions should be taken in the first place. The research conducted so far has not allowed us to indicate this cause as the main cause of the problem. The model allowed us to determine the ranking of other improvement actions. The initial test of the proposed model shows the possibilities of its application to improve the quality of the product in casting processes. It turned out that visualization of incompatibility causes supported by additional decision tools could be the right way to improve the quality of products that occur in special processes. The developed model allows us to show that the ranking of identified causes has essential practical meaning because, according to rule 20/80, reducing the most important causes results in a significant improvement of product quality. This also translates into savings related to, for example, materials. To confirm the results obtained in the preliminary test, it will be necessary to carry out a more extensive investigation for other products and other processes.

The limitation of the proposed model is the need to verify the problem based on the knowledge and experience of the team of experts. This applies to the need for a thoughtful verification of the incompatibility by a properly selected team of experts [9]. Furthermore, the ranking of causes of the problem resulted from individual analysis of problem. Therefore, it may be different in other cases. Additionally, it should be mentioned that the proposed method ignores the statistical dispersion of the quality parameters. The method is complex, although the instruments that can be used in this process support a precise way of right identifying causes of incompatibility. Therefore, it is possible to reduce errors. In future research, they plan to make a computer implementation of this model, and this interface could be relatively simple to use. This study has the objective of showing that it is possible to develop a model to support the process of identifying causes of incompatibility of products.

The verification of the model shows its practical potential for use. In the future, a comparable verification of the model is planned to be performed on various products (including services). Carrying out a large number of verifications (tests) is likely to highlight its advantages and possibly show its limitations. Therefore, future research will focus on developing a computer program for this model. In addition, a dynamic decision-making platform is planned to be developed to make decisions on various types of incompatibilities. As part of future research, it is planned to do more extensive research on the model because the developed model needs tests to show its accuracy in predicting/determining the errors and incompatibilities. However, the results of the model will be influenced by many factors, e.g., the selection of a team of experts, which may make it difficult to compare the results.

5. Conclusions

Improving the quality of industrial products requires thoughtful and standardized actions. Therefore, the aim of the article was to develop a new model that supports the stabilization quality of industrial products. It refers to the sequential and coherent way of determining the causes of problems with the quality of products, where the number of verified causes can be equal to even 4. Moreover, it is possible to reduce the causes of the problem from potential causes to the most important causes, which possibly have the most degree of influence on the occurrence of the problem. The model was developed by integrating and used in a sequential way by the selected techniques. Those techniques were quality management tools and decision methods in the fuzzy area, that is, SMART(-ER) method, method of selecting a team of experts, brainstorming (BM), Ishikawa diagram with the 5M rule, technique of importance in the Likert scale, arithmetic average, and Grey Relational Analysis (GRA).

The model was carried out by searching and ranking incompatibilities to often identify incompatibility of cracks on the outer hull of the four-point bearing of alloy 418 alloy (CPW-S 5616). Incompatibility was identified by non-destructive testing (magnetic powder method) in the Polish industry. After testing the model, it was concluded that this model can help support the process of stability, quality of materials, the processes of its formation, and industrial product processes by determining the classification of causes that have an impact on the incompatibility of the product; it is realized by determining all potential causes and then their sequential and constructive reduction to reduce the main causes. It was concluded that the application of the model may help enterprises stabilize the quality of products. Additionally, the model can be used for any type of product, for any incompatibilities, and it can be combined with any quality control. The proposed model can be widely used in many industries. It seems particularly advantageous to use it for special processes. Other areas of application may be industries where a particularly high quality of products is required. It is connected with the necessity of continuous improvement based on identifying and then reducing the causes, and not the effects of noncompliance. Examples of such industries may be the automotive, aviation, etc., industries. The use of the model in relation to various industries requires the execution of appropriate tests.

Author Contributions: Conceptualization, A.P. and D.S.; methodology, A.P. and D.S.; formal analysis, D.S.; writing—original draft preparation, D.S.; writing—review and editing, A.P.; visualization, A.P. and D.S.; supervision, A.P.; project administration, A.P. and D.S. All authors have read and agreed to the published version of the manuscript.

Funding: This research received no external funding.

Institutional Review Board Statement: Not applicable.

Informed Consent Statement: Not applicable.

Data Availability Statement: Not applicable.

Conflicts of Interest: The authors declare no conflict of interest.

References

1. Lehner, P.; Hrabová, K. Relationship of Time-Dependent Parameters from Destructive and Non-Destructive Tests of Structural Concrete. *Mathematics* **2022**, *10*, 460. [CrossRef]
2. Chokkalingram, B.; Raja, V.; Anburaj, J.; Immanual, R.; Dhineshkumar, M. Investigation of Shrinkage Defect in Castings by Quantitative Ishikawa Diagram. *Arch. Foundry Eng.* **2017**, *17*, 174–178. [CrossRef]
3. Pacana, A.; Siwiec, D.; Bednarova, L. Analysis of the incompatibility of the product with fluorescent method. *Metalurgija* **2019**, *58*, 337–340.
4. Midor, K. An analysis of the causes of product defects using quality management tools. *Manag. Syst. Prod. Eng.* **2014**, *16*, 162–167. [CrossRef]
5. Siwiec, D.; Pacana, A. A Pro-Environmental Method of Sample Size Determination to Predict the Quality Level of Products Considering Current Customers' Expectations. *Sustainability* **2021**, *13*, 5542. [CrossRef]
6. Markulik, Š.; Petrík, J.; Šolc, M.; Blaško, P.; Palfy, P.; Sütőová, A.; Girmanová, L. Analysis of Fault Conditions in the Production of Prestressed Concrete Sleepers. *Appl. Sci.* **2022**, *12*, 928. [CrossRef]
7. Pacana, A.; Bednarova, L.; Pacana, J.; Liberko, I.; Wozny, A.; Malindzak, D. Effect of selected factors of the production process of stretch film for its resistance to puncture. *Przem. Chem.* **2014**, *12*, 2263–2264.
8. Midor, K.; Zarnovsky, J. Innovative use of quality management methods for product improvement. *Manag. Syst. Prod. Eng.* **2016**, *24*, 264–267. [CrossRef]
9. Goel, P.; Kumar, R.; Banga, H.K.; Kaur, S.; Kumar, R.; Pimenov, D.Y.; Giasin, K. Deployment of Interpretive Structural Modeling in Barriers to Industry 4.0: A Case of Small and Medium Enterprises. *J. Risk Financ. Manag.* **2022**, *15*, 171. [CrossRef]
10. Hofele, J.; Link, G.; Jelonnek, J. Reaction Kinetics and Process Model of the Polyacrylonitrile Fibers Stabilization Process Based on Dielectric Measurements. *Materials* **2022**, *15*, 1222. [CrossRef]
11. Jonšta, P.; Jonšta, Z.; Brožová, S.; Ingaldi, M.; Pietraszek, J.; Klimecka-Tatar, D. The Effect of Rare Earth Metals Alloying on the Internal Quality of Industrially Produced Heavy Steel Forgings. *Materials* **2021**, *14*, 5160. [CrossRef] [PubMed]
12. Pacana, A.; Siwiec, D. Universal Model to Support the Quality Improvement of Industrial Products. *Materials* **2021**, *14*, 7872. [CrossRef] [PubMed]
13. Zimon, G.; Zimon, D. Quality Management Systems and Working Capital SMEs in GPO—A Case of Poland. *Adm. Sci.* **2020**, *10*, 76. [CrossRef]
14. Pacana, A.; Siwiec, D. Model to Predict Quality of Photovoltaic Panels Considering Customers' Expectations. *Energies* **2022**, *15*, 1101. [CrossRef]
15. Pinho, T.; Zhiltsova, T.; Oliveira, M.; Costa, A. Computer-Aided Reengineering towards Plastic Part Failure Minimization. *Materials* **2021**, *14*, 6303. [CrossRef]
16. Saroso, D.S. Decision Making Models for Quality Improvement. *Int. J. Technol.* **2012**, *3*, 169–181. [CrossRef]
17. Shinde, D.D.; Ahirrao, S.; Prasad, R. Fishbone Diagram: Application to Identify the Root Causes of Student–Staff Problems in Technical Education. *Wirel. Pers. Commun.* **2018**, *100*, 653–664. [CrossRef]
18. Rusiecki, A. The analysis of quality management tools used for improving the production process based on a chosen foundry. *Qual. Prod. Improv.* **2019**, *2*, 113–124. [CrossRef]
19. Szczucka-Lasota, B.; Wolniak, R.; Fryc, M. The identification of incompatibility in the process of laser cutting of the semi-finding. *Sil. Univ. Technol. Sci. Pap. Organ. Manag.* **2017**, *105*, 429–438.
20. Coccia, M. The Fishbone diagram to identify, systematize and analyze the sources of general purpose technologies. *J. Soc. Adm. Sci.* **2017**, *4*, 291–303. [CrossRef]
21. Luca, L. A new model of Ishikawa diagram for quality assessment. *IOP Conf. Ser. Mater. Sci. Eng.* **2016**, *161*, 012099.

22. Raman, R.; Basavaraj, Y. Quality Improvement of Capacitor through Fishbone and Pareto techniques. *Int. J. Recent Technol. Eng.* **2019**, *8*, 2248–2252. [CrossRef]
23. Liu, T.T.; Duan, G.J. A quality improvement model for the development of complex products based on data mining and mechanism analysis. *J. Chin. Inst. Eng.* **2020**, *43*, 763–774. [CrossRef]
24. Soares, J.C.; Tereso, A.P.; Sousa, S.D. A decision-making model for the rework of defective products. *Int. J. Qual. Reliab. Manag.* **2021**, *38*, 68–97. [CrossRef]
25. Siwiec, D.; Pacana, A. Method of improve the level of product quality. *Prod. Eng. Arch.* **2021**, *27*, 1–7. [CrossRef]
26. Siwiec, D.; Pacana, A. Model Supporting Development Decisions by Considering Qualitative–Environmental Aspects. *Sustainability* **2021**, *13*, 9067. [CrossRef]
27. Siwiec, D.; Pacana, A. Identifying the source of the problem by using implemented the FAHP method in the selected quality management techniques. *Prod. Eng. Arch.* **2020**, *26*, 5–10. [CrossRef]
28. Pavletic, D.; Sokovic, M.; Paliska, G. Practical Application of Quality Tools. *Int. J. Qual. Res.* **2008**, *2*, 197–205.
29. Pacana, A.; Czerwinska, K.; Bednarova, L. Comprehensive improvement of the surface quality of the diesel engine piston. *Metalurgija* **2019**, *58*, 329–332.
30. Lawlor, K.B.; Hornyak, M.J. Smart Goals: How The Application Of Smart Goals Can Contribute To Achievement Of Student Learning Outcomes. *Dev. Bus. Simul. Exp. Learn.* **2012**, *39*, 259–267.
31. Kupraszewicz, W.; Żółtowski, B. The selection of the experts team to diagnose the condition of the machines. *Diagn. Artyk. Główne* **2002**, *26*, 94–100.
32. Kohn, N.; Smith, S. Collaborative Fixation: Effexts of Others' Ideas on Brainstorming. *Appl. Cogn. Psychol.* **2011**, *25*, 359–371. [CrossRef]
33. Putman, V.; Paulus, P. Brainstorming, Brainstorming. Rules and Decision Making. *J. Creat. Behav.* **2009**, *43*, 29–40. [CrossRef]
34. Ertuğrul, İ.; Öztaş, T.; Özçil, A.; Öztaş, G.Z. Grey Relational Analysis Approach In Academic Performance Comparison Of University: A Case Study Of Turkish Universities. *Eur. Sci. J.* **2016**, *12*, 128–139.
35. Javed, S.A.; Khan, A.M.; Dong, W.; Raza, A.; Liu, S. Systems Evaluation through New Grey Relational Analysis Approach: An Application on Thermal Conductivity—Petrophysical Parameters' Relationships. *Processes* **2019**, *7*, 348. [CrossRef]
36. Liu, A.; Zhu, Q.; Ji, X.; Lu, H.; Tsai, S.B. Novel method for perceiving key requirements of customer collaboration low-carbon product design. *Int. J. Environ. Res. Public Health* **2018**, *15*, 1446. [CrossRef]
37. Wang, F.; Li, H.; Dong, M. Capturing the key customer requirements for complex equipment design using Grey Relational Analysis. *J. Grey Syst.* **2015**, *27*, 51–70.
38. Wang, T. A Novel Approach of Integrating Natural Language Processing Techniques with Fuzzy TOPSIS for Product Evaluation. *Symmetry* **2022**, *14*, 120. [CrossRef]
39. Lu, H.; Zhao, Y.; Zhou, X.; Wei, Z. Selection of Agricultural Machinery Based on Improved CRITIC-Entropy Weight and GRA-TOPSIS Method. *Processes* **2022**, *10*, 266. [CrossRef]
40. Fabis-Domagala, J.; Domagala, M.; Momeni, H. A Matrix FMEA Analysis of Variable Delivery Vane Pumps. *Energies* **2021**, *14*, 1741. [CrossRef]
41. Sutrisno, A.; Hunawan, I.; Tangkuman, S. Modified Failure Mode and Effect Analysis (FMEA) Model for Accessing the Risk of Maintenance Waste. *Procedia Manuf.* **2015**, *4*, 23–29. [CrossRef]
42. Rachwał, A.; Wolniak, R. The Influence Of Nonconformities On Welding Process Performance. *Zesz. Nauk. Politech. Śląskiej Organ. I Zarządzanie* **2018**, *118*, 491–501. [CrossRef]
43. Rowiński, A. Dynamiczne obciążenia kadłubów silników turbinowych i ich zamocowań na płatowcach. *Pr. Inst. Lotnictwa* **2011**, *211*, 245–251.
44. Four-Point Ball Bearings. Available online: https://www.nskeurope.pl/pl/nsk-innovative-products/contact-ball-bearings-4-points.html# (accessed on 2 February 2022).
45. Greek Ascoloy (Alloy 418). Available online: https://www.neonickel.com/generate-alloy-pdf/?id=13027 (accessed on 2 February 2022).
46. Ostasz, G.; Siwiec, D.; Pacana, A. Universal Model to Predict Expected Direction of Products Quality Improvement. *Energies* **2022**, *15*, 1751. [CrossRef]
47. Wang, L.; Shi, H.B.; Yu, S.; Li, H.; Liu, L.; Bi, Z.M.; Fu, L.L. An application of enterprise systems in quality management of products. *Inf. Technol. Manag.* **2012**, *13*, 389–402. [CrossRef]
48. Ostasz, G.; Czerwinska, K.; Pacana, A. Quality management of aluminum pistons with the use of quality control points. *Manag. Syst. Prod. Eng.* **2020**, *28*, 29–33. [CrossRef]
49. Bilsel, R.U.; Lin, D.K.J. Ishikawa Cause and Effect Diagrams Using Capture Recapture Techniques. *Qual. Technol. Quant. Manag.* **2012**, *9*, 137–152. [CrossRef]
50. Pacana, A.; Siwiec, D.; Bednárová, L. Method of Choice: A Fluorescent Penetrant Taking into Account Sustainability Criteria. *Sustainability* **2020**, *12*, 5854. [CrossRef]

materials

MDPI

Review

An Overview of Recycling Wastes into Graphene Derivatives Using Microwave Synthesis; Trends and Prospects

Nuralmeera Balqis [1], Badrul Mohamed Jan [1,*], Hendrik Simon Cornelis Metselaar [2], Akhmal Sidek [3], George Kenanakis [4,*] and Rabia Ikram [1,*]

1 Department of Chemical Engineering, Faculty of Engineering, Universiti Malaya, Kuala Lumpur 50603, Malaysia; nuralmeerabalqis@gmail.com
2 Centre of Advanced Materials, Department of Mechanical Engineering, Universiti Malaya, Kuala Lumpur 50603, Malaysia; h.metselaar@um.edu.my
3 Petroleum Engineering Department, School of Chemical and Energy Engineering, Faculty of Engineering, Universiti Teknologi Malaysia, Johor Bahru 81310, Malaysia; akhmalsidek@utm.my
4 Institute of Electronic Structure and Laser, Foundation for Research and Technology-Hellas, N. Plastira 100, Vasilika Vouton, GR-700 13 Heraklion, Crete, Greece
* Correspondence: badrules@um.edu.my (B.M.J.); gkenanak@iesl.forth.gr (G.K.); raab@um.edu.my or raabch123@gmail.com (R.I.)

Abstract: It is no secret that graphene, a two-dimensional single-layered carbon atom crystal lattice, has drawn tremendous attention due to its distinct electronic, surface, mechanical, and optoelectronic properties. Graphene also has opened up new possibilities for future systems and devices due to its distinct structure and characteristics which has increased its demand in a variety of applications. However, scaling up graphene production is still a difficult, daunting, and challenging task. Although there is a vast body of literature reported on the synthesis of graphene through conventional and eco-friendly methods, viable processes for mass graphene production are still lacking. This review focuses on the variety of unwanted waste materials, such as biowastes, coal, and industrial wastes, for producing graphene and its potential derivatives. Among the synthetic routes, the main emphasis relies on microwave-assisted production of graphene derivatives. In addition, a detailed analysis of the characterization of graphene-based materials is presented. This paper also highlights the current advances and applications through the recycling of waste-derived graphene materials using microwave-assisted technology. In the end, it would alleviate the current challenges and forecast the specific direction of waste-derived graphene future prospects and developments.

Keywords: unwanted wastes; nanomaterials; graphene; microwave; batteries; sensors; supercapacitors

1. Introduction

Graphite has been utilized as the essential raw material in the production of graphene since its discovery. Graphite has an exceptionally anisotropic construction which leads to its in-plane and out-of-plane surface properties being very different [1]. Graphene is a layer of graphite. It is a solitary atom thick sheet of sp^2 hybridized carbon atoms organized in a hexagonal grid structure with extraordinary properties, such as high surface area, high electrical conductivity, and excellent mechanical strength [2]. Due to its exceptional physical characteristics, such as its ultra-thin properties, significant nonlinearity, and electrical tunability, graphene is frequently used in combination with other materials to create tunable optical and other electronic devices [3]. Since each carbon particle has an unhybridized single bond, graphene has high native flexibility and electronic conductivity. Recently, 3D structures of graphene honeycombs have been studied through large-scale molecular dynamics simulations for mechanistic understanding and deformation behaviors as displayed in Figure 1 [4].

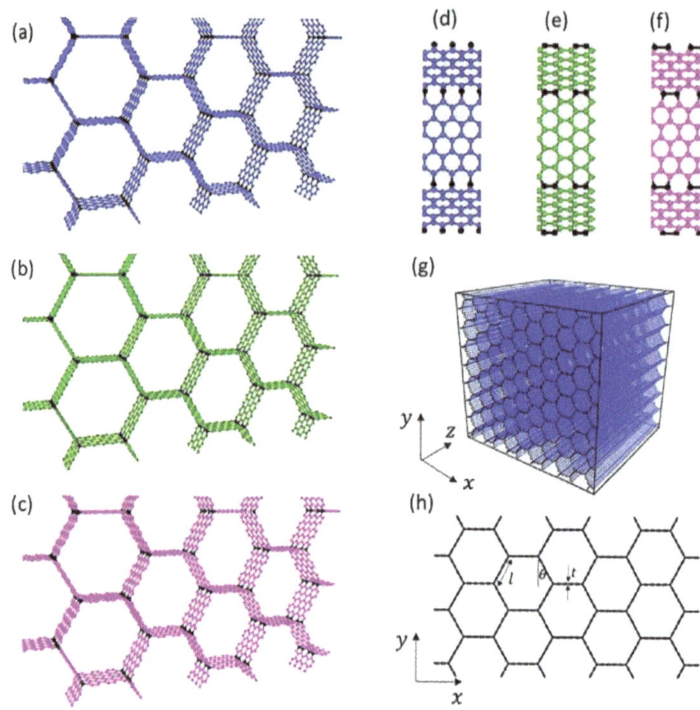

Figure 1. A variety of Graphene honeycomb 3D structures (**a–h**) [4].

Graphene oxide (GO) is not a conductor. However, it can be reduced by heat processes into conductive reduced GO (rGO) [5]. rGO is conveyed by disposing of the oxygenated groups of GO, where GO is a variant of graphene adorned with functional groups [6]. Despite the fact that rGO is a derivative of graphene, the rigorous process of oxidation and reduction familiarizes harmed areas with the rGO sheets. There are unreacted functional groups attached to the rGO plane (Figure 2) [7].

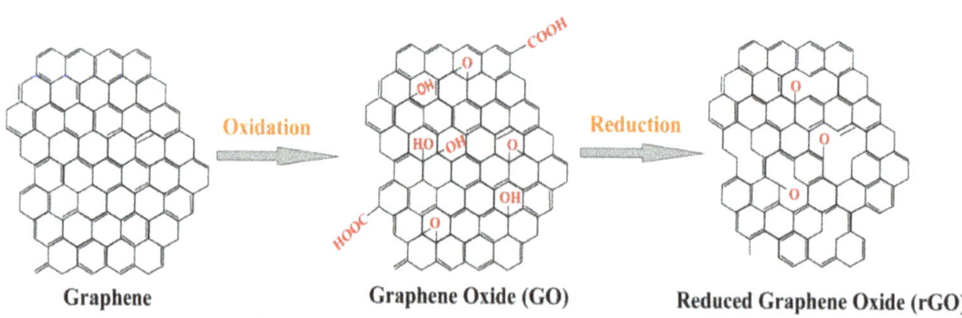

Figure 2. The structure of GO and rGO [6].

Graphene, both single layer and multilayer, can now be manufactured in a variety of ways. The layers of graphene union are fabricated through a hierarchical or base methodology [8]. Graphite is composed of graphene layers. The graphene layers have two types of bond structures. The weak Van der Waal interactions hold the graphene bond layers together with a distance of around 0.341 nm between the adjoining graphene layers [9]. The Van der Waal interactions have a significant impact on the frequency of modes with relative movement between the layers in vibrational dispersion [10]. Some of the phenomena related to Van der Waal interactions include friction, surface tension, viscosity, adhesion, cohesion, and wetting [11]. The assembly of carbon atoms into a graphene arrangement is the bottom-up method of synthesis. The two methodologies have advantages and downsides that have been explored in literature [12,13]. In this study, we aim to review the available literature on the synthesis of graphene and graphene-based materials derived from wastes in the last decade. The focus of waste is biowaste, coal, and industrial waste as source materials. The specific synthesis method is microwave synthesis. Moreover, numerous characterization techniques have been discussed along with the emerging future prospects and recommendations.

2. Synthesis of Waste into Graphene Derivatives

2.1. *Biowastes*

It has become challenging in the 21st century to obtain clean, affordable, and reliable energy sources which are essential from both a financial and natural outlook. Biomass has been identified as one of the most favorable sustainable sources of energy [14]. Biomass is standard and normal material derived from plants and animals (microorganisms) and it contains stored energy from the sun [15]. Since plants and animals are classified as sustainable, the word "renewable" is applicable to both. Moreover, biomass is often time obtained from forestry, agricultural, industrial, household, and municipal solid wastes (MSW) [16,17]. Every year, various bio-waste from large-scale livestock or agricultural sources are dumped into the environment [18]. Biomass is mostly comprised of long chains of carbon, hydrogen, and oxygen compounds with a carbon fixation as high as 55% by weight [19]. The carbon content of biomass should be concentrated before it can be converted completely to graphene. The industry has utilized this strategy to make biochar. Biochemicals, biofuels, and even bio-vehicles are created from biomass utilizing heat treatment methods, such as gasification, carbonization, liquefaction, and pyrolysis [20]. Carbonization is a pyrolysis process that converts biomass into a carbonaceous, charcoal-like material [21]. On the other hand, graphitization is a method in which amorphous carbon is heated before being converted into three-layered graphite [22]. It ought to be noticed that the carbonization cycle habitually brings about amorphous carbon instead of graphite-like carbon. Pyrolyzed carbon exists in two forms which are hard and soft carbon. In fact, despite being heated to extremely high temperatures, hard carbon graphitization is yet to be achieved [23]. In the meantime, heat treatment readily converts soft carbon into graphite. In spite of the way that the properties of the converted carbon structures are similar to graphene, they are not unmodified graphene because of the presence of extra carbon components [24].

Thermal exfoliation and carbon growth are two methods for the thermal degradation of biomass [25]. The exfoliation technique with graphitized biomass incorporates breaking up the carbon structure by overcoming the Van der Waals forces, resulting in graphene sheets (GSs). This process is similar to the conversion of graphite into graphene with graphitized biomass substituted for graphite [26,27]. Table 1 shows examples of methods for the conversion of biomass into graphene derivatives.

Table 1. Methods for the conversion of biomass into graphene derivatives.

Waste Sources	Methods	Temperature	Atmosphere	Time	Graphene Derivatives	References
Petals of lotus and hibiscus flowers	Chemical vapor deposition (CVD)	800–1600 °C	Air	0.5 h	graphene	[28]
Newspaper	Carbonization	450 °C	Air	5 days	graphene	[29]
Chitosan	Pyrolysis, Chemical activation	800 °C 900 °C	N_2 gas	3 h 2 h	graphene	[30]
Camphor leaves	Pyrolysis	1200 °C	Nitrogen gas	4 min	graphene	[31]
Wheat straw	Hydrothermal, Pyrolysis, Pyrolysis	150 °C 800 °C 2600 °C	Air N_2 gas Ar gas	6 h 3 h	graphene	[26]
Oil palm leaves & Palm kernel shell	Pyrolysis	700 °C	N_2 gas	3 h	GO	[32]
Oil palm fiber	CVD & Pyrolysis	1020 °C	Ar and H_2 gas	30 min	graphene	[33]
Rice husks	Chemical activation	400 °C 800 °C	Air	2 h	graphene	[34]
Palm oil	Pyrolysis	900 °C	Ar gas	10 min	GO	[35]
Spruce bark	Hydrothermal Pyrolysis	180 °C 1000 °C	Air N_2 gas	12 h 2 h	graphene	[36]
Mango peel	Pyrolysis	750 °C	H_2 gas Ar gas	15 min	graphene	[37]
Macademia nut shell	Hydrothermal Pyrolysis	180 °C 800 °C	Air Argon gas	12 h 2 h	graphene	[38]
Soybeans	Pyrolysis	800 °C	Nitrogen gas	2 h	graphene	[39]
Empty fruit brunch	Pyrolysis Graphitization	350 °C 900 °C	N_2 gas	2 h	graphene	[40]
Bengal gram bean husk	Pyrolysis	400 °C 850 °C	Nitrogen gas	2 h	graphene	[41]
Populus wood	Pyrolysis	950 °C	Nitrogen gas	1 h	graphene	[42]
Lignin biomass	Hydrothermal	180 °C	Air	12 h	graphene	[43]
Walnut shell	Pyrolysis	700 °C	Argon gas	4 h	graphene	[44]
Coconut shells, Oil palm empty fruit bunches (OPEFB), Rice husks	Carbonization	250, 300, 350, 400, 450 °C 105 °C 250, 300, 350 °C	NaOH NaOH Air	2 h 24 h 2.5 h	GO	[45]

2.2. Coal Waste

Coal is a unique carbon material that can be subdivided into lignite, bituminous coal, and anthracite [46]. Lignite and sub-bituminous coal are classified as inferior coal because of their high moisture content, high impurity, highly volatile matter substance, and low quantitative worth [47]. Coal is generally converted into fuel through various cycles, such as ignition, pyrolysis, gasification, and liquefaction [48]. The traditional methods have drawbacks which include a lack of energy efficiency and ecological contamination [49,50]. Subsequently, a high-esteem and earth-manageable technique in using coal is required [51]. Coal particles, in contrast with normal pieces of graphite and other precursors, contain a number of aromatic units as well as short aliphatic and ether bonds [52]. It is believed that coal might be a good option for creating carbon nanomaterials because of its staggered nanoarchitecture and explicit capabilities. Savitskii et al. [53] utilized anthracite coal and a thermo-oxidative technique to produce colloidal GO nanoparticle scatterings in size range from 122 nm to 190 nm. Pakhira et al. [54] showed that GO can be synthesized from low-grade coal. It was molded from the natural coalification of plant metabolites isolated by chemical exfoliation of cold HNO_3. However, such GO sheets are bound to break into negligible round shapes of nanometers. It is striking that there is an expansion in the utilization of coal-derived nanomaterials for a variety of industrial applications [52].

Currently, the strategy for reprocessing coal into graphene is to initially convert huge molecule coal into an antecedent carbon source prior to synthesizing graphene. The precursor carbon source can be gaseous or a particular form [55]. Primer screening of crude coal, debasement expulsion, pyrolysis (dry refining), gasification, and liquefaction of coal steps in the preparation of a precursor carbon source [56]. Zhou et al. utilized a reactant graphitization-helped dielectric barrier discharge (DBD) plasma strategy to make Taixi anthracite-based synthetically inferred graphene as well as metallic nanoparticle-enhanced graphene sheets [57]. In this method, crude coal was graphitized at 2400 °C for 2 h (under Ar) directly with $Fe_2(SO_4)_3$ as a catalyst followed by Hummers' method oxidation into the corresponding graphite-like carbon oxides (TX-NC-GO and TX-C-GO, separately) [57,58].

2.3. *Industrial Wastes*

Malaysia is an emerging nation that relies on modern efficiency as one of its monetary donors. Different types of wastes are produced in industrial processes, including chemical effluents, industrial plants waste, paper waste, metals, concrete, sludge, electronic devices wastes, etc. [59]. A number of significant materials (e.g., graphite, Cu, Fe, and Zn) from industrial waste can be recuperated utilizing a hydrometallurgical technique called leaching [60]. The commercialization of graphite-based products has immensely improved during the twenty-first century [61]. It is due to their unique physical and manufactured properties, such as high chemical resistance, heat capacity, high electrical conductivity, and lubricity. These unique properties are suitable for various modern applications, such as contraptions, oils, and metallurgy [62].

A modified Hummers method was utilized to prepare to GO from graphite obtained from modern waste filtering [63]. Concentrated sulfuric acid (H_2SO_4) and graphite (30 mL & 1 g) were mixed homogeneously in an ice bath for 30 min during the synthesis cycle. A total of 5 g of potassium permanganate ($KMnO_4$) was added and mixed for another 15 min at temperatures below 10 °C [64]. The extent of $KMnO_4$ was subsequently increased from 1:3 to 1:5 to speed up the oxidation rate. From that point onward, 8 mL of ultrapure water was added dropwise for 15 min, and the temperature of the mixture was kept under 98 °C for around 60 min. Finally, the oxidation reaction was obtained by adding 60 mL ultrapure water followed by 1 mL H_2O_2 [65].

3. Microwave Synthesis of Graphene Nanomaterials from Waste Materials

Microwave radiation is electromagnetic radiation with wavelengths ranging from 0.01 to 1 m and frequencies ranging from 300 MHz to 300 GHz [66]. Modern microwaves have two frequencies, 915 MHz and 2.45 GHz, while the consumed microwave only has one frequency, 2.45 GHz, and a wavelength of 12.25 cm [67]. Microwaves are widely used to heat materials that can absorb and convert microwave radiation to heat [68]. These dipolar particles that are changed can quickly rearrange toward the electric field, leading to expanded inward atomic contact, and volumetric warming of the whole substance [69]. As a result, microwave-assisted technology is able to provide a quick and efficient method of evenly heating the material or system from within. The conventional heating system, on the other hand, is relatively slow and ineffective [70].

Graphite or GO, is a typical wellspring of GSs, which are made from a conventional or modified Hummer's method [71]. Hummers' method is the most widely used method in the synthesis of GO through a mixture of concentrated H_2SO_4 and $KMnO_4$ [72]. Since then, numerous modified versions have been developed. However, the experimental procedures are mainly very similar to the original Hummers method. Oxidation is achieved using

$KMnO_4$ and the reaction is stabilized by adding hydrogen peroxide into the solution [73]. A few hazardous reducing agents, such as hydrazine (N_2H_4) and $NaBH_4$, are normally utilized in substance methodology to reduce GO. Thermal treatment, on the other hand, does not require the utilization of hazardous reducing agents making it a more attractive option [74]. The microwave-assisted technique has acquired ubiquity as an alternative to conventional graphene preparation. It treats GO or normal graphite in a microwave or microwave plasma-assisted chemical vapor deposition (MPCVD) framework which utilizes microwave-assisted solvothermal/aqueous strategies [75]. Microwave radiation provides a quick and uniform heating rate that leads to fast particle nucleation and growth which may reduce the reaction time that eventually led to significant energy saving [76]. Figure 3 portrays one potential microwave-assisted strategy for graphene synthesis. Microwave illumination produces very high temperatures and tensions, and energy is transferred directly into the GO [77]. Furthermore, the interaction of polar solvents with the surface oxides on GO sheets is the key factor in determining deposit regularity [77].

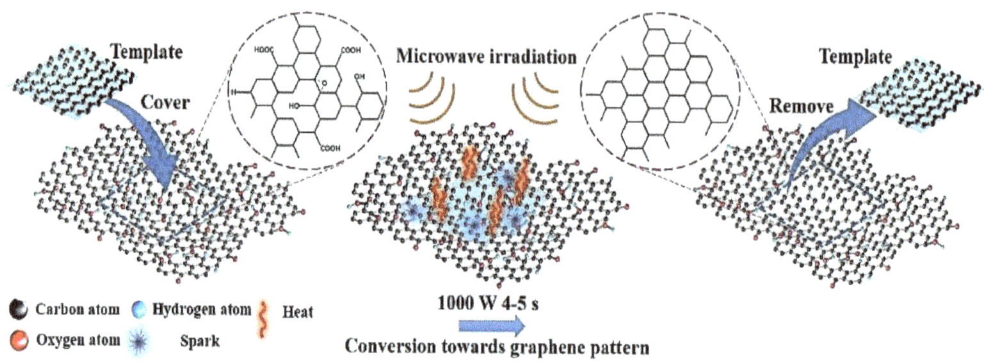

Figure 3. Schematic illustration of the synthesis of graphene and graphene-based composites with the assistance of microwave irradiation [77].

Furthermore, the reduction degree of GSs was further enhanced, and the functional groups on the surface of GO are successfully lowered [78]. There are several obvious advantages to producing graphene using microwave technology. Firstly, the advantage of microwave-assisted heating over traditional heating methods is its uniform and rapid heating of the reaction mixture [79]. In addition, microwave-assisted heating can significantly improve the transfer of energy directly to the reactants, resulting in an instantaneous internal temperature rise [80]. Furthermore, microwave technology enables the use of environmentally friendly solvents, resulting in cleaner products that do not require additional purification steps [81]. Since it involves a quick warming and very fast rate of crystallization to create the ideal nanocrystalline items, microwave illumination has recently been proposed as a valuable procedure for delivering carbon-related composites with uniform scattering as well as size and morphology control [82]. Table 2 shows examples of waste materials in graphene derivatives by using the microwave method.

Table 2. Examples of waste materials into graphene derivatives by using microwave method.

Types of Waste	Microwave Experimental Parameters				Characterizations	Applications	References
	Power	Time	Reagents	Frequency			
Graphite powder	700 W	60 s			XPS, XRD and TEM	Fuel cell catalysts	[83]
Cellulose	950 W	2 h	H_2SO_4	-	XRD	Biobased GO Quantum dots (GOQD)	[84]
Sugarcane bagasse (dried) & bulk	700 W 800 W	2 min 10 min	H_2SO_4 Argon gas	2450 MHz	FESEM, XRD, XPS and Raman spectroscopy SEM, Raman spectroscopy	Li-ion battery (LIB)	[85] [86]
Betalain from dragon fruit	100 W	10 min		-	ANOVA and BBD design matrix	Coloring food product	[87]
Spent tea waste	100-900W	15-180 min		-	TEM, XPS and FTIR	Graphene quantum dots (GQDs)	[88]
Waste palm	700 W	5 min		-	FESEM, XRD, XPS, TEM and Raman spectroscopy	Supercapacitor	[89]
Poly (Ethylene terephthalate)	700 W	300 s	Iron nano-particles	2450 MHz	XPS, Raman spectroscopy, FESEM, SEM, HRTEM and EDX	Bisphenol-A removal from contaminated water	[90]
Coconut shells	800 W	10, 20, 30, 40 min	L-ascorbic acid	2.45 MHz	FTIR, SEM, EDAX, XRD, LCR-Meters.	Effects of microwave irradiation	[91]
Sorghum stalk	700 W	3 min		-	SEM, XRD, XPS, TEM and EDS	Supercapacitors	[92]
Coconut coir and coconut shell		12 h		-	XRD, XPS, TEM and SEM	Electrical transportation system	[93]
Bovine blood waste	700 W	10 min		-	XPS and Raman spectroscopy	Food industry	[94]
Coconut shells	900 W	15 min		-	XRD and Raman spectroscopy	A hybrid gas sensor from room temperature	[95]
Waste PET bottle	600 W	2 min		-	EDX, FTIR XRD and SEM.	Tetracyclines removal	[96]
Disposable mask				-	SEM microscopy	Composite materials	[97]
Grass waste		5 h		-	FTIR, TEM, Raman, AFM, XPS, UV-Vis and HRTEM	Nonlinear optical applications	[98]
Styrofoam waste	1100 W	30 min		2.45GHz	TEM, Raman, XRD, FTIR and SAED	Nonpolar GQDs-based hydrophobic coating	[99]

Table 2. *Cont.*

Types of Waste	Microwave Experimental Parameters				Characterizations	Applications	References
	Power	Time	Reagents	Frequency			
Bamboo waste	2000 W	25 min	K_2CO_3	-	XRD, TEM, SEM and XPS	Biochar containing graphene (BCG)	[100]
Melamine sponge and arjuna bark	700 W	10 min		-	FTIR, XPS and TEM	Cell imaging and H_2O_2 sensing	[101]
Toner powder waste	350 W	30s		-	Raman, FTIR, UV-Vis spectrometer and FETEM	Color converting film	[102]

4. Characterization Techniques

4.1. X-ray Diffraction (XRD) and X-ray Photoelectron Spectroscopy (XPS)

Firstly, XRD is a reliable technique for the structural analysis of GO. This analysis can be used to assess the pattern/shape and crystallinity of GO [103]. XRD also is comparable to a fingerprint that is unique for each sample or species. This is due to the evaluation of achieved data which can be compared with the database results to identify that material [104]. In spite of the fact that XRD is certainly not an optimal device for recognizing single-layer graphene, it can be used to recognize graphite and graphene tests. In the XRD design, the unblemished graphite has a basal reflection (002) peak at $2\theta = 26.6°$ (d spacing = 0.335 nm). Later, the oxidation of graphite into graphite oxide shows middle basal (002) reflection peak moves to 11.2°, corresponding to a d spacing of 0.79 nm [105]. The increase in interlayer space is due to water atoms intercalating between the oxidized graphene layers. The presence of metallic mixtures in graphene structures was analyzed utilizing XRD examination. In addition, an x-beam connection with a graphitic translucent stage produces a diffraction design [106]. Non-covalent functionalization of rGO with two poly ionic fluids (PIL), poly (1-vinylimidazole) (PVI), and 2-bromopropionyl bromide resulted in the disappearance of a sharp GO diffraction peak at $2\theta = 11.8°$ in PIL-rGO diffractograms [107]. This trend is predictable with the detailed information and a slight expansion in the power of the GO trademark top in $2\theta = 44.5°$ (101) which relates to the basal reflexing plane of the tri-layered graphite [108]. Figure 4 displays the XRD profiles of graphite, GO-I, GO-II, and rGO. The formation of GO was confirmed by the diffraction peak at $2\theta = 11.01°$ at a reflection plane (001). A diffraction peak that appeared at $2\theta = 26.8°$ at a reflection plane (002) after the thermochemical treatment confirmed the reduction of GO [109]. This diffractogram demonstrated the disappearance of the GO peak, providing evidence that GO was converted into rGO. In addition, GO was prepared using different ratios of acids (I and II) as shown in Figure 4 [109,110].

XPS is one of the most common techniques used to study the relative amount of carbon, oxygen, and functional groups present in GO and electrochemically rGO (ErGO). It is an accurate technique to determine the amount of carbon and oxygen compared to elemental analysis because it is difficult to fully dehydrate a GO sample [111]. This is a quantitative and reliable technique in removing electrons from the C 1s and O 1s levels of graphene using X-rays and the energies of the emitted electrons are determined by the atomic composition of the material [112]. XPS can quantify the different types of carbon functionalities present and indicate the formation of chemical bonds, and evaluate the physisorption of molecules through the O/C ratio [113]. This quantification is critical to correlate the graphene-based materials' chemical properties versus their performance, for example, in permeability [114], water purification [115], or bio-sensing [116]. Furthermore, the surface chemistry and binding sites of both electrically conducting and non conducting materials are also studied by XPS. It is possible to characterize the networks and bonds in the material sample. The photoelectric effect serves as the basis for the theory. Additionally, XPS can shed light on the atomic composition's percentage. Figure 5 displays the GO and rGO of the XPS spectra that exhibit distinctive patterns which reveal their chemical composition [117].

The C(1 s) and (O1 s) peaks, which are located at about 285 and 532 eV, respectively, in the XPS full scan spectra of GO and rGO, are discernible [118]. The bonding involved is further highlighted by the deconvolution of the core orbitals of C(1 s) and O(1 s) [119]. Peaks for C=O, O=C-OH, C=C and C-C bonds are respectively visible in the C(1 s) deconvolution for GO at binding energies of 287, 289, 284, and 285 eV. The C-OH and C-O-C groups have peaks on the O(1 s) deconvolution curve for GO at 532 and 533 eV, respectively [120].

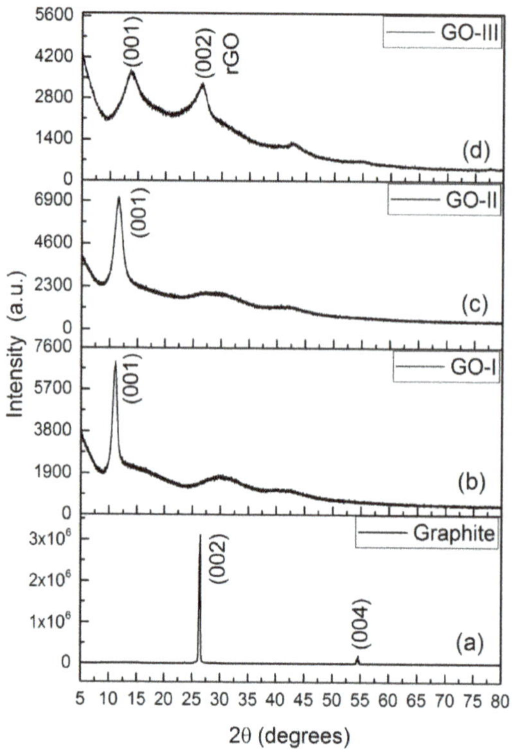

Figure 4. XRD patterns of graphite (**a**), GO-I (**b**), GO-II (**c**) and rGO (**d**) [109].

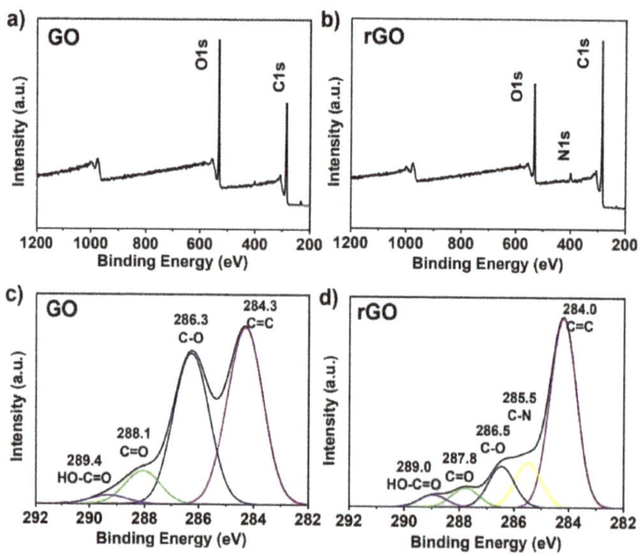

Figure 5. XPS spectra of (**a**) GO and (**b**) rGO. C1s XPS spectra of (**c**) GO and (**d**) rGO [117].

Table 3 has highlighted the GO and rGO binding energy values. In the case of rGO, these peaks show up with low intensity, confirming the reduction of GO. As a result, the peaks in rGO become narrower when GO is reduced to rGO. Additionally, it appears intense to restore the π-conjugation in the rGO peak at 284 eV which corresponds to C=C. The deconvoluted peaks in rGO shift to a binding energy value greater than GO for O(1 s).

Table 3. Binding energy values of GO and rGO in (eV) from the XPS plot [117,119].

Bond	GO		rGO	
	C (1 s)	O (1 s)	C (1 s)	O (1 s)
C=C	284	-	Increase in intensity	-
C=O	287	-	Decrease in intensity	-
C-O-H	285	532	Decrease in intensity	Narrowing of peaks
C-O-C	-	533	-	Decrease in intensity

The appearance of distinctive peaks in XPS can be used to verify that graphene has been successfully non-covalently functionalized [121]. According to Khan et al., two distinct peaks at 729 and 715.3 eV can be used to identify the presence of magnetic nanoparticles anchored on the GO surface [122]. Furthermore, it is noted that the XPS C1 spectrum after Fe3O4-functionalization shows peaks associated with C=O (285 eV), C=C (286.2 eV), and C-O-O (289 eV) bonds. XPS can also be used to identify active sites and further illuminate associated reaction mechanisms in graphene-based catalytic materials. This is best illustrated by the direct observation of active sites during the oxygen reduction reaction (ORR) over nitrogen-doped graphene (NG) catalysts [123]. Even though many simulation results showed various reaction pathways and adsorption sites for ORR over NG, the actual mechanism is still in dispute, primarily because there is not any direct evidence of the detection of intermediate species or active sites [124].

4.2. Other Characterization Methods

4.2.1. Raman Spectroscopy and Fourier-Transform Infrared Spectroscopy (FTIR)

Raman spectroscopy detects the transformation in energy connected with the Stokes and anti-Stokes transitions between the scattered photons. It is a non-destructive technique that provides information on chemical structure and molecular interactions by the combination of light within the bond of material [125]. Moreover, Raman spectroscopy is one of the most useful assets for concentrating on the construction and nature of carbon-based materials, for example, graphene [126]. It is a powerful, quick, delicate, and logical technique for giving subjective and quantitative information to graphene-based materials [127]. Raman spectroscopy is a significant instrument for deciding the quantity of graphite layers and the level of graphitization [128]. Graphene shows D, G, and 2D bands for the most parts in Raman analysis [129]. The D band is commonly situated around 1350 cm^{-1} and addresses the level of defects in the graphite. The higher the D band, the more defects in the graphite are observed [130]. The G band is linked to the in-plane vibration of sp^2 hybridization of carbon atoms which is located near 1580 cm^{-1}. The 2D peak, also known as G′, represents the number of graphene layers and is observed at 2700 cm^{-1} [131]. Figure 6 depicts the Raman spectra of graphene reduced with various reduction conditions which reflect the significant structural changes that occur during each stage of the electro and thermal processing [132].

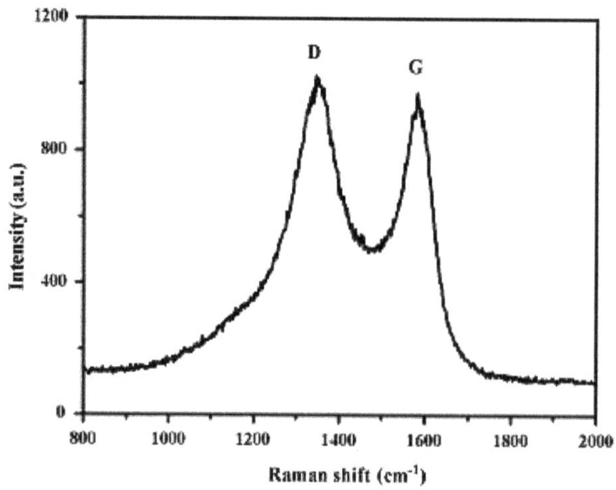

Figure 6. Raman spectra of samples at various stages of processing [132].

Pristine graphene is a carbon allotrope, and no signal can be collected using FTIR. Graphite oxide exfoliation is one of the primary routes for preparing practical graphene which supports catalytic research, and the oxidation step is critical [133]. As a result, many functional groups may remain in graphene-based catalysts even after being "completely removed", having a significant impact on catalytic performance [134]. Therefore, it is important to evaluate the reduction level. FTIR is one of the most efficient and simple methods for investigating residual functional groups [135]. Other than that, the method to determine the bonding configuration of different types of oxygen is FTIR analysis. Additionally, FTIR is a tool that complements Raman spectroscopy. The identifiable functional groups do not show any distinctive peaks in the pristine graphite FTIR spectrum [136]. It only shows two peaks at about 1610 and 450 cm^{-1} which are attributed to the vibration of adsorbed water molecules (the O-H stretching) and the skeletal vibrations from graphite domains, respectively (the sp^2 aromatic C=C) [137]. The oxygenated GSs may exhibit a variety of absorption bands or characteristic peaks ranging from 900 to 3500 cm1 following treatment with oxidizing agents [138]. These include the stretching vibrations of epoxy C-O groups (1000–1280 cm^{-1}), alkoxy stretching vibrations (1040–1170 cm^{-1}), O-H stretching vibrations (3300–3500 cm^{-1}), O-H deformation peaks (1300–1400 cm^{-1}), and carboxyl peaks (1700–1750 cm^{-1}) [139]. Notably, between 1600 and 1650 cm^{-1}, the aromatic C=C peak was visible. This peak is a result of the sp^2 domains in the unoxidized region of the graphite, and the vibration that is produced there is known as skeletal vibration [140].

4.2.2. Atomic Force Microscopy (AFM)

As a result of the limits of scanning tunneling microscopy (STM), such as the requirement for conductive examples, atomic force microscopy (AFM) was created in 1985 [141]. AFM is a multifunctional instrument that can envision the topography of a sample, measure its roughness, and distinguish the various periods of a composite [142]. It is widely used to measure the adhesive strength and mechanical properties of materials. It requires the utilization of conductive tips that act as top terminals as well as related to programming. Furthermore, nanoindentation can be utilized to quantify mechanical properties, such as Young's modulus and hardness [143]. AFM is broadly utilized in materials science [144], life science, and other disciplines [145]. As AFM innovation progresses, perception goal improves, and application scope extends and also more quantitative investigation of noticed pictures has started [146]. For instance, in the field of biomedicine, most exploratory examinations have zeroed in on the connection between the design and related elements of natural macromolecules, especially nucleic acids and proteins [147]. AFM in materials science can provide data related to the three-layered morphology and surface roughness of a material surface, as well as the distinction in the distribution of actual properties on the material surface, for example, morphological analysis [148] and dielectric constant [149]. A modified Langmuir–Schaefer deposition method was used to create a thin monolayer film suitable for imaging in the samples for AFM measurements. Figure 7 shows a representative AFM image of the GO monolayer deposited on the Si substrate as well as the corresponding size distribution of the GO sheets [150].

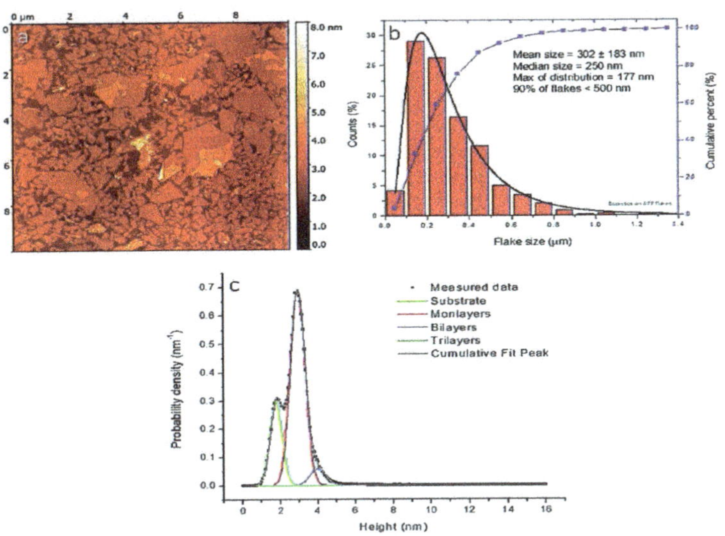

Figure 7. Purified GO: (**a**) atomic force microscopy (AFM) image, (**b**) GO size distribution, and (**c**) AFM scan height analysis [150].

4.2.3. Scanning Electron Microscopy-Energy Dispersive X-ray Spectroscopy (SEM-EDS)

Since it can rapidly examine/imagine the morphology of a huge sample, electron microscopy is broadly utilized in everyday schedule examinations [151]. A potential difference accelerates thermionic electrons transmitted by a tungsten fiber (cathode) close to the anode (1.0–30.0 kV). A condenser and objective electromagnetic focal points are utilized to adjust the bar to the example under vacuum (105 Dad) [152]. Secondary and backscattered electrons are transmitted during the output, as well as Auger electrons and X-rays, and their interaction with electrons which changes them completely to grayscale pictures. Pictures of the sample are given by secondary and backscattered electron identifiers, while compositional data is given by the X-ray spectrometer [153]. Secondary electrons are fundamentally created by the outer shell's inelastic scattering, while backscattered electrons are delivered by the primary electrons [154]. To avoid surface and underlying damage from the rays, delicate examples, such as polymers, need to be treated carefully. Nonconductive examples require surface pre-treatment and the sample is normally covered with a gold or carbon overlayer [155]. Due to the oxygenated epoxy groups of GO, it shows multilayers with some wrinkles [156]. SEM images provide 3D visualization of nanoparticles morphology, dispersion in cells, and other matrices. Lateral dimension and rapid analysis of nanoparticles element composition and surface flaws, such as cracks, etching residues, differential swelling, and holes can also be seen [157]. Figure 8 shows SEM images of protruded GNP produced by GNP debonding from the polymer matrix upon failure as indicated by circles when GNP loading is increased to 10% and 20%, respectively. It has been observed that while GNP loading is increased to 10% and 20% (Figure 8c,d), the fractured surfaces become much coarser [158].

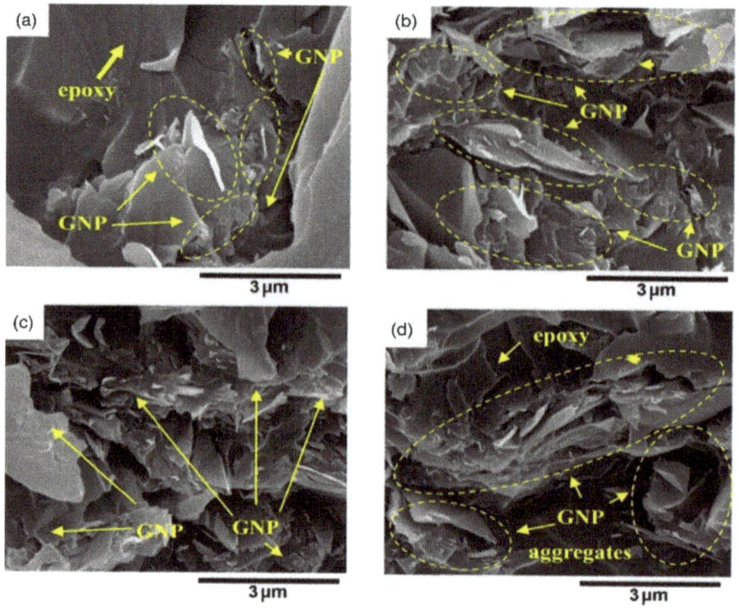

Figure 8. SEM images of GNP fracture surfaces in epoxy at loadings of (**a**) fGNP = 1%, (**b**) fGNP = 2%, (**c**) fGNP = 10%, and (**d**) fGNP = 20% [158].

4.2.4. Transmission Electron Microscopy (TEM) and High-Resolution Transmission Electron Microscopy (HRTEM)

TEM is best known for imaging a specimen's morphology, a wide variety of other combined techniques are also available in TEM to extract chemical, electrical, and structural data. For instance, local diffraction patterns can be measured using the parallel electron beam of the TEM which can offer precise measurements of the crystal system and parameters [159]. Furthermore, the transparent, corrugated, or wrinkled structure of the two-dimensional (2D) GO and rGO nanosheets is visible under the TEM [160]. It is also described as having the morphology of an ultrathin silk veil with folds and scrolls along its edges and it is attributed to graphene's inherent properties [161]. A highly effective method for characterizing the structure of graphene is HRTEM. It is a special tool for describing graphene's atomic structures and interfaces. It has been used to observe graphene flakes in a fraction of a micron and to reveal the fine chemical structure of GO [162]. Based on a TEM image of the folds formed at the edge, HRTEM also provides data on the number of graphene layers. Graphene's electron diffraction pattern can also be used by HRTEM to identify its crystalline nature [163]. It is noteworthy that HRTEM can reveal the quantity of layers present in various areas of the sheets [164]. The measured lattice spacing of single-layer graphene using this method is 0.236 nm [165]. Figure 9 shows TEM and HRTEM images of rGO.

4.2.5. Field Emission Scanning Electron Microscopy (FESEM)

The image of the materials' microstructure is captured using the cutting-edge technology known as FESEM. Gas molecules have a tendency to disturb the electron beam and the emitted secondary and backscattered electrons used for imaging and FESEM is typically carried out in a high vacuum [166]. The difference between the surface morphology of GO and rGO was further demonstrated by FESEM analysis [167]. It has been demonstrated that the rGO's FESEM image from Figure 10 has more wrinkles than GO [168]. The removal of oxygenated functional groups from the GO surface during the reduction process was supposed to be the cause of the corrugations on the rGO surface [169].

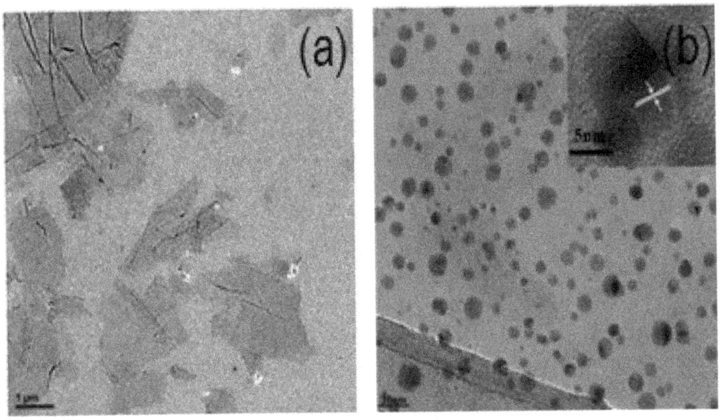

Figure 9. TEM image of bare GO (**a**) TEM and HRTEM image of rGO-Au (**b**) [165].

Figure 10. FESM micrograph of GO (**a**) and rGO (**b**) [168].

5. Future Prospects

Even though scientific interest in graphene has increased for a variety of applications, there are still several significant obstacles and challenges that need to be addressed and overcome. One of the critical issues is the reproducibility of waste materials into graphenaceous materials. Improved morphological properties should be combined with procedures that are both scalable and affordable. Excitedly, there is a sustained interest in the synthesis of materials based on graphene and the evaluation of their production and fusion with other materials. Although waste precursors have been the subject of recent studies, none of them have yet been able to be marked into commercially available products. Figure 11 shows the future prospects in graphene synthesis from a variety of wastes. Noteworthy future prospects include;

- Optimization of process variables and techniques to regulate the size, quality, and morphology of graphene-derived materials from waste materials.
- Improved synthetic concepts and methods are highly inspiring and necessitate commercial research involving renewable and biodegradable waste materials.
- Well-ordered oxidation/decrease and functionalization are expected for calibrating material properties, for example, band hole, electrical conductivity, and mechanical properties [170].
- Controlled graphite, GO, and rGO adjustment is in this way basic for widening the utilizations of graphene-based materials.

- To survey the wellbeing risk related with graphene and its subsidiaries, the poisonousness and biocompatibility of these unique carbon structures and their subordinates should be examined [171].
- Due to its extensive property, graphene preparation is a crucial area for material scientists. As a result, the scientific community should focus on advanced and novel microwave instruments which would be a great substitute of toxic and harsh chemicals
- To explore more variations that involving novel synthetic techniques, high purity GO for its mass production.
- There should be more consideration to lessen the cost effects of graphene derivatives.
- There should be more emphasis on the high yield and purity of graphene derivatives using a variety of wastes through microwave synthesis.
- This may also lead towards the excellence of functionalization, such as ID, 2D, and 3D graphene members, to fabricate waste materials into graphene-based structures with enhanced functionalities and high surface areas [172].
- Improving synthetic ideas and microwave approaches are remarkably motivating and requires further investigations by recycling waste materials for the optimization of parameters, such as time, power, and frequency.
- Further analysis of microwave synthesis and applications should be explored where the waste-based graphene derivatives can be utilized and, thus, the structures and properties can be modified as per the industrial demands.

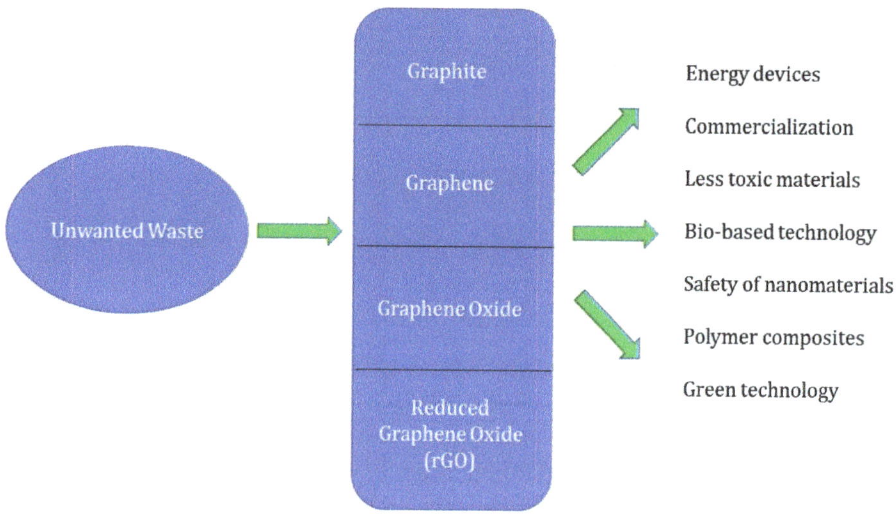

Figure 11. Future prospects in graphene synthesis from a variety of wastes.

6. Conclusions

Due to graphene's industrial significance, there is great concern about its sources and synthesis methods. These variables affect the price of graphene, and its industrial applications are constrained. The current study presents an overview of various types of wastes followed for the synthesis of graphene-based materials. Graphene creates wonders with its intriguing qualities and great attention from researchers all over the world. GO isolation has been established for more than ten years. However, the process is a continual exploration of variations involving novel synthesis techniques, and highly pure GO for its mass production and commercialization. The major attention is to have a process that is cost effective and economical. The literature is rich with important process parameters, their optimizations, and the synthesis of GO from a variety of waste which is useful for a wide range of applications. We have narrated literature that lists the synthetic routes for GO, particularly microwave synthesis, as well as different characterization methods. Although waste biomass-inferred graphene is one more encouraging material with various applications, its synthesis method is still open to be explored. Accordingly, more examination is expected to exhibit the best strategy for creating graphene with the best properties and optimization. In addition, successful and cost-effective planning may lead to the use of graphene in a wide range of applications from energy to the environment. Furthermore, material progress always demonstrates a superior effect in any field. Due to its diverse properties, graphene preparation is an important area for material scientists. As a result, the scientific community will always give attention to effective and efficient graphene preparation. The improvement of graphene through green combination addresses a huge progression in graphene innovation. The cost of producing graphene in large quantities could be reduced in alternative ways using carbonaceous wastes as raw materials. The production of graphene for industrial applications should successfully utilize a variety of environmentally hazardous solid waste precursors. Since waste-derived graphene might have impurities, additional purification procedures are needed. Future research is therefore required to increase graphene production with better yield and properties.

Author Contributions: N.B., original concept and initial draft of the paper, R.I., G.K. and A.S., processing, data analysis, and validation, B.M.J. and H.S.C.M., supervised and coordinated the work, G.K. and R.I., funding acquisition. All authors have read and agreed to the published version of the manuscript.

Funding: The authors would like to thank the Malaysia–Thailand Joint Authority (MTJA) under grant number IF062-2019 and RU Geran-Fakulti Program GPF054B-2020 from University of Malaya for providing funds during the course of this study. This work was also supported by proposal number 101092339 (acronym Exploit4InnoMat), under the Call: HORIZON-CL4-2022-RESILIENCE-01 (Topic: HORIZON-CL4-2022-RESILIENCE-01-20; Type of Action: HORIZON-IA).

Institutional Review Board Statement: Not applicable.

Informed Consent Statement: Not applicable.

Data Availability Statement: Not applicable.

Conflicts of Interest: The authors declare that there are no conflicts of interest regarding the publication of this manuscript.

Abbreviations

GO	Graphene oxide
rGO	Reduced graphene oxide
MSW	Metropolitan strong squanders
GSs	Graphene sheets
MPCVD	Microwave plasma assisted chemical vapor deposition
PIL-rGO	Poly ionic liquid reduced graphene oxide
ErGO	Electrochemically reduced graphene oxide
DBD	Dielectric barrier discharge
TX-NC-GO and TX-C-GO	Coal-based graphite-like carbons oxides
XRD	X-ray diffraction
XPS	X-ray photoelectron spectroscopy
TEM	Transmission electron microscopy
FESEM	Field emission scanning electron microscopy
ANOVA	Analysis of variance
BBD Design	Box-behnken design
SEMEDAX	Scanning electron microscopy energy dispersive X-Ray
EDS	Energy dispersive spectroscopy
SEM	Scanning electron microscopy

FTIR	Fourier transform infrared spectroscopy
AFM	Atomic force microscopy
SAED	Selected area (electron) diffraction
FETEM	Field-emission transmission electron microscope

References

1. Ke, Q.; Wang, J. Graphene-based materials for supercapacitor electrodes–A review. *J. Mater.* **2016**, *2*, 37–54. [CrossRef]
2. Luo, B.; Liu, S.; Zhi, L. Chemical approaches toward graphene-based nanomaterials and their applications in energy-related areas. *Small* **2012**, *8*, 630–646. [CrossRef] [PubMed]
3. Wu, Y.; Zhao, X.; Hu, J.; Xu, H. Low threshold optical bistability based on coupled graphene Tamm states. *Results Phys.* **2021**, *21*, 103824. [CrossRef]
4. Chen, Y.; Meng, F.; Bie, X.; Ou, P.; Song, J. Atomistic and continuum modeling of 3D graphene honeycombs under uniaxial in-plane compression. *Comput. Mater. Sci.* **2021**, *197*, 110646. [CrossRef]
5. Safian, M.T.U.; Umar, K.; Ibrahim, M.N.M. Synthesis and scalability of graphene and its derivatives: A journey towards sustainable and commercial material. *J. Clean. Prod.* **2021**, *318*, 128603. [CrossRef]
6. Priyadarsini, S.; Mohanty, S.; Mukherjee, S.; Basu, S.; Mishra, M. Graphene and graphene oxide as nanomaterials for medicine and biology application. *J. Nanostructure Chem.* **2018**, *8*, 123–137. [CrossRef]
7. Somanathan, T.; Prasad, K.; Ostrikov, K.K.; Saravanan, A.; Krishna, V.M. Graphene oxide synthesis from agro waste. *Nanomaterials* **2015**, *5*, 826–834. [CrossRef]
8. Bhuyan, M.S.A.; Uddin, M.N.; Islam, M.M.; Bipasha, F.A.; Hossain, S.S. Synthesis of graphene. *Int. Nano Lett.* **2016**, *6*, 65–83. [CrossRef]
9. Cai, M.; Thorpe, D.; Adamson, D.H.; Schniepp, H.C. Methods of graphite exfoliation. *J. Mater. Chem.* **2012**, *22*, 24992–25002. [CrossRef]
10. Tan, P.H.; Han, W.P.; Zhao, W.J.; Wu, Z.H.; Chang, K.; Wang, H.; Wang, Y.F.; Bonini, N.; Marzari, N.; Pugno, N.; et al. The shear mode of multilayer graphene. *Nat. Mater.* **2012**, *11*, 294–300. [CrossRef]
11. Vasu, K.S.; Prestat, E.; Abraham, J.; Dix, J.; Kashtiban, R.J.; Beheshtian, J.; Sloan, J.; Carbone, P.; Neek-Amal, M.; Haigh, S.J.; et al. Van der Waals pressure and its effect on trapped interlayer molecules. *Nat. Commun.* **2016**, *7*, 1–6. [CrossRef]
12. Avouris, P.; Dimitrakopoulos, C. Graphene: Synthesis and applications. *Mater. Today* **2012**, *15*, 86–97. [CrossRef]
13. Dasari, B.L.; Nouri, J.M.; Brabazon, D.; Naher, S. Graphene and derivatives–Synthesis techniques, properties and their energy applications. *Energy* **2017**, *140*, 766–778. [CrossRef]
14. Junginger, H.M.; Jonker, J.G.G.; Faaij, A.P.C.; Cocchi, M.; Schouwenberg, P. *Summary, Synthesis and Conclusions from IEA Bioenergy Task 40 Country Reports on International Bioenergy Trade*; Utrecht University: Utrecht, The Netherlands, 2011.
15. Lehr, J.H.; Keeley, J. (Eds.) *Alternative Energy and Shale Gas Encyclopedia*; John Wiley & Sons: Hoboken, NJ, USA, 2016.
16. Hakeem, K.R.; Jawaid, M.; Alothman, O.Y. (Eds.) *Agricultural Biomass Based Potential Materials*; Springer: Berlin/Heidelberg, Germany, 2015.
17. Tripathi, N.; Hills, C.D.; Singh, R.S.; Atkinson, C.J. Biomass waste utilisation in low-carbon products: Harnessing a major potential resource. *NPJ Clim. Atmos. Sci.* **2019**, *2*, 35. [CrossRef]
18. Ikram, R.; Jan, B.M.; Ahmad, W. Advances in synthesis of graphene derivatives using industrial wastes precursors; prospects and challenges. *J. Mater. Res. Technol.* **2020**, *9*, 15924–15951. [CrossRef]
19. Xie, X.; Goodell, B. Thermal degradation and conversion of plant biomass into high value carbon products. In *Deterioration and Protection of Sustainable Biomaterials*; American Chemical Society: Washington, DC, USA, 2014; pp. 147–158.
20. He, L.; Guo, S.; Lei, J.; Sha, Z.; Liu, Z. The effect of Stone–Thrower–Wales defects on mechanical properties of graphene sheets—A molecular dynamics study. *Carbon* **2014**, *75*, 124–132. [CrossRef]
21. Ronsse, F.; Nachenius, R.W.; Prins, W. Carbonization of biomass. In *Recent Advances in Thermo-Chemical Conversion of Biomass*; Elsevier: Amsterdam, The Netherlands, 2015; pp. 293–324.
22. Zhang, C.; Guimin, L.U.; Ze, S.U.N.; Jianguo, Y.U. Catalytic graphitization of carbon/carbon composites by lanthanum oxide. *J. Rare Earths* **2012**, *30*, 128–132. [CrossRef]
23. Pan, G.; Liang, W.; Liang, P.; Chen, Q. Effect of vacuum-carbonization treatment of soft carbon anodes derived from coal-based mesophase pitch for lithium-ion batteries. *Clean Energy* **2019**, *3*, 211–216. [CrossRef]
24. Ikram, R.; Mohamed Jan, B.; Nagy, P.B.; Szabo, T. Recycling waste sources into nanocomposites of graphene materials: Overview from an energy-focused perspective. *Nanotechnol. Rev.* **2023**, *12*, 20220512. [CrossRef]
25. Safian, M.T.U.; Haron, U.S.; Ibrahim, M.M. A review on bio-based graphene derived from biomass wastes. *BioResources* **2020**, *15*, 9756. [CrossRef]
26. Chen, F.; Yang, J.; Bai, T.; Long, B.; Zhou, X. Facile synthesis of few-layer graphene from biomass waste and its application in lithium ion batteries. *J. Electroanal. Chem.* **2016**, *768*, 18–26. [CrossRef]
27. Kumar, R.; Joanni, E.; Singh, R.K.; Singh, D.P.; Moshkalev, S.A. Recent advances in the synthesis and modification of carbon-based 2D materials for application in energy conversion and storage. *Prog. Energy Combust. Sci.* **2018**, *67*, 115–157. [CrossRef]
28. Ray, A.K.; Sahu, R.K.; Rajinikanth, V.; Bapari, H.; Ghosh, M.; Paul, P. Preparation and characterization of graphene and Ni-decorated graphene using flower petals as the precursor material. *Carbon* **2012**, *50*, 4123–4129. [CrossRef]

29. Akhavan, O.; Bijanzad, K.; Mirsepah, A. Synthesis of graphene from natural and industrial carbonaceous wastes. *RSC Adv.* **2014**, *4*, 20441–20448. [CrossRef]
30. Hao, P.; Zhao, Z.; Leng, Y.; Tian, J.; Sang, Y.; Boughton, R.I.; Wong, C.P.; Liu, H.; Yang, B. Graphene-based nitrogen self-doped hierarchical porous carbon aerogels derived from chitosan for high performance supercapacitors. *Nano Energy* **2015**, *15*, 9–23. [CrossRef]
31. Shams, S.S.; Zhang, L.S.; Hu, R.; Zhang, R.; Zhu, J. Synthesis of graphene from biomass: A green chemistry approach. *Mater. Lett.* **2015**, *161*, 476–479. [CrossRef]
32. Nasir, S.; Hussein, M.Z.; Yusof, N.A.; Zainal, Z. Oil palm waste-based precursors as a renewable and economical carbon sources for the preparation of reduced graphene oxide from graphene oxide. *Nanomaterials* **2017**, *7*, 182. [CrossRef]
33. Tahir, N.A.M.; Abdollah, M.F.B.; Tamaldin, N.; Amiruddin, H.; Tokoroyama, T.; Umehara, N. Potential of growing graphene from solid waste products. In Proceedings of the SAKURA Symposium on Mechanical Science and Engineering, Nagoya, Japan, 12 September 2017; pp. 26–28.
34. Ismail, M.S.; Yusof, N.; Yusop, M.Z.M.; Ismail, A.F.; Jaafar, J.; Aziz, F.; Karim, Z.A. Synthesis and characterization of graphene derived from rice husks. *Malays. J. Fundam. Appl. Sci.* **2019**, *15*, 516–521. [CrossRef]
35. Mamat, R.H.; Hamzah, F.; Hashim, A.; Abdullah, S.; Alrokayan, S.A.; Khan, H.A.; Safiay, M.; Jafar, S.M.; Asli, A.; Khusaimi, Z.; et al. Influence of volume variety of waste cooking palm oil as carbon source on graphene growth through double thermal chemical vapor deposition. In Proceedings of the 2018 IEEE International Conference on Semiconductor Electronics (ICSE), Kuala Lumpur, Malaysia, 15–17 August 2018; pp. 53–56.
36. Sun, Z.; Zheng, M.; Hu, H.; Dong, H.; Liang, Y.; Xiao, Y.; Lei, B.; Liu, Y. From biomass wastes to vertically aligned graphene nanosheet arrays: A catalyst-free synthetic strategy towards high-quality graphene for electrochemical energy storage. *Chem. Eng. J.* **2018**, *336*, 550–561. [CrossRef]
37. Shah, J.; Lopez-Mercado, J.; Carreon, M.G.; Lopez-Miranda, A.; Carreon, M.L. Plasma synthesis of graphene from mango peel. *ACS Omega* **2018**, *3*, 455–463. [CrossRef]
38. Lu, X.; Xiang, K.; Zhou, W.; Zhu, Y.; He, Y.; Chen, H. Graphene-like carbon derived from macadamia nut shells for high-performance supercapacitor. *Russ. J. Electrochem.* **2019**, *55*, 242–246. [CrossRef]
39. Sha, T.; Liu, J.; Sun, M.; Li, L.; Bai, J.; Hu, Z.; Zhou, M. Green and low-cost synthesis of nitrogen-doped graphene-like mesoporous nanosheets from the biomass waste of okara for the amperometric detection of vitamin C in real samples. *Talanta* **2019**, *200*, 300–306. [CrossRef] [PubMed]
40. Widiatmoko, P.; Sukmana, I.F.; Nurdin, I.; Prakoso, T.; Devianto, H. Increasing yield of graphene synthesis from oil palm empty fruit bunch via two-stages pyrolysis. In *IOP Conference Series: Materials Science and Engineering*; IOP Publishing: Bristol, UK, 2019; Volume 543, p. 012032.
41. Gupta, K.; Gupta, D.; Khatri, O.P. Applied Surface Science Graphene-like porous carbon nanostructure from Bengal gram bean husk and its application for fast and efficient adsorption of organic dyes. *Appl. Surf. Sci.* **2019**, *476*, 647–657. [CrossRef]
42. Ekhlasi, L.; Younesi, H.; Rashidi, A.; Bahramifar, N. Populus wood biomass-derived graphene for high CO2 capture at atmospheric pressure and estimated cost of production. *Process Saf. Environ. Prot.* **2018**, *113*, 97–108. [CrossRef]
43. Ding, Z.; Yuan, T.; Wen, J.; Cao, X.; Sun, S.; Xiao, L.P.; Shi, Q.; Wang, X.; Sun, R. Green synthesis of chemical converted graphene sheets derived from pulping black liquor. *Carbon* **2020**, *158*, 690–697. [CrossRef]
44. Roquia, A.; khalfan hamed Alhashmi, A.; hamed Abdullah alhasmi, B. Synthesis and characterisation of carbon nanotubes from waste of Juglans regia (walnut) shells. *Fuller. Nanotub. Carbon Nanostructures* **2021**, *29*, 860–867. [CrossRef]
45. Ramli, R.; Hidayat, R. *Graphene Oxide Based on Biomass Waste: Synthesis and Applications*; IntechOpen: London, UK, 2022. [CrossRef]
46. Lee, S.Y.; Mahajan, R.L. A facile method for coal to graphene oxide and its application to a biosensor. *Carbon* **2021**, *181*, 408–420. [CrossRef]
47. Xu, Y.; Zhang, Y.; Zhang, G.; Guo, Y.; Zhang, J.; Li, G. Pyrolysis characteristics and kinetics of two Chinese low-rank coals. *J. Therm. Anal. Calorim.* **2015**, *122*, 975–984. [CrossRef]
48. Xu, T.; Srivatsa, S.C.; Bhattacharya, S. In-situ synchrotron IR study on surface functional group evolution of Victorian and Thailand low-rank coals during pyrolysis. *J. Anal. Appl. Pyrolysis* **2016**, *122*, 122–130. [CrossRef]
49. Zhu, Z.; Zuo, H.; Li, S.; Tu, J.; Guan, W.; Song, W.L.; Zhao, J.; Tian, D.; Jiao, S. A green electrochemical transformation of inferior coals to crystalline graphite for stable Li-ion storage. *J. Mater. Chem. A* **2019**, *7*, 7533–7540. [CrossRef]
50. Kaklidis, N.; Kyriakou, V.; Marnellos, G.E.; Strandbakke, R.; Arenillas, A.; Menéndez, J.A.; Konsolakis, M. Effect of fuel thermal pretreament on the electrochemical performance of a direct lignite coal fuel cell. *Solid State Ion.* **2016**, *288*, 140–146. [CrossRef]
51. Zhang, X.; Sun, B.; Fan, X.; Liang, P.; Zhao, G.; Saikia, B.K.; Wei, X. Hierarchical porous carbon derived from coal and biomass for high performance supercapacitors. *Fuel* **2022**, *311*, 122552. [CrossRef]
52. Li, K.K.; Liu, G.Y.; Zheng, L.S.; Jia, J.; Zhu, Y.Y.; Zhang, Y.T. Coal-derived carbon nanomaterials for sustainable energy storage applications. *New Carbon Mater.* **2021**, *36*, 133–154. [CrossRef]
53. Savitskii, D.P. Preparation and characterization of colloidal dispersions of graphene-like structures from different ranks of coals. *J. Fuel Chem. Technol.* **2017**, *45*, 897–907. [CrossRef]
54. Pakhira, B.; Ghosh, S.; Maity, S.; Sangeetha, D.N.; Laha, A.; Allam, A.; Sarkar, S. Extraction of preformed graphene oxide from coal: Its clenched fist form entrapping large molecules. *RSC Adv.* **2015**, *5*, 89076–89082. [CrossRef]

55. Tran, V.T.; Saint-Martin, J.; Dollfus, P. Electron transport properties of graphene nanoribbons with Gaussian deformation. *Phys. Rev. B* **2020**, *102*, 075425. [CrossRef]
56. Canel, M.; Mısırlıoğlu, Z.; Canel, E.; Bozkurt, P.A. Distribution and comparing of volatile products during slow pyrolysis and hydropyrolysis of Turkish lignites. *Fuel* **2016**, *186*, 504–517. [CrossRef]
57. Zhou, Q.; Zhao, Z.; Zhang, Y.; Meng, B.; Zhou, A.; Qiu, J. Graphene sheets from graphitized anthracite coal: Preparation, decoration, and application. *Energy Fuels* **2012**, *26*, 5186–5192. [CrossRef]
58. Abidi, I.H.; Liu, Y.; Pan, J.; Tyagi, A.; Zhuang, M.; Zhang, Q.; Cagang, A.A.; Weng, L.T.; Sheng, P.; Goddard, W.A., III; et al. Regulating Top-Surface Multilayer/Single-Crystal Graphene Growth by "Gettering" Carbon Diffusion at Backside of the Copper Foil. *Adv. Funct. Mater.* **2017**, *27*, 1700121. [CrossRef]
59. Jadhav, U.U.; Hocheng, H. A review of recovery of metals from industrial waste. *J. Achiev. Mater. Manuf. Eng.* **2012**, *54*, 159–167.
60. Li, H.; Eksteen, J.; Oraby, E. Hydrometallurgical recovery of metals from waste printed circuit boards (WPCBs): Current status and perspectives–A review. *Resour. Conserv. Recycl.* **2018**, *139*, 122–139. [CrossRef]
61. Simandl, G.J.; Paradis, S.; Akam, C. Graphite deposit types, their origin, and economic significance. *Br. Columbia Minist. Energy Mines Br. Columbia Geol. Surv.* **2015**, *3*, 163–171.
62. Zhang, Z.J.; Simionesie, D.; Schaschke, C. Graphite and hybrid nanomaterials as lubricant additives. *Lubricants* **2014**, *2*, 44–65. [CrossRef]
63. Xuan, Y.; Jiang, G.; Li, Y. Nanographite oxide as ultrastrong fluid-loss-control additive in water-based drilling fluids. *J. Dispers. Sci. Technol.* **2014**, *35*, 1386–1392. [CrossRef]
64. Singh, R.K.; Kumar, R.; Singh, D.P. Graphene oxide: Strategies for synthesis, reduction and frontier applications. *RSC Adv.* **2016**, *6*, 64993–65011. [CrossRef]
65. Siaw, W.C.; Tsuji, T.; Manaf, N.A.; Patah, M.A.; Jan, B.M. Synthesis of graphene oxide from industrial waste. In *IOP Conference Series: Materials Science and Engineering*; IOP Publishing: Bristol, UK, 2020; Volume 778, p. 012050.
66. Wojnarowicz, J.; Chudoba, T.; Lojkowski, W. A review of microwave synthesis of zinc oxide nanomaterials: Reactants, process parameters and morphologies. *Nanomaterials* **2020**, *10*, 1086. [CrossRef]
67. Chandrasekaran, S.; Ramanathan, S.; Basak, T. Microwave food processing—A review. *Food Res. Int.* **2013**, *52*, 243–261. [CrossRef]
68. Nakamura, N.; Reeja-Jayan, B. Synchrotron X-ray characterization of materials synthesized under microwave irradiation. *J. Mater. Res.* **2019**, *34*, 194–205. [CrossRef]
69. Xia, X.; Zhang, Y.; Chao, D.; Guan, C.; Zhang, Y.; Li, L.; Ge, X.; Bacho, I.M.; Tu, J.; Fan, H.J. Solution synthesis of metal oxides for electrochemical energy storage applications. *Nanoscale* **2014**, *6*, 5008–5048. [CrossRef]
70. Wong, C.H.A.; Jankovský, O.; Sofer, Z.; Pumera, M. Vacuum-assisted microwave reduction/exfoliation of graphite oxide and the influence of precursor graphite oxide. *Carbon* **2014**, *77*, 508–517. [CrossRef]
71. Wang, C.; Wang, Y.; Graser, J.; Zhao, R.; Gao, F.; O'Connell, M.J. Solution-based carbohydrate synthesis of individual solid, hollow, and porous carbon nanospheres using spray pyrolysis. *ACS Nano* **2013**, *7*, 11156–11165. [CrossRef]
72. Yu, H.; Zhang, B.; Bulin, C.; Li, R.; Xing, R. High-efficient synthesis of graphene oxide based on improved hummers method. *Sci. Rep.* **2016**, *6*, 36143. [CrossRef]
73. Alam, S.N.; Sharma, N.; Kumar, L. Synthesis of graphene oxide (GO) by modified hummers method and its thermal reduction to obtain reduced graphene oxide (rGO). *Graphene* **2017**, *6*, 1–18. [CrossRef]
74. Ikram, R.; Mohamed Jan, B.; Abdul Qadir, M.; Sidek, A.; Stylianakis, M.M.; Kenanakis, G. Recent advances in chitin and chitosan/graphene-based bio-nanocomposites for energetic applications. *Polymers* **2021**, *13*, 3266. [CrossRef]
75. Tatarova, E.; Dias, A.; Henriques, J.; do Rego, A.B.; Ferraria, A.M.; Abrashev, M.V.; Luhrs, C.C.; Phillips, J.; Dias, F.M.; Ferreira, C.M. Microwave plasmas applied for the synthesis of free standing graphene sheets. *J. Phys. D Appl. Phys.* **2014**, *47*, 385501. [CrossRef]
76. Zheng, Y.; Lv, K.; Wang, Z.; Deng, K.; Li, M. Microwave-assisted rapid synthesis of anatase TiO2 nanocrystals with exposed {0 0 1} facets. *J. Mol. Catal. A Chem.* **2012**, *356*, 137–143. [CrossRef]
77. Zhao, Y.; He, J. Novel template-assisted microwave conversion of graphene oxide to graphene patterns: A reduction transfer mechanism. *Carbon* **2019**, *148*, 159–163. [CrossRef]
78. Wang, Z.; Yu, J.; Zhang, X.; Li, N.; Liu, B.; Li, Y.; Wang, Y.; Wang, W.; Li, Y.; Zhang, L.; et al. Large-scale and controllable synthesis of graphene quantum dots from rice husk biomass: A comprehensive utilization strategy. *ACS Appl. Mater. Interfaces* **2016**, *8*, 1434–1439. [CrossRef]
79. Van Khai, T.; Kwak, D.S.; Kwon, Y.J.; Cho, H.Y.; Huan, T.N.; Chung, H.; Ham, H.; Lee, C.; Van Dan, N.; Tung, N.T.; et al. Direct production of highly conductive graphene with a low oxygen content by a microwave-assisted solvothermal method. *Chem. Eng. J.* **2013**, *232*, 346–355. [CrossRef]
80. Yan, Q.; Liu, Q.; Wang, J. A simple and fast microwave assisted approach for the reduction of graphene oxide. *Ceram. Int.* **2016**, *42*, 3007–3013. [CrossRef]
81. Sengupta, J. Different synthesis routes of graphene-based metal nanocomposites. *arXiv* **2019**, arXiv:1911.01720.
82. Thomas, R.; Rao, K.Y.; Rao, G.M. Morphology and electrochemical performance of graphene nanosheet array for Li-ion thin film battery. *Electrochim. Acta* **2013**, *108*, 458–464. [CrossRef]
83. Xin, Y.; Liu, J.G.; Jie, X.; Liu, W.; Liu, F.; Yin, Y.; Gu, J.; Zou, Z. Preparation and electrochemical characterization of nitrogen doped graphene by microwave as supporting materials for fuel cell catalysts. *Electrochim. Acta* **2012**, *60*, 354–358. [CrossRef]

84. Adolfsson, K.H.; Hassanzadeh, S.; Hakkarainen, M. Valorization of cellulose and waste paper to graphene oxide quantum dots. *RSC Adv.* **2015**, *5*, 26550–26558. [CrossRef]
85. Agarkar, S.; Yadav, P.; Fernandes, R.; Kothari, D.; Suryawanshi, A.; Ogale, S. Minute-made activated porous carbon from agro-waste for Li-ion battery anode using a low power microwave oven. *Electrochim. Acta* **2016**, *212*, 535–544. [CrossRef]
86. Wu, J.; Zhao, J.; Vaidhyanathan, B.; Zhang, H.; Anshuman, A.; Nare, A.; Saremi-Yarahmadi, S. Rapid microwave-assisted bulk production of high-quality reduced graphene oxide for lithium ion batteries. *Materialia* **2020**, *13*, 100833. [CrossRef]
87. Thirugnanasambandham, K.; Sivakumar, V. Microwave assisted extraction process of betalain from dragon fruit and its antioxidant activities. *J. Saudi Soc. Agric. Sci.* **2017**, *16*, 41–48. [CrossRef]
88. 88. Abbas, A.; Tabish, T.A.; Bull, S.J.; Lim, T.M.; Phan, A.N. High yield synthesis of graphene quantum dots from biomass waste as a highly selective probe for Fe3+ sensing. *Sci. Rep.* **2020**, *10*, 1–16. [CrossRef]
89. Liu, C.; Chen, W.; Hong, S.; Pan, M.; Jiang, M.; Wu, Q.; Mei, C. Fast microwave synthesis of hierarchical porous carbons from waste palm boosted by activated carbons for supercapacitors. *Nanomaterials* **2019**, *9*, 405. [CrossRef]
90. Sridhar, V.; Park, H. Transforming waste poly (ethylene terephthalate) into nitrogen doped carbon nanotubes and its utility in oxygen reduction reaction and bisphenol-a removal from contaminated water. *Materials* **2020**, *13*, 4144. [CrossRef]
91. Yanti, D.R.; Hikmah, U.; Prasetyo, A.; Hastuti, E. The effect of microwave irradiation on reduced graphene oxide from coconut shells. In *IOP Conference Series: Earth and Environmental Science*; IOP Publishing: Bristol, UK, 2020; Volume 456, p. 012008.
92. Fu, M.; Zhu, Z.; Zhang, Z.; Zhuang, Q.; Chen, W.; Liu, Q. Microwave deposition synthesis of Ni (OH) 2/sorghum stalk biomass carbon electrode materials for supercapacitors. *J. Alloy. Compd.* **2020**, *846*, 156376. [CrossRef]
93. Tamilselvi, R.; Ramesh, M.; Lekshmi, G.S.; Bazaka, O.; Levchenko, I.; Bazaka, K.; Mandhakini, M. Graphene oxide–Based supercapacitors from agricultural wastes: A step to mass production of highly efficient electrodes for electrical transportation systems. *Renew. Energy* **2020**, *151*, 731–739. [CrossRef]
94. Chalmpes, N.; Asimakopoulos, G.; Baikousi, M.; Moschovas, D.; Avgeropoulos, A.; Bourlinos, A.B.; Sedajova, V.; Bakandritsos, A.; Gournis, D.; Karakassides, M.A. Fast and direct microwave synthesis of carbon from bovine blood waste: A feedstock material for extractive metallurgy, carbon dots production and graphite synthesis. *J. Nanotechnol. Res.* **2021**, *3*, 011–028.
95. Parveen, R.A.; Rakkesh, R.A.; Durgalakshmi, D.; Balakumar, S. Graphene-Ag2S hybrid nanostructures: A hybrid gas sensor for room temperature hydrogen sensing application. *Mater. Lett.* **2021**, *303*, 130470. [CrossRef]
96. Ersan, M.; Dogan, H. Development of new adsorbents via microwave treatment magnetic PET synthesis from waste PET and investigation of TC removal. *Colloid Interface Sci. Commun.* **2021**, *42*, 100416. [CrossRef]
97. Irez, A.B.; Okan, C.; Kaya, R.; Cebe, E. Development of recycled disposable mask based polypropylene matrix composites: Microwave self-healing via graphene nanoplatelets. *Sustain. Mater. Technol.* **2022**, *31*, e00389. [CrossRef]
98. Danial, W.H.; Abdullah, M.; Bakar, M.A.A.; Yunos, M.S.; Ibrahim, A.R.; Iqbal, A.; Adnan, N.N. The valorisation of grass waste for the green synthesis of graphene quantum dots for nonlinear optical applications. *Opt. Mater.* **2022**, *132*, 112853. [CrossRef]
99. Kumar, D.; Rani, S.; Nandan, B.; Srivastava, R.K. Nonpolar Graphene Quantum Dot-Based Hydrophobic Coating from Microwave-Assisted Treatment of Styrofoam Waste. *ACS Sustain. Chem. Eng.* **2022**, *10*, 1070–1077. [CrossRef]
100. Wang, Y.; Srinivasakannan, C.; Wang, H.; Xue, G.; Wang, L.; Wang, X.; Duan, X. Preparation of novel biochar containing graphene from waste bamboo with high methylene blue adsorption capacity. *Diam. Relat. Mater.* **2022**, *125*, 109034. [CrossRef]
101. Khose, R.V.; Bangde, P.; Bondarde, M.P.; Dhumal, P.S.; Bhakare, M.A.; Chakraborty, G.; Ray, A.K.; Dandekar, P.; Some, S. Waste derived approach towards wealthy fluorescent N-doped graphene quantum dots for cell imaging and H2O2 sensing applications. *Spectrochim. Acta Part A Mol. Biomol. Spectrosc.* **2022**, *266*, 120453. [CrossRef]
102. Hong, W.T.; Moon, B.K.; Yang, H.K. Microwave irradiation and color converting film application of carbon dots originated from wasted toner powder. *Mater. Res. Bull.* **2022**, *156*, 111999. [CrossRef]
103. Smith, A.T.; LaChance, A.M.; Zeng, S.; Liu, B.; Sun, L. Synthesis, properties, and applications of graphene oxide/reduced graphene oxide and their nanocomposites. *Nano Mater. Sci.* **2019**, *1*, 31–47. [CrossRef]
104. Zhang, R.; Thiyagarajan, P.; Tiede, D.M. Probing protein fine structures by wide angle solution X-ray scattering. *J. Appl. Crystallogr.* **2020**, *33*, 565–568. [CrossRef]
105. Spyrou, K.; Rudolf, P. An introduction to graphene. In *Functionalization of Graphene*; Wiley: Hoboken, NJ, USA, 2014; pp. 1–20.
106. Yousefi, N.; Lu, X.; Elimelech, M.; Tufenkji, N. Environmental performance of graphene-based 3D macrostructures. *Nat. Nanotechnol.* **2019**, *14*, 107–119. [CrossRef] [PubMed]
107. Zhao, W.; Tang, Y.; Xi, J.; Kong, J. Functionalized graphene sheets with poly (ionic liquid) s and high adsorption capacity of anionic dyes. *Appl. Surf. Sci.* **2015**, *326*, 276–284. [CrossRef]
108. Mu, S.J.; Su, Y.C.; Xiao, L.H.; Liu, S.D.; Hu, T.; Tang, H.B. X-ray difraction pattern of graphite oxide. *Chin. Phys. Lett.* **2013**, *30*, 096101. [CrossRef]
109. Al-Gaashani, R.; Najjar, A.; Zakaria, Y.; Mansour, S.; Atieh, M.A. XPS and structural studies of high quality graphene oxide and reduced graphene oxide prepared by different chemical oxidation methods. *Ceram. Int.* **2019**, *45*, 14439–14448. [CrossRef]
110. Ikram, R.; Jan, B.M.; Ahmad, W. An overview of industrial scalable production of graphene oxide and analytical approaches for synthesis and characterization. *J. Mater. Res. Technol.* **2020**, *9*, 11587–11610. [CrossRef]
111. Shao, G.; Lu, Y.; Wu, F.; Yang, C.; Zeng, F.; Wu, Q. Graphene oxide: The mechanisms of oxidation and exfoliation. *J. Mater. Sci.* **2012**, *47*, 4400–4409. [CrossRef]

112. Buchner, F.; Forster-Tonigold, K.; Bozorgchenani, M.; Gross, A.; Behm, R.J. Interaction of a self-assembled ionic liquid layer with graphite (0001): A combined experimental and theoretical study. *J. Phys. Chem. Lett.* **2016**, *7*, 226–233. [CrossRef]
113. Zaldivar, R.J.; Adams, P.M.; Nokes, J.; Kim, H.I. Surface functionalization of graphene like materials by carbon monoxide atmospheric plasma treatment for improved wetting without structural degradation. *J. Vac. Sci. Technol. B Nanotechnol. Microelectron. Mater. Process. Meas. Phenom.* **2012**, *30*, 03D107.
114. Amadei, C.A.; Arribas, P.; Vecitis, C.D. Graphene oxide standardization and classification: Methods to support the leap from lab to industry. *Carbon* **2018**, *133*, 398–409. [CrossRef]
115. Rosenzweig, S.; Sorial, G.A.; Sahle-Demessie, E.; McAvoy, D.C. Optimizing the physical-chemical properties of carbon nanotubes (CNT) and graphene nanoplatelets (GNP) on Cu (II) adsorption. *J. Hazard. Mater.* **2014**, *279*, 410–417. [CrossRef]
116. Maccaferri, G.; Zanardi, C.; Xia, Z.Y.; Kovtun, A.; Liscio, A.; Terzi, F.; Palermo, V.; Seeber, R. Systematic study of the correlation between surface chemistry, conductivity and electrocatalytic properties of graphene oxide nanosheets. *Carbon* **2017**, *120*, 165–175. [CrossRef]
117. Shin, Y.E.; Sa, Y.J.; Park, S.; Lee, J.; Shin, K.H.; Joo, S.H.; Ko, H. An ice-templated, pH-tunable self-assembly route to hierarchically porous graphene nanoscroll networks. *Nanoscale* **2014**, *6*, 9734–9741. [CrossRef]
118. Huang, H.H.; De Silva, K.K.H.; Kumara, G.R.A.; Yoshimura, M. Structural evolution of hydrothermally derived reduced graphene oxide. *Sci. Rep.* **2018**, *8*, 6849. [CrossRef]
119. Zhu, Y.; Huang, H.; Zhou, W.; Li, G.; Liang, X.; Guo, J.; Tang, S. Low temperature reduction of graphene oxide film by ammonia solution and its application for high-performance supercapacitors. *J. Mater. Sci. Mater. Electron.* **2017**, *28*, 10098–10105. [CrossRef]
120. Muralikrishna, S.; Sureshkumar, K.; Varley, T.S.; Nagaraju, D.H.; Ramakrishnappa, T. In situ reduction and functionalization of graphene oxide with L-cysteine for simultaneous electrochemical determination of cadmium (II), lead (II), copper (II), and mercury (II) ions. *Anal. Methods* **2014**, *6*, 8698–8705. [CrossRef]
121. Feng, J.; Hou, Y.; Wang, X.; Quan, W.; Zhang, J.; Wang, Y.; Li, L. In-depth study on adsorption and photocatalytic performance of novel reduced graphene oxide-$ZnFe_2O_4$-polyaniline composites. *J. Alloy. Compd.* **2016**, *681*, 157–166. [CrossRef]
122. Khan, A.J.; Khan, A.; Javed, M.S.; Arshad, M.; Asim, S.; Khalid, M.; Siyal, S.H.; Hussain, S.; Hanif, M.; Liu, Z. Surface assembly of Fe_3O_4 nanodiscs embedded in reduced graphene oxide as a high-performance negative electrode for supercapacitors. *Ceram. Int.* **2020**, *46*, 19499–19505. [CrossRef]
123. Xing, T.; Zheng, Y.; Li, L.H.; Cowie, B.C.; Gunzelmann, D.; Qiao, S.Z.; Huang, S.; Chen, Y. Observation of active sites for oxygen reduction reaction on nitrogen-doped multilayer graphene. *Acs Nano* **2014**, *8*, 6856–6862. [CrossRef] [PubMed]
124. Ikram, R.; Mohamed Jan, B.; Atif Pervez, S.; Papadakis, V.M.; Ahmad, W.; Bushra, R.; Kenanakis, G.; Rana, M. Recent advancements of N-doped graphene for rechargeable batteries: A review. *Crystals* **2020**, *10*, 1080. [CrossRef]
125. Cao, X.; Hong, T.; Yang, R.; Tian, J.H.; Xia, C.; Dong, J.C.; Li, J.F. Insights into the catalytic activity of barium carbonate for oxygen reduction reaction. *J. Phys. Chem. C* **2016**, *120*, 22895–22902. [CrossRef]
126. Whelan, P.R.; Jessen, B.S.; Wang, R.; Luo, B.; Stoot, A.C.; Mackenzie, D.M.; Braeuninger-Weimer, P.; Jouvray, A.; Prager, L.; Camilli, L.; et al. Raman spectral indicators of catalyst decoupling for transfer of CVD grown 2D materials. *Carbon* **2017**, *117*, 75–81. [CrossRef]
127. Barbon, A.; Tampieri, F. Identification of slow relaxing spin components by pulse EPR techniques in graphene-related materials. *AIMS Mater Sci.* **2017**, *4*, 147–157. [CrossRef]
128. Qi, B.; Ren, K.; Lin, Y.; Zhang, S.; Wei, T.; Fan, Z. Design of layered-stacking graphene assemblies as advanced electrodes for supercapacitors. *Particuology* **2022**, *60*, 1–13. [CrossRef]
129. Kumar, T.N.; Vardhan, K.V.; Krishna, M.H.; Nagaraja, P.V. Effect of graphene oxide on strength properties of cementitious materials: A review. *Mater. Today Proc.* **2021**, *46*, 2157–2160. [CrossRef]
130. Liu, G.; Xiong, Z.; Yang, L.; Shi, H.; Fang, D.; Wang, M.; Shao, P.; Luo, X. Electrochemical approach toward reduced graphene oxide-based electrodes for environmental applications: A review. *Sci. Total Environ.* **2021**, *778*, 146301. [CrossRef]
131. Goodwin, S.; Coldrick, Z.; Heeg, S.; Grieve, B.; Vijayaraghavan, A.; Hill, E.W. Fabrication and electrochemical response of pristine graphene ultramicroelectrodes. *Carbon* **2021**, *177*, 207–215. [CrossRef]
132. He, D.; Wang, W.; Fu, Y.; Zhao, R.; Xue, W.; Hu, W. Formation of three-dimensional honeycomb-like nitrogen-doped graphene for use in energy-storage devices. *Compos. Part A Appl. Sci. Manuf.* **2016**, *91*, 140–144. [CrossRef]
133. Hu, M.; Wang, X. Effect of N3− species on selective acetylene hydrogenation over Pd/SAC catalysts. *Catal. Today* **2016**, *263*, 98–104. [CrossRef]
134. Lin, C.; Wei, W.; Hu, Y.H. Catalytic behavior of graphene oxide for cement hydration process. *J. Phys. Chem. Solids* **2016**, *89*, 128–133. [CrossRef]
135. Yao, H.; Li, F.; Lutkenhaus, J.; Kotaki, M.; Sue, H.J. High-performance photocatalyst based on nanosized ZnO-reduced graphene oxide hybrid for removal of Rhodamine B under visible light irradiation. *AIMS Mater Sci.* **2016**, *3*, 1410–1425.
136. Geng, X.; Guo, Y.; Li, D.; Li, W.; Zhu, C.; Wei, X.; Chen, M.; Gao, S.; Qiu, S.; Gong, Y.; et al. Interlayer catalytic exfoliation realizing scalable production of large-size pristine few-layer graphene. *Sci. Rep.* **2013**, *3*, 1–6. [CrossRef]
137. Li, C.; Wang, X.; Liu, Y.; Wang, W.; Wynn, J.; Gao, J. Using glucosamine as a reductant to prepare reduced graphene oxide and its nanocomposites with metal nanoparticles. *J. Nanoparticle Res.* **2012**, *14*, 1–11. [CrossRef]
138. Chua, C.K.; Ambrosi, A.; Pumera, M. Graphene oxide reduction by standard industrial reducing agent: Thiourea dioxide. *J. Mater. Chem.* **2012**, *22*, 11054–11061. [CrossRef]

139. Zhang, X.; Li, K.; Li, H.; Lu, J.; Fu, Q.; Chu, Y. Graphene nanosheets synthesis via chemical reduction of graphene oxide using sodium acetate trihydrate solution. *Synth. Met.* **2014**, *193*, 132–138. [CrossRef]
140. Ahmed, M.S.; Han, H.S.; Jeon, S. One-step chemical reduction of graphene oxide with oligothiophene for improved electrocatalytic oxygen reduction reactions. *Carbon* **2013**, *61*, 164–172. [CrossRef]
141. Lin, L.; Peng, H.; Liu, Z. Synthesis challenges for graphene industry. *Nat. Mater.* **2019**, *18*, 520–524. [CrossRef]
142. Chen, H.; Du, W.; Liu, J.; Qu, L.; Li, C. Efficient room-temperature production of high-quality graphene by introducing removable oxygen functional groups to the precursor. *Chem. Sci.* **2019**, *10*, 1244–1253. [CrossRef]
143. Saeed, M.; Alshammari, Y.; Majeed, S.A.; Al-Nasrallah, E. Chemical vapour deposition of graphene—Synthesis, characterisation, and applications: A review. *Molecules* **2020**, *25*, 3856. [CrossRef] [PubMed]
144. Li, N.W.; Shi, Y.; Yin, Y.X.; Zeng, X.X.; Li, J.Y.; Li, C.J.; Wan, L.J.; Wen, R.; Guo, Y.G. A flexible solid electrolyte interphase layer for long-life lithium metal anodes. *Angew. Chem.* **2018**, *130*, 1521–1525. [CrossRef]
145. Baldassarre, L.; Giliberti, V.; Rosa, A.; Ortolani, M.; Bonamore, A.; Baiocco, P.; Kjoller, K.; Calvani, P.; Nucara, A. Mapping the amide I absorption in single bacteria and mammalian cells with resonant infrared nanospectroscopy. *Nanotechnology* **2016**, *27*, 075101. [CrossRef] [PubMed]
146. Jayaseelan, C.; Rahuman, A.A.; Kirthi, A.V.; Marimuthu, S.; Santhoshkumar, T.; Bagavan, A.; Gaurav, K.; Karthik, L.; Rao, K.B. Novel microbial route to synthesize ZnO nanoparticles using Aeromonas hydrophila and their activity against pathogenic bacteria and fungi. *Spectrochim. Acta Part A Mol. Biomol. Spectrosc.* **2012**, *90*, 78–84. [CrossRef] [PubMed]
147. Rana, A.; Yadav, K.; Jagadevan, S. A comprehensive review on green synthesis of nature-inspired metal nanoparticles: Mechanism, application and toxicity. *J. Clean. Prod.* **2020**, *272*, 122880. [CrossRef]
148. Pyne, A.L.; Noy, A.; Main, K.H.; Velasco-Berrelleza, V.; Piperakis, M.M.; Mitchenall, L.A.; Cugliandolo, F.M.; Beton, J.G.; Stevenson, C.E.; Hoogenboom, B.W.; et al. Base-pair resolution analysis of the effect of supercoiling on DNA flexibility and major groove recognition by triplex-forming oligonucleotides. *Nat. Commun.* **2021**, *12*, 1053. [CrossRef]
149. Alemu, D.; Wei, H.Y.; Ho, K.C.; Chu, C.W. Highly conductive PEDOT: PSS electrode by simple film treatment with methanol for ITO-free polymer solar cells. *Energy Environ. Sci.* **2012**, *5*, 9662–9671. [CrossRef]
150. Bugárová, N.; Špitálsky, Z.; Mičušík, M.; Bodík, M.; Šiffalovič, P.; Koneracká, M.; Závišová, V.; Kubovčíková, M.; Kajanová, I.; Zaťovičová, M.; et al. A multifunctional graphene oxide platform for targeting cancer. *Cancers* **2019**, *11*, 753. [CrossRef]
151. Lee, X.J.; Hiew, B.Y.Z.; Lai, K.C.; Lee, L.Y.; Gan, S.; Thangalazhy-Gopakumar, S.; Rigby, S. Review on graphene and its derivatives: Synthesis methods and potential industrial implementation. *J. Taiwan Inst. Chem. Eng.* **2019**, *98*, 163–180. [CrossRef]
152. Ossonon, B.D.; Bélanger, D. Synthesis and characterization of sulfophenyl-functionalized reduced graphene oxide sheets. *RSC Adv.* **2017**, *7*, 27224–27234. [CrossRef]
153. Dimiev, A.M.; Tour, J.M. Mechanism of graphene oxide formation. *ACS Nano* **2014**, *8*, 3060–3068. [CrossRef]
154. Trikkaliotis, D.G.; Mitropoulos, A.C.; Kyzas, G.Z. Low-cost route for top-down synthesis of over-and low-oxidized graphene oxide. *Colloids Surf. A Physicochem. Eng. Asp.* **2020**, *600*, 124928. [CrossRef]
155. Gurzęda, B.; Buchwald, T.; Nocuń, M.; Bąkowicz, A.; Krawczyk, P. Graphene material preparation through thermal treatment of graphite oxide electrochemically synthesized in aqueous sulfuric acid. *RSC Adv.* **2017**, *7*, 19904–19911. [CrossRef]
156. Li, K.; Han, Z.; Wang, L.; Wang, J.; Zhang, C.; Lin, J.; Luo, S.; Peng, L.; Fang, W.; Liu, Y.; et al. Wrinkling modes of graphene oxide assembled on curved surfaces. *Nano Res.* **2023**, *16*, 1801–1809. [CrossRef]
157. Mourdikoudis, S.; Pallares, R.M.; Thanh, N.T. Characterization techniques for nanoparticles: Comparison and complementarity upon studying nanoparticle properties. *Nanoscale* **2018**, *10*, 12871–12934. [CrossRef]
158. Ravindran, A.R.; Feng, C.; Huang, S.; Wang, Y.; Zhao, Z.; Yang, J. Effects of graphene nanoplatelet size and surface area on the AC electrical conductivity and dielectric constant of epoxy nanocomposites. *Polymers* **2018**, *10*, 477. [CrossRef]
159. Wang, W.; Wu, X.; Zhang, J. Graphene and other 2D material components dynamic characterization and nanofabrication at atomic scale. *J. Nanomater.* **2015**, *16*, 81. [CrossRef]
160. Štengl, V.; Henych, J.; Bludská, J.; Ecorchard, P.; Kormunda, M. A green method of graphene preparation in an alkaline environment. *Ultrason. Sonochem.* **2015**, *24*, 65–71. [CrossRef]
161. Khanra, P.; Kuila, T.; Kim, N.H.; Bae, S.H.; Yu, D.S.; Lee, J.H. Simultaneous bio-functionalization and reduction of graphene oxide by baker's yeast. *Chem. Eng. J.* **2012**, *183*, 526–533. [CrossRef]
162. Ibrahim, A.; Klopocinska, A.; Horvat, K.; Abdel Hamid, Z. Graphene-based nanocomposites: Synthesis, mechanical properties, and characterizations. *Polymers* **2021**, *13*, 2869. [CrossRef]
163. Priya, B.; Shandilya, P.; Raizada, P.; Thakur, P.; Singh, N.; Singh, P. Photocatalytic mineralization and degradation kinetics of ampicillin and oxytetracycline antibiotics using graphene sand composite and chitosan supported BiOCl. *J. Mol. Catal. A Chem.* **2016**, *423*, 400–413. [CrossRef]
164. Bose, S.; Kuila, T.; Mishra, A.K.; Kim, N.H.; Lee, J.H. Dual role of glycine as a chemical functionalizer and a reducing agent in the preparation of graphene: An environmentally friendly method. *J. Mater. Chem.* **2012**, *22*, 9696–9703. [CrossRef]
165. Yang, H.; Zhou, W.; Yu, B.; Wang, Y.; Cong, C.; Yu, T. Uniform decoration of reduced graphene oxide sheets with gold nanoparticles. *J. Nanotechnol.* **2012**, 2012. [CrossRef]
166. Jaya, R.P. Porous concrete pavement containing nanosilica from black rice husk ash. In *New Materials in Civil Engineering*; Butterworth-Heinemann: Oxford, UK, 2020; pp. 493–527.

167. Shahriary, L.; Athawale, A.A. Graphene oxide synthesized by using modified hummers approach. *Int. J. Renew. Energy Environ. Eng.* **2014**, *2*, 58–63.
168. Sharma, N.; Sharma, V.; Vyas, R.; Kumari, M.; Kaushal, A.; Gupta, R.; Sharma, S.K.; Sachdev, K. A new sustainable green protocol for production of reduced graphene oxide and its gas sensing properties. *J. Sci. Adv. Mater. Devices* **2019**, *4*, 473–482. [CrossRef]
169. Gupta, V.; Sharma, N.; Singh, U.; Arif, M.; Singh, A. Higher oxidation level in graphene oxide. *Optik* **2017**, *143*, 115–124. [CrossRef]
170. Li, M.; Ma, Q.; Luo, A.; Hong, W. Multiple toroidal dipole symmetry-protected bound states in the continuum in all-dielectric metasurfaces. *Opt. Laser Technol.* **2022**, *154*, 108252. [CrossRef]
171. Xu, B.; Zhao, X.; Li, G.; Zhang, P.; Zhao, D.; Kong, X.; Hua, R. Large spatial Goos-Hänchen shifts from quasicrystals with graphene. *Results Phys.* **2020**, *19*, 103349. [CrossRef]
172. Zhao, D.; Wang, L.; Liu, F.; Zhong, D.; Wu, M. Photonic Stopband Filters Based on Graphene-Pair Arrays. *Appl. Sci.* **2021**, *11*, 11557. [CrossRef]

Disclaimer/Publisher's Note: The statements, opinions and data contained in all publications are solely those of the individual author(s) and contributor(s) and not of MDPI and/or the editor(s). MDPI and/or the editor(s) disclaim responsibility for any injury to people or property resulting from any ideas, methods, instructions or products referred to in the content.

MDPI AG
Grosspeteranlage 5
4052 Basel
Switzerland
Tel.: +41 61 683 77 34

Materials Editorial Office
E-mail: materials@mdpi.com
www.mdpi.com/journal/materials

Disclaimer/Publisher's Note: The title and front matter of this reprint are at the discretion of the Guest Editors. The publisher is not responsible for their content or any associated concerns. The statements, opinions and data contained in all individual articles are solely those of the individual Editors and contributors and not of MDPI. MDPI disclaims responsibility for any injury to people or property resulting from any ideas, methods, instructions or products referred to in the content.

www.ingramcontent.com/pod-product-compliance
Lightning Source LLC
Chambersburg PA
CBHW061559210926
51850CB00009B/2150